MATERIALS SCIENCE OF POLYMERS

Plastics, Rubber, Blends, and Composites

MATERIALS SCIENCE OF POLYMERS

Plastics, Rubber, Blends, and Composites

Edited by
A. K. Haghi, PhD, Eduardo A. Castro, PhD,
Sabu Thomas, PhD, P. M. Sivakumar, PhD, and
Andrew G. Mercader, PhD

Apple Academic Press Inc. | Apple Academic Press Inc.
3333 Mistwell Crescent | 9 Spinnaker Way
Oakville, ON L6L 0A2 | Waretown, NJ 08758
Canada | USA

©2015 by Apple Academic Press, Inc.

Exclusive worldwide distribution by CRC Press, a member of Taylor & Francis Group

No claim to original U.S. Government works

Printed in the United States of America on acid-free paper

International Standard Book Number-13: 978-1-77188-066-4 (Hardcover)

All rights reserved. No part of this work may be reprinted or reproduced or utilized in any form or by any electric, mechanical or other means, now known or hereafter invented, including photocopying and recording, or in any information storage or retrieval system, without permission in writing from the publisher or its distributor, except in the case of brief excerpts or quotations for use in reviews or critical articles.

This book contains information obtained from authentic and highly regarded sources. Reprinted material is quoted with permission and sources are indicated. Copyright for individual articles remains with the authors as indicated. A wide variety of references are listed. Reasonable efforts have been made to publish reliable data and information, but the authors, editors, and the publisher cannot assume responsibility for the validity of all materials or the consequences of their use. The authors, editors, and the publisher have attempted to trace the copyright holders of all material reproduced in this publication and apologize to copyright holders if permission to publish in this form has not been obtained. If any copyright material has not been acknowledged, please write and let us know so we may rectify in any future reprint.

Trademark Notice: Registered trademark of products or corporate names are used only for explanation and identification without intent to infringe.

Library and Archives Canada Cataloguing in Publication

Materials science of polymers : plastics, rubber, blends, and composites / edited by A.K. Haghi, PhD, Eduardo A. Castro, PhD, Sabu Thomas, PhD, P.M. Sivakumar, PhD, and Andrew G. Mercader, PhD.

Includes bibliographical references and index.
ISBN 978-1-77188-066-4 (bound)

1. Polymers. 2. Polymerization. 3. Plastics. 4. Rubber. 5. Composite materials. 6. Materials science. I. Castro, E. A. (Eduardo Alberto), 1944-, editor II. Thomas, Sabu, editor III. Haghi, A. K., editor IV. Mercader, Andrew G., editor V. Sivakumar, P. M., editor

TA455.P58M38 2015   620.1'92   C2015-902228-2

CIP data on file with US Library of Congress

Apple Academic Press also publishes its books in a variety of electronic formats. Some content that appears in print may not be available in electronic format. For information about Apple Academic Press products, visit our website at **www.appleacademicpress.com** and the CRC Press website at **www.crcpress.com**

ABOUT THE EDITORS

A. K. Haghi, PhD

A. K. Haghi, PhD, holds a BSc in urban and environmental engineering from University of North Carolina (USA); a MSc in mechanical engineering from North Carolina A&T State University (USA); a DEA in applied mechanics, acoustics and materials from Université de Technologie de Compiègne (France); and a PhD in engineering sciences from Université de Franche-Comté (France). He is the author and editor of 165 books as well as 1,000 published papers in various journals and conference proceedings. Dr. Haghi has received several grants, consulted for a number of major corporations, and is a frequent speaker to national and international audiences. Since 1983, he served as a professor at several universities. He is currently Editor-in-Chief of the *International Journal of Chemoinformatics and Chemical Engineering* and *Polymers Research Journal* and on the editorial boards of many international journals. He is a member of the Canadian Research and Development Center of Sciences and Cultures (CRDCSC), Montreal, Quebec, Canada.

Eduardo A. Castro, PhD

Eduardo A. Castro, PhD, is a Superior Researcher at the Argentina National Research Council. He is a full professor of theoretical chemistry at the Universidad Nacional de La Plata and a career investigator with the Consejo Nacional de Investigaciones Cientificas y Tecnicas, both based in Buenos Aires, Argentina. He is the author of nearly 1,000 academic papers in theoretical chemistry and other topics, and he has published several books. He serves on the editorial advisory boards of several chemistry journals and is often an invited speaker at international conferences in South America and elsewhere.

Sabu Thomas, PhD

Dr. Sabu Thomas is the Director of the School of Chemical Sciences, Mahatma Gandhi University, Kottayam, India. He is also a full professor of polymer science and engineering and the Director of the International and Inter University Centre for Nanoscience and Nanotechnology of the same university. He is a fellow of many professional bodies. Professor Thomas has authored or co-authored many papers in international peer-reviewed journals in the area of polymer processing. He has organized several international conferences and has more than 420 publications, 11 books, and two patents to his credit. He has been involved in a number of books both as author and editor. He is a reviewer to many international journals and has received many awards for his excellent work in polymer processing. His h-Index is 42. Professor Thomas is listed as the 5th position in the list of Most Productive Researchers in India, in 2008.

P. M. Sivakumar, PhD

P. M. Sivakumar, PhD, is a Foreign Postdoctoral Researcher (FPR) at RIKEN, Wako Campus, in Japan. RIKEN is Japan's largest comprehensive research institution renowned for high-quality research in a diverse range of scientific disciplines. He received his PhD from the Department of Biotechnology, Indian Institute of Technology Madras, India. He is a member of the editorial boards of several journals and has published papers in international peer-reviewed journals and professional conferences. His research interests include bionanotechnology and biomaterials.

Andrew G. Mercader, PhD

Andrew G. Mercader, PhD, studied physical chemistry at the Faculty of Chemistry of La Plata National University (UNLP), Buenos Aires, Argentina, from 1995 to 2001. Afterwards he joined Shell Argentina to work as Luboil, Asphalts and Distillation Process Technologist, as well as Safeguarding and Project Technologist from 2001 to 2006. His PhD work, on the development and applications of QSAR/QSPR theory, was performed at the Theoretical and Applied Research Institute located at La Plata National University (INIFTA), from 2006 to 2009. After that he obtained a postdoctoral scholarship to work on theoretical-experimental studies of biflavonoids at IBIMOL (ex PRALIB), Faculty of Pharmacy and Biochemistry, University of Buenos Aires (UBA), from 2009 to 2012. He is currently appointed as a member of the Scientific Researcher Career in the Argentina National Research Council at INIFTA.

CONTENTS

List of Contributors .... *ix*
List of Abbreviations .... *xiii*
List of Symbols .... *xv*
Preface .... *xvii*

1. **A Detailed Review on Characteristics, Application and Limitation of Amorphous Glassy Polymers as Natural Nanocomposites** 1
   G. V. Kozlov, I. V. Dolbin, Jozef Richert, O. V. Stoyanov, and G. E. Zaikov

2. **Structure of Graphitic Carbons: A Comprehensive Review** 51
   Heinrich Badenhorst

3. **Radiation Crosslinking of Acrylonitrile-Butadiene Rubber** 81
   Katarzyna Bandzierz, Dariusz M. Bielinski, Adrian Korycki, and Grazyna Przybytniak

4. **Rubber Vulcanizates Containing Plasmochemically Modified Fillers** 91
   Dariusz M. Bieliński, Mariusz Siciński, Jacek Grams, and Michał Wiatrowski

5. **Modification of the Indian Rubber in the Form of Latex with Ozone** ... 103
   L. A. Vlasova, P. T. Poluektov, S. S. Nikulin, and V. M. Misin

6. **Influence of the Structure of Polymer Material on Modification of the Surface Layer of Iron Counterface in Tribological Contact** 111
   Dariusz M. Bieliński, Mariusz Siciński, Jacek Grams, and Michał Wiatrowski

7. **Boron Oxide as a Fluxing Agent for Silicone Rubber-Based Ceramizable Composites** 125
   R. Anyszka, D. M. Bieliński, and Z. Pędzich

8. **Application of Micro-dispersed Silicon Carbide Along with Slurries as a Functional Fin Fire and Heat Resistant Elastomer Compositions** 139
   V. S. Liphanov, V. F. Kablov, S.V. Lapin, V. G. Kochetkov, O. M. Novopoltseva, and G. E. Zaikov

9. **Thermal Stability of Elastic Polyurethane** 145
   I. A. Novakov, M. A. Vaniev, D.V. Medvedev, N. V. Sidorenko, G. V. Medvedev, and D. O. Gusev

10. **PAN/Nano–TiO$_2$–S Composites: Physico–Chemical Properties** 155
   M. M. Yatsyshyn, A. S. Kun'ko, and O. V. Reshetnyak

11. **Viscoelastic Properties of the Polystyrene** 173
   Yu. G. Medvedevskikh, O. Yu. Khavunko, L. I. Bazylyak, and G. E. Zaikov

12. **Nanostructured Polymeric Composites Filled with Nanoparticles** 211
   A. K. Mikitaev, A. Yu. Bedanokov, and M. A. Mikitaev

13. **Structure, Properties and Application of Dendritic Macromolecules in Various Fields: Molecular Simulation Techniques in Hyperbranched Polymer and Dendrimers** 241
   M. Hasanzadeh and B. Hadavi Moghadam

14. **A Study on Influence of Electrospinning Parameters on the Contact Angle of the Electrospun PAN Nanofiber Mat Using Response Surface Methodology (RSM) and Artificial Neural Network (ANN)** 261
   B. Hadavi Moghadam and M. Hasanzadeh

15. **Fabrication and Characterization of the Metal Nano-Sized Branched Structures and the Composite Nanostructures Grown on Insulator Substrates by the EBID Process** 279
   Guoqiang Xie, Minghui Song, Kazuo Furuya, and Akihisa Inoue

16. **A Case Study on Hyperbranched Polymers** 297
   Ramin Mahmoodi, Tahereh Dodel, Tahereh Moieni, and Mahdi Hasanzadeh

17. **A Study on Network of Sodium Hyaluronate with Nano-Knots Junctions** 307
   Shin-ichi Hamaguchi and Toyoko Imae

18. **The Magnetic Photocatalyst Conversion to the Magnetic Dye-Adsorbent Catalyst via Hydrothermal Followed by Typical Washing and Thermal Treatments** 323
   Satyajit Shukla

19. **Solid Polymer Fuel Cell: A Three-Dimensional Computation Model and Numerical Simulations** 339
   Mirkazem Yekani, Meysam Masoodi, Nima Ahmadi, Mohamad Sadeghi Azad, and Khodadad Vahedi

Index 363

LIST OF CONTRIBUTORS

Nima Ahmadi
Mechanical Engineering Department, Urmia University, of Technology, Iran

R. Anyszka
Lodz University of Technology, Faculty of Chemistry, Institute of Polymer and Dye Technology, 90-924, Lodz, Poland

Mohamad Sadeghi Azad
Mechanical Engineering Department, Urmia University, of Technology, Iran

Heinrich Badenhorst
SARChI Chair in Carbon Materials and Technology, Department of Chemical Engineering, University of Pretoria, Pretoria, Gauteng, 0002, Gauteng, 0169, South Africa, Email: heinrich.badenhorst@up.ac.za

Katarzyna Bandzierz
Faculty of Chemistry, Lodz University of Technology, Poland

L. I. Bazylyak
Physical Chemistry of Combustible Minerals Department, Institute of Physical–Organic Chemistry and Coal Chemistry named after L. M. Lytvynenko, National Academy of Sciences of Ukraine, 79053, Lviv, Ukraine

D. M. Bieliński
Institute for Engineering of Polymer Materials and Dyes, Division of Elastomers and Rubber Technology, 05-820 Piastow, Poland

Tahereh Dodel
Amirkabir University of Technology, Iran

I. V. Dolbin
Kabardino-Balkarian State University, Nal'chik, 360004, Russian Federation

Kazuo Furuya
High Voltage Electron Microscopy Station, National Institute for Materials Science, Tsukuba, 305-0003, Japan

Jacek Grams
Institute of General and Ecological Chemistry, Technical University of Łódź, Łódź, Poland.

D. O. Gusev
Volgograd State Technical University, 400005, Volgograd, Russia

Shin-ichi Hamaguchi
Graduate School of Science, Nagoya University, Chikusa, Nagoya 464-8602, Japan

M. Hasanzadeh
Department of Textile Engineering, University of Guilan, Rasht, Iran

Toyoko Imae
Graduate Institute of Engineering, National Taiwan University of Science and Technology, Taipei 10607, Taiwan; Graduate School of Science, Nagoya University, Chikusa, Nagoya 464-8602, Japan

Akihisa Inoue
WPI Advanced Institute for Materials Research, Tohoku University, Sendai, 980-8577, Japan

V. F. Kablov
Volzhsky Polytechnical Institute (branch) Volgograd State Technical University, Volzhsky, Volgograd Region, 404121, Russian Federation, E-mail: nov@volpi.ru; www.volpi.ru

O. Yu. Khavunko
Physical Chemistry of Combustible Minerals Department, Institute of Physical–Organic Chemistry and Coal Chemistry named after L. M. Lytvynenko, National Academy of Sciences of Ukraine, 79053, Lviv, Ukraine, e-mail: hop_vfh@ukr.net

V. G. Kochetkov
Volzhsky Polytechnical Institute (branch) Volgograd State Technical University, Volzhsky, Volgograd Region, 404121, Russian Federation, E-mail: nov@volpi.ru; www.volpi.ru

Adrian Korycki
Faculty of Chemistry, Lodz University of Technology, Poland

G. V. Kozlov
Kabardino-Balkarian State University, Nal'chik, 360004, Russian Federation, E-mail: I_dolbin@mail.ru

A. S. Kun'ko
Department of Physical and Colloidal Chemistry, Ivan Franko National University of L'viv, 79005, L'viv, Ukraine, e-mail: m_yatsyshyn@franko.lviv.ua

S. V. Lapin
Volzhsky Polytechnical Institute (branch) Volgograd State Technical University, Volzhsky, Volgograd Region, 404121, Russian Federation, E-mail: nov@volpi.ru; www.volpi.ru

V. S. Liphanov
Volzhsky Polytechnical Institute (branch) Volgograd State Technical University, Volzhsky, Volgograd Region, 404121, Russian Federation, E-mail: nov@volpi.ru; www.volpi.ru

Ramin Mahmoodi
Amirkabir University of Technology, Iran

Meysam Masoodi
Department of Chemical Engineering-Faculty of Engineering- Imam Hossein University, Tehran, Iran

D. V. Medvedev
Elastomer Limited Liability Company, 400005, Volgograd, Russia

G. V. Medvedev
Volgograd State Technical University, 400005, Volgograd, Russia

Yu. G. Medvedevskikh
Physical Chemistry of Combustible Minerals Department, Institute of Physical–Organic Chemistry and Coal Chemistry named after L. M. Lytvynenko, National Academy of Sciences of Ukraine, 79053, Lviv, Ukraine, e-mail: hop_vfh@ukr.net

A. K. Mikitaev
Kabardino Balkarian State University, Nalchik, Russia

List of Contributors

M. A. Mikitaev
L.Ya.Karpov Research Institute, Moscow, Russia

V. M. Misin
Voronezh State University of the engineering technologies, N. M. Emanuel Institute of Biochemical Physics, Russian Academy of Sciences, Moscow

B. Hadavi Moghadam
Department of Textile Engineering, University of Guilan, Rasht, Iran

Tahereh Moieni
Amirkabir University of Technology, Iran

S. S. Nikulin
Voronezh State University of the Engineering Technologies, N. M. Emanuel Institute of Biochemical Physics, Russian Academy of Sciences, Moscow

I. A. Novakov
Volgograd State Technical University, 400005, Volgograd, Russia

O. M. Novopoltseva
Volzhsky Polytechnical Institute (branch) Volgograd State Technical University, Volzhsky, Volgograd Region, 404121, Russian Federation, E-mail: nov@volpi.ru; www.volpi.ru

Z. Pędzich
AGH University of Science and Technology, Faculty of Materials Science and Ceramics, Department of Ceramics and Refractory Materials, Al., 30-045, Krakow, Poland

P. T. Poluektov
Voronezh State University of the Engineering Technologies, N. M. Emanuel Institute of Biochemical Physics, Russian Academy of Sciences, Moscow

Grazyna Przybytniak
Faculty of Chemistry, Lodz University of Technology, Poland

O. V. Reshetnyak
Department of Physical and Colloidal Chemistry, Ivan Franko National University of L'viv, 79005, L'viv, Ukraine; Department of Chemistry, Army Academy named after hetman Petro Sahaydachnyi, 79012, L'viv, Ukraine

Jozef Richert
Institut Inzynierii Materialow Polimerowych I Barwnikow, 87-100 Torun, Poland, E-mail: j.richert@impib.pl

Satyajit Shukla
Ceramic Technology Department, Materials and Minerals Division (MMD), National Institute for Interdisciplinary Science and Technology (NIIST), Council of Scientific and Industrial Research (CSIR), Thiruvananthapuram 695019, Kerala, India

Mariusz Siciński
Institute of Polymer and Dye Technology, Technical University of Łódź, 90-924, Łódź, Poland

N. V. Sidorenko
Volgograd State Technical University, 400005, Volgograd, Russia

Minghui Song
High Voltage Electron Microscopy Station, National Institute for Materials Science, Tsukuba 305-0003, Japan

O. V. Stoyanov
Kazan National Research Technological University, Kazan, Tatarstan, Russia, E-mail: OV_Stoyanov@mail.ru

khodadad Vahedi
Mechanical Engineering Department, Imam Hosein University, Tehran, Iran

M. A. Vaniev
Volgograd State Technical University, 400005, Volgograd, Russia

L. A. Vlasova
Voronezh State University of the Eengineering Technologies, N. M. Emanuel Institute of Biochemical Physics, Russian Academy of Sciences, Moscow

Michał Wiatrowski
Department of Molecular Physics, Technical University of Łódź, Łódź, Poland

Guoqiang Xie
Institute for Materials Research, Tohoku University, Sendai 980-8577, Japan

M. M. Yatsyshyn
Department of Physical and Colloidal Chemistry, Ivan Franko National University of L'viv, 79005, L'viv, Ukraine, e–mail: m_yatsyshyn@franko.lviv.ua

Mirkazem Yekani
Aerospace Engineering Department, Imam Hosein University, Tehran, Iran, Email:meyekani@yahoo.com

Bedanokov A. Yu
D.I.Mendeleev Russian University for Chemical Technology, Moscow, Russia

G. E. Zaikov
Kinetics of Chemical and Biological Processes Division, Institute of Biochemical Physics named after N. N. Emanuel, Russian Academy of Sciences, 119991, Moscow, RUSSIA, e–mail: chembio@sky.chph.ras.ru

G. E. Zaikov
N. M. Emanuel Institute of Biochemical Physics of Russian Academy of Sciences, Moscow 119334, Russian Federation, E-mail: Chembio@sky.chph.ras.ru

LIST OF ABBREVIATIONS

AFM	atomic force microscopic
ANN	artificial neural network
ASA	active surface area
CA	contact angle
CCBB	continuous configurational boltzmann biased
CCD	central composite design
CP	conducting polymer
CVD	chemical vapor deposition
DB	degree of branching
DFT	density functional theory
DTA	differential thermal analysis
HBPs	hyperbranched polymers
EB	electron beam
EBID	electron-beam-induced deposition
EDS	X-ray energy dispersive spectroscopy
EDX	energy dispersive X–ray
EHT	energy of electrons
FTIR	fourier transform infrared spectroscopy
HA	hyaluronic acid
HBPs	hyperbranched polymers
LDA	local density approximation
MC	monte carlo
MD	molecular dynamics
NEMD	non-equilibrium molecular dynamics
NNG	natural source
OIT	oxidation induction time
PEMFCs	proton exchange membrane fuel cells
PET	poly(ethylene terephthalate)
PUE	polyurethane elastomers
RSM	response surface methodology
SAD	selected-area diffraction
SBR	styrene-butadiene rubber
SEM	scanning electron microscope
SFE	surface free energy
SiC	silicon carbide
TBMD	tight bonding molecular dynamics

TEM	transmission electron microscope
TES	tear strength
TGA	thermogravimetric analyzer
TS	tensile strength

LIST OF SYMBOLS

a	water activity
C	molar concentration (mol/m3)
D	mass diffusion coefficient (m2/s)
F	faraday constant (C/mol)
I	local current density (A/m2)
J	exchange current density (A/m2)
K	permeability (m2)
M	molecular weight (kg/mol)
n_d	electro-osmotic drag coefficient
P	pressure (Pa)
R	universal gas constant (J/mol-K)
T	temperature (K)
t	thickness
$\vec{u}$	velocity vector
Vcell	cell voltage
Voc	open-circuit voltage
W	width
X	mole fraction

Greek letters

ρ	water transfer coefficient
ρ	effective porosity
ρ	density (kg/m3)
μ	viscosity (kg/m-s)
σ_e	membrane conductivity (1/ohm-m)
λ	water content in the membrane
η	stoichiometric ratio
η	over potential (v)
λ_{eff}	effective thermal conductivity (w/m-k)
ϕ_e	electrolyte phase potential (v)

SUBSCRIPTS AND SUPERSCRIPTS

a	anode
c	cathode

ch	channel
k	chemical species
m	membrane
MEA	membrane electrolyte assembly
ref	reference value
sat	saturated
w	water

PREFACE

This book skillfully blends and integrates polymer science, plastic technology and rubber technology. The fundamentals of polymerization, polymer characteristics, rheology and morphology as well as the composition, technology, testing and evaluation of various plastics, rubbers, fibers, adhesives, coatings and composites are comprehensively presented. The book is highly suitable for all entrepreneurs and professionals engaged in production of as well as research and development in polymers. It will also be found immensely useful by advanced- level research students of physics, chemistry, and materials science, specializing in polymers, as well as students of chemical and metallurgical engineering having courses in polymer technology/materials science and technology.

This volume highlights the latest developments and trends in advanced poly-blends and their structures. It presents the developments of advanced poly-blends and respective tools to characterize and predict the material properties and behavior. The book provides important original and theoretical experimental results that use nonroutine methodologies often unfamiliar to many readers. Furthermore chapters on novel applications of more familiar experimental techniques and analyses of composite problems are included, which indicate the need for the new experimental approaches that are presented.

Technical and technological development demands the creation of new materials that are stronger, more reliable, and more durable, i.e. materials with new properties. Up-to-date projects in creation of new materials go along the way of nanotechnology.

With contributions from experts from both industry and academia, this book presents the latest developments in the identified areas. This book incorporates appropriate case studies, explanatory notes, and schematics for more clarity and better understanding. This book will be useful for chemists, chemical engineers, technologists, and students interested in advanced nanopolymers with complex behavior and their applications.

This new book:
- Gives an up-to-date and thorough exposition of the present state of the art of polyblends and composites.
- Familiarizes the reader with new aspects of the techniques used in the examination of polymers, including chemical, physicochemical, and purely physical methods of examination.
- Describes the types of techniques now available to the polymer chemist and technician and discusses their capabilities, limitations, and applications.

- Provides a balance between materials science and mechanics aspects, basic and applied research, and high-technology and high-volume (low cost) composite development.

CHAPTER 1

A DETAILED REVIEW ON CHARACTERISTICS, APPLICATION, AND LIMITATION OF AMORPHOUS GLASSY POLYMERS AS NATURAL NANOCOMPOSITES

G. V. KOZLOV, I. V. DOLBIN, JOZEF RICHERT, O. V. STOYANOV, G. E. ZAIKOV

1.1 INTRODUCTION

The stated results in the present article give purely practical aspect of such theoretical concepts as the cluster model of polymers amorphous state stricture and fractal analysis application for the description of structure and properties of polymers, treated as natural nanocomposites. The necessary nanostructure goal-directed making will allow to obtain polymers, not yielding (and even exceeding), by their properties to the composites, produced on their basis. Structureless (defect-free) polymers are imagined to be the most perspective in this respect. Such polymers can be natural replacement for a large number of elaborated at present polymer nanocomposites. The application of structureless polymers as artificial nanocomposites polymer matrix can give much larger effect. Such approach allows to obtain polymeric materials, comparable by their characteristics with metals (e.g., with aluminum).

The idea of different classes of polymer representation as composites is not new. Even 35 years ago, Kardos and Raisoni [1] offered to use composite models for the description of semicrystalline polymer properties number and obtained prediction of the indicated polymer stiffness and thermal strains to a precision of ±20 percent. They considered semicrystalline polymer as composite, in which matrix is the amorphous and the crystallites are the filler. The authors [1] also supposed that other polymers, for example, hybrid polymer systems, in which two components with different mechanical properties were present obviously, can be simulated by a similar method.

In paper [2] it has been pointed out, that the most important consequence from works by supramolecular formation study is the conclusion that physical-mechanical properties depend in the first place on molecular structure, but are realized through supramolecular formations. At scales interval and studies methods resolving ability of polymers structure, the nanoparticle size can be changed within the limits of 1 100 and more nanometers. The polymer crystallite size makes up 10 20 nm. The macromolecule can be included in several crystallites, since at molecular weight of order of 6 10^4 its length makes up more than 400 nm. These reasonings point out that macromolecular formations and polymer systems in virtue of their structure features are always nanostructural systems.

However, in the cited above works the amorphous glassy polymers consideration as natural composites (nanocomposites) is absent, although they are one of the most important classes of polymeric materials. This gap reason is quite enough (i.e., polymers amorphous state quantitative model absence). However, such model appearance lately [3–5] allows to consider the amorphous glassy polymers (both linear and cross-linked ones) as natural nanocomposites, in which local order regions (clusters) are nanofiller and surrounded by loosely packed matrix of amorphous polymers structure which is matrix of nanocomposite. Proceeding from the said above, in the present chapter description of amorphous glassy polymers as natural nanocomposites, their limiting characteristics determination, and practical recommendation by the indicated polymers properties improvement will be given.

1.1.1 NATURAL NANOCOMPOSITES STRUCTURE

The synergetics principles revealed structure adaptation mechanism to external influence and are universal ones for self-organization laws of spatial structures in dynamical systems of different nature. The structure adaptation is the reformation process of structure, which loses stability, with the new, much stable structure self-organization. The fractal (multifractal) structure, which is impossible to describe within the framework of Euclidean geometry, is formed in reformation process. A wide spectrum of natural and artificial topological forms, the feature of which is self-similar hierarchically organized structure and which amorphous glassy polymers possessed [6], belongs to fractal structures.

The authors [7, 8] considered the typical amorphous glassy polymer (polycarbonate) structure change within the frameworks of solid-body synergetics.

The local order region, consisting of several densely packed collinear segments of various polymer chains (for more details see refrerencesRefs. [3–8]) according to a sign number should be attributed to the nanoparticles (nanoclusters) [9]:

1. Their size makes up 2–5 nm;
2. They are formed by self-assemble method and adapted to the external influence (e.g., temperature change results to segments number per one nanocluster change);
3. Each statistical segment represents an atoms group and boundaries between these groups are coherent owing to collinear arrangement of one segment relative to another.

The main structure parameter of cluster model nanoclusters relative fraction φ_{cl}, which is polymers structure order parameter in strict physical sense of this term, can be calculated according to the equation (see previous paper). In its turn, the polymer structure fractal dimension d_f value is determined according to the equations (see referenceRef. [9]).

In Figure 1.1, the dependence of φ_{cl} on testing temperature T for PC is shown, which can be approximated by the broken line, where points of folding (bifurcation points) correspond to energy dissipation mechanism change, coupling with the threshold values φ_{cl} reaching. So, in Figure 1.1, T_1 corresponds to structure "freezing" temperature T_0 [4], T_2 to loosely packed matrix glass transition temperature T_g' [10-11], and T_3 to polymer glass transition temperature T_g.

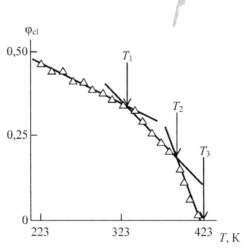

FIGURE 1.1 The dependence of nanoclusters relative fraction φ_{cl} on testing temperature T for PC. The critical temperatures of bifurcation points are indicated by arrows (explanations are given in the text) [18].

Within the frameworks of solid-body synergetics, it has been shown [12] that at structures self-organization, the adaptation universal algorithm [12] is realized at transition from previous point of structure instability to subsequent one. The value $m = 1$ corresponds to structure minimum adaptivity and $m = m^*$ to maximum one. In paper [12], the table is adduced, in which values A_m, m, and Δ_i are given, determined by the gold proportion rule and corresponding to spectrum of structure stability measure invariant magnitudes for the alive and lifeless nature systems. The indicated table usage facilitates determination are interconnected by the power law stability and adaptivity of structure to external influence [12].

Using as the critical magnitudes of governing parameter, the values φ_{cl} in the indicated bifurcation points T_0, T_g', and T_g (ϕ_{cl}' and T_{cl}^*, accordingly) together with the above mentioned table data [12], values A_m, Δ_i, and for PC can be obtained, which are adduced in Table 1.1. As it follows from the data of this table, systematic reduction of parameters A_m and Δ_i at the condition $m = 1 = const$ is observed. Hence, within the frameworks of solid-body synergetics temperature T_g' can be characterized as bifurcation point ordering-degradation of nanostructure and T_g – as nanostructure degradation-chaos [12].

It is easy to see that Δ_i decrease corresponds to bifurcation point critical temperature increase.

TABLE 1.1 The critical parameters of nanoclaster structure state for PC [8]

The Temperature Range	ϕ_{cl}'	ϕ_{cl}^*	A_m	D_i	m	m^*
213 , 333 K	0,528	0,330	0,623	0,618	1	1
333 , 390 K	0,330	0,153	0,465	0,465	1	2
390 , 425 K	0,153	0,049	0,324	0,324	1	8

Therefore, critical temperatures T_{cr} (T_0, T_g' and T_g) values increase should be expected at nanocluster structure stability measure Δ_i reduction. In Figure 1.2, the dependence of T_{cr} in Δ_i reciprocal value for PC is adduced, on which corresponding values for polyarylate (PAr) are also plotted. This correlation proved to be linear one and has two characteristic points. At $\Delta_i = 1$ the linear dependence $T_{cr}(\Delta_i^{-1})$ extrapolates to $T_{cr} = 293K$, i.e., this means that at the indicated Δ_i value, glassy polymer turns into rubber-like state at the used testing temperature $T = 293K$. From [12], $\Delta_i = 0,213$ at $m = 1$. In the plot of Figure 1.2, the greatest for polymers critical temperature $T_{cr} = T_{ll}$ (T_{ll} is the temperature of "liquid 1 to liquid 2" transition), defining the transition to "structureless liquid" [13], corresponds to this minimum Δmagnitude. For polymers this means the absence of even dynamical short-lived local order [13].

Hence, the above stated results allow to give the following interpretation of critical temperatures T_g' and T_g of amorphous glassy polymers structure within the frameworks of solid-body synergetics. These temperatures correspond to governing parameter (nanocluster contents) φ_{cl} critical values, at which reaching one of the main principles of synergetics is realized-subordination principle, when a variables set is controlled by one (or several) variable, which is an order parameter. Let us also note that reformations number $m = 1$ corresponds to structure formation mechanism particle-cluster [4, 5].

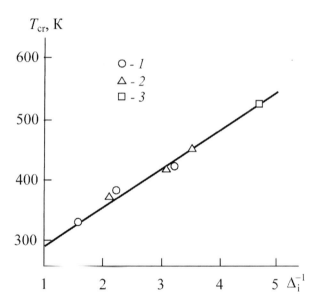

FIGURE 1.2 The dependence of critical temperatures T_{cr} on reciprocal value of nanocluster structure stability measure Δ_i for PC (1) and PAr(2), $3 - T_{ll}$ value for PC [19].

The authors [14, 15] considered synergetics principles application for the description of behavior of separate nanocluster structure, characterized by the integral parameter φ_{cl} nanoclusters in the system for the same amorphous glassy polymers. This aspect is very important, since, as it will be shown is subsequent sections, just separate nanoclusters characteristics define natural nanocomposites properties by critical mode. One from the criterions of nanoparticle definition has been obtained in paper [16]: atoms number N_{at} in it should not exceed 10^3–10^4. In paper [15], this criterion was applied to PC local order regions, having the greatest number of statistical segments $n_{cl} = 20$. Since nanocluster is amorphous analog of crystallite with the stretched chains and at its functionality F a number of chains emerging from it is accepted, then the value n_{cl} is determined as follows [4]:

$$n_{cl} = \frac{F}{2}, \quad (1.1)$$

where the value F was calculated according to the Eq. (1.7) in previous publication.

The statistical segment volume simulated as a cylinder is equal to $l_{st}S$ and further the volume per one atom of substance (PC) a^3 can be calculated according to the equation [17]:

$$a^3 = \frac{M}{\rho N_A p}, \quad (1.2)$$

where M is repeated link molar mass, ρ is polymer density, N_A is Avogadro number, and p is atoms number in a repeated link.

For PC $M = 264$ g/mole, $\rho = 1,200$ kg/m³, and $p = 37$. Then $a^3 = 9,54$ Å³ and the value N_{at} can be estimated according to the following simple equation [17]:

$$N_{at} = \frac{l_{st} \cdot S \cdot n_{cl}}{a^3}. \quad (1.3)$$

For PC $N_{at} = 193$ atoms per one nanocluster (for $n_{cl} = 20$) is obtained. It is obvious that the indicated value N_{at} corresponds well to the adduced above nanoparticle definition criterion ($N_{at} = 10^3$–10^4) [9, 17].

Let us consider synergetics of nanoclusters formation in PC and PAr. Using Eq. (1.3) as governing parameter critical magnitudes n_{cl} values at testing temperature T consecutive change and the above indicated table determined by gold proportion law values A_m, m, and Δ_i, the dependence $\Delta(T)$ can be obtained, which is adduced in Figure 1.3. As it follows from this figure data, the nanoclusters stability within the temperature range of 313–393K is approximately constant and small ($\Delta_i \approx 0,232$ at minimum value $\Delta_i \approx 0,213$) and at $T > 393$K, fast growth Δ_i (nanoclusters stability enhancement) begins for both considered polymers.

This plot can be explained within the frameworks of a cluster model [3–5]. In Figure 1.3, glass transition temperatures of loosely packed matrix T'_g, which are approximately 50 K lower than polymer macroscopic glass transition temperature T_g, are indicated by vertical shaded lines. At T'_g instable nanoclusters (i.e., having small n_{cl}), decay occurs. At the same time, stable and, hence, more steady nanoclusters remain as a structural element, that results in Δ_i growth [14].

A Detailed Review on Characteristics, Application

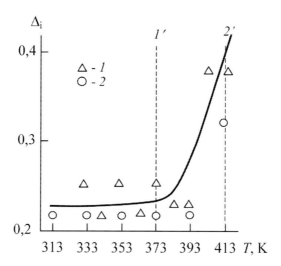

FIGURE 1.3 The dependence of nanoclusters stability measure Δ_i on testing temperature T for PC(1) and PAR(2). The vertical shaded lines indicate temperature T'_g for PC (1') and PAR (2') [14].

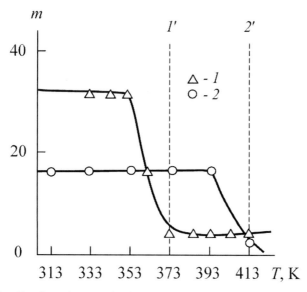

FIGURE 1.4 The dependences of reformations number m for nanoclusters on testing temperature T. The designations are the same as in Figure 1.3 [14].

In Figure 1.4, the dependences of reformations number m on testing temperature T for PC and PAR are adduced. At relatively low temperatures ($T < T'_g$), segments number in nanoclusters is large and segment joining (separation) to nanoclusters occurs easily enough, that explains large values of m. At $T \to T'_g$, reformations number reduces sharply and at $T > T'_g$, $m \approx 4$. Since at $T > T'_g$ in the system only stable clusters remain, then it is necessary to assume that large m at $T < T'_g$ are due to reformation of just instable nanoclusters [15].

In Figure 1.5 the dependence of n_{cl} on m is adduced. As one can see, even small m enhancement within the range of 2–16 results in sharp increase in segments number per one nanocluster. At $m \geq 32$, the dependence $n_{cl}(m)$ attains asymptotic branch for both studied polymers. This supposes that $n_{cl} \geq 16$ is the greatest magnitude for nanoclusters and for $m \geq 32$, this term belongs equally to both joining and separation of such segment from nanocluster.

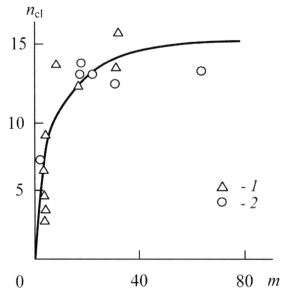

FIGURE 1.5 The dependence of segments number per one nanocluster n_{cl} on reformations number m for PC (1) and PAR (2) [14].

In Figure 1.6, the relationship of stability measure Δ_i and reformations number m for nanoclusters in PC and PAR is adduced. As it follows from the data of this figure, at $m \geq 16$ (or, according to the data of Figure 1.5, $n_{cl} \geq 12$), Δ_i value attains its minimum asymptotic magnitude $\Delta_i = 0.213$ [12]. This means that for the indicated n_{cl} values, nanoclusters in PC and PAR structure are adopted well to the external influence change ($A_m \geq 0,91$).

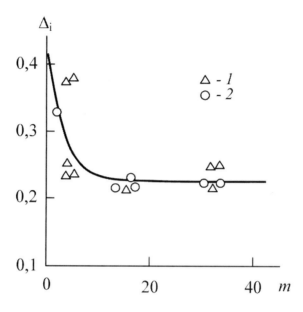

FIGURE 1.6 The dependence of stability measure Δ_i on reformation number m for PC (1) and PAR (2) [14].

Nanoclusters formation synergetics is directly connected with the studied polymers structure macroscopic characteristics. As it has been noted above, the fractal structure, characterized by the dimension d_f, is formed as a result of nanoclusters reformations. In Figure 1.7 the dependence $d_f(\Delta_i)$ for the considered polymers is adduced, from which d_f increase at Δ_i growth follows. This means that the increasing of possible reformations number m, resulting to Δ_i reduction (Figure 1.6), defines the growth of segments number in nanoclusters, the latter relative fraction φ_{cl} enhancement and, as consequence, d_f reduction [3–5].

And let us note in conclusion the following aspect, obtaining from the plot $\Delta_i(T)$ (Figure 1.3) extrapolation to maximum magnitude $\Delta_i \approx 1,0$. The indicated Δ_i value is reached approximately at $T \approx 458$ K that corresponds to mean glass transition temperature for PC and Par. Within the frameworks of the cluster model T_g reaching means polymer nanocluster structure decay [3–5] and, in its turn, realization at T_g of the condition $\Delta_i \approx 1,0$ means that the "degenerated" nanocluster, consisting of one statistical segment or simply statistical segment, possesses the greatest stability measure. Several such segments joining up in nanocluster maintains its stability reduction (see Figures 1.5 and 1.6), which is the cause of glassy polymers structure thermodynamical nonequilibrium [14].

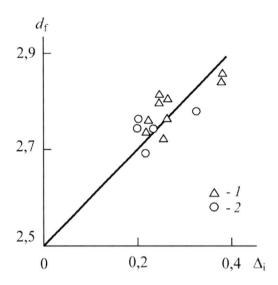

FIGURE 1.7 The dependence of structure fractal dimension d_f on stability measure of nanoclusters Δ_i for PC (1) and PAR (2) [14].

Therefore, the above stated results showed synergetics principles applicability for the description of association (dissociation) processes of polymer segments in local order domains (nanoclusters) in case of amorphous glassy polymers. Such conclusion can be a priori, since nanoclusters are dissipative structures [6]. Testing temperature increase raises nanoclusters stability measure at the expense of possible reformations number reduction [14, 15].

As it has been shown lately, the notion "nanoparticle" (nanocluster) gets well over the limits of purely dimensional definition and means substance state specific character in sizes nanoscale. The nanoparticles, sizes of which are within the range of order of 1–100 nm, are already not classical macroscopic objects. They represent themselves the boundary state between macro- and microworld and in virtue of this they have specific features number, to which the following ones are attributed:

1. Nanoparticles are self-organizing nonequilibrium structures, which submit to synergetics laws;
2. They possess very mature surface;
3. Nanoparticles possess quantum (wave) properties.

For the nanoworld structures in the form of nanoparticles (nanoclusters), their size, defining the surface energy critical level, is the information parameter of feedback [19].

The first from the indicated points was considered in detail above. The authors [20, 21] showed that nanoclusters surface fractal dimension changes within the range of 2,15–2,85, which is their well-developed surface sign. And at last, let us consider quantum (wave) aspect of nanoclusters nature on the example of PC [22]. Structural levels hierarchy formation and development "scenario" in this case can be presented with the aid of iterated process [23]:

$$l_k = \langle a \rangle B_\lambda^k; \quad \lambda_k = \langle a \rangle B_\lambda^{k+1}; \quad k = 0, 1, 2, \ldots, \tag{1.4}$$

where l_k is specific spatial scale of structural changes; λ_k is length of irradiation sequence, which is due to structure reformation; k is structural hierarchy sublevel number; $B_\lambda = \lambda_b/\langle a \rangle = 2{,}61$ is discretely wave criterion of microfracture; and λ_b is the smallest length of acoustic irradiation sequence.

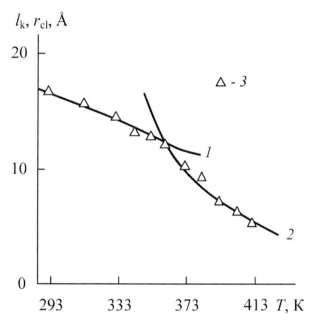

FIGURE 1.8 The dependences of structural changes at specific spatial scale l_k at $B_\lambda = 1{,}06$ (1) and 1,19 (2) and nanoclusters radius r_{cl} (3) on testing temperature T for PC [22].

In Figure 1.8, the dependences of l_k and nanoclusters radius r_{cl} on T are adduced, where l_k was determined according to the Eq. (1.4) and the value r_{cl} was calculated according to the formula (in previous paper). As it follows from the data of Figure 1.8, the values l_k and r_{cl} agree within the whole studied temperatures range. Let us note that if in paper [23] the value $B_\lambda = 2{,}61$, then for PC the above indicated agreement was obtained at $B_\lambda = 1{,}19$ and 1,06. This distinction confirms the thesis about distinction of synergetics laws in reference to nano-microworld objects (let us remind that the condition $B_\lambda = 2{,}61$ is valid even in the case of earthquakes [14]). It is interesting to note that B_λ change occurs at glass transition temperature of loosely packed matrix (i.e., approximately at $T_g - 50$ K) [11].

Hence, the above stated results demonstrated that the nanocluster possessed all nanoparticles properties (i.e., they belonged to substance intermediate state—nanoworld).

And in completion of the present section, let us note one more important feature of natural nanocomposites structure. In papers [24, 25], the absence of interfacial regions in amorphous glassy polymers, treated as natural nanocomposites, was shown. This means that such nanocomposites structure represents a nanofiller (nanoclusters), immersed in matrix (loosely packed matrix of amorphous polymer structure), that is, unlike polymer nanocomposites with inorganic nanofiller (artificial nanocomposites), they have only two structural components.

1.1.2 THE NATURAL NANOCOMPOSITES REINFORCEMENT

As it is well-known [26], very often a filler introduction in polymer matrix is carried out for the last stiffness enhancement. Therefore the reinforcement degree of polymer composites, defined as a composite and matrix polymer elasticity moduli ratio, is one of their most important characteristics.

Amorphous glassy polymers as natural nanocomposites treatment, the estimation of filling degree or nanoclusters relative fraction φ_{cl} has an important significance. Therefore the authors [27] carried out the comparison of the indicated parameter estimation different methods, one of which is Electron paramagnetic resonance EPR-spectroscopy (the method of spin probes). The indicated method allows to study amorphous polymer structural heterogeneity, using radicals distribution character. As it is known [28], the method, based on the parameter d_1/d_c—the ratio of spectrum extreme components total intensity to central component intensity-measurement (this is the simplest and most suitable method of nitroxil radicals local concentrations determination). The value of dipole-dipole interaction ΔH_{dd} is directly proportional to spin probes concentration C_w [29]:

$$\Delta H_{dd} = A \times C_w, \tag{1.5}$$

where $A = 5 \times 10^{-20}$ Ersted×cm^3 in the case of radicals chaotic distribution.

On the basis of Eq. (1.5) the relationship was obtained, which allows to calculate the average distance r between two paramagnetic probes [29]:

$$r = 38(\Delta H_{dd})^{-1/3}, \text{Å} \tag{1.6}$$

where ΔH_{dd} is given in Ersteds.

In Figure 1.9, the dependence of d_1/d_c on mean distance r between chaotically distributed and amorphous PC radicals-probes is adduced. For PC at $T = 77$K, the values of $d_1/d_c = 0{,}38$–$0{,}40$ were obtained. One can make an assumption about vol-

A Detailed Review on Characteristics, Application   13

ume fractions relation for the ordered domains (nanoclusters) and loosely packed matrix of amorphous PC. The indicated value d_1/d_c means that in PC, probes statistical distribution 0,40 of its volume is accessible for radicals and approximately 0,60 of volume remains unoccupied by spin probes (i.e., the nanoclusters relative fraction φ_{cl} according to the EPR method makes up approximately 0,60–0,62).

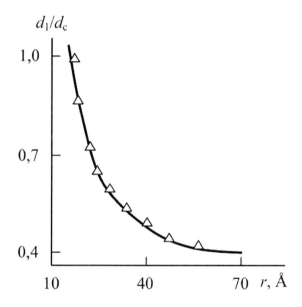

FIGURE 1.9 The dependence of parameter d_1/d_c of EPR spectrum on the value of mean distance r between radicals for PC [27].

This result corresponds well to the experimental data of Yech [30] and Perepechko [31], who obtained the values 0,60 and 0,63 for densely packed regions relative fraction in amorphous polymers.

The authors of paper [11] fulfilled φ_{cl} estimation with the aid of reversed gas chromatography and obtained the following magnitudes of this parameter for PC, poly (methyl methacrylate), and polysulfone: 0,70; 0,60; and 0,65, accordingly.

Within the frameworks of the cluster model φ_{cl}, estimation can be fulfilled by the percolation relationship (in previous paper) usage. Let us note that in the given case, the temperature of polymers structure quasi-equilibrium state attainment, lower of which φ_{cl} value does not change (i.e., T_0) [32], is accepted as testing temperature T. The calculation φ_{cl} results according to the equation (in previous paper) for the mentioned above polymers. Proceeding from the circumstance, that radicals-probes are concentrated mainly in intercluster regions, the nanocluster size can be estimated, which in amorphous PC should be approximately equal to mean distance r between two paramagnetic probes (i.e., ~50 Å) (Figure 1.9). This value corresponds well to the experimental data, obtained by dark-field electron microscopy method (30–100 Å) [33].

Within the frameworks of the cluster model, the distance between two neighboring nanoclusters can be estimated according to the equation (in previous paper) as $2R_{cl}$. The estimation $2R_{cl}$ by this mode gives the value 53,1 Å (at F = 41), that corresponds excellently to the method EPR data.

Thus, the paper [27] results showed, that the obtained by EPR method natural nanocomposites (amorphous glassy polymers) structure characteristics corresponded completely to both the cluster model theoretical calculations and other authors estimations. In other words, EPR data are experimental confirmation of the cluster model of polymers amorphous state structure.

The treatment of amorphous glassy polymers as natural nanocomposites allows to use for their elasticity modulus E_p (and hence, the reinforcement degree $E_p/E_{l.m.}$, where $E_{l.m.}$ is loosely packed matrix elasticity modulus) description theories, developed for polymer composites reinforcement degree description [9, 17]. The authors [34] showed correctness of particulate-filled polymer nanocomposites reinforcement of two concepts on the example of amorphous PC. For theoretical estimation of particulate-filled polymer nanocomposites reinforcement degree E_n/E_m two equations can be used. The first of them looks like this [35]:

$$\frac{E_n}{E_m} = 1 + \phi_n^{1.7}, \quad (1.7)$$

where E_n and E_m are elasticity moduli of nanocomposites and matrix polymer, accordingly, and φ_n is nanofiller volume contents.

The second equation offered by the authors of paper [36] is:

$$\frac{E_n}{E_m} = 1 + \frac{0,19 W_n l_{st}}{D_p^{1/2}}, \quad (1.8)$$

where W_n is nanofiller mass contents in mas .% and D_p is nanofiller particles diameter in nm.

Let us consider Eqs. (1.7) and (1.8) parameters estimation methods. It is obvious that in the case of natural nanocomposites, one should accept $E_n = E_p$, $E_m = E_{l.m.}$, and $\varphi_n = \varphi_{cl}$, the value of the latter can be estimated according to the equation (in previous paper).

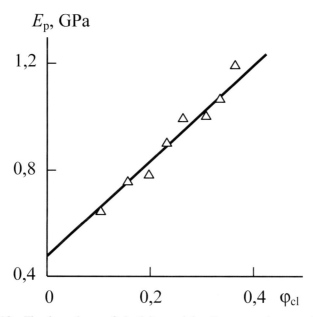

FIGURE 1.10 The dependence of elasticity modulus E_p on nanoclusters relative fraction φ_{cl} for PC [34].

The mass fraction of nanoclusters W_{cl} can be calculated as follows [37]:

$$W_{cl} = \rho \varphi_{cl}, \qquad (1.9)$$

where ρ is nanofiller (nanoclusters) density which is equal to 1,300 kg/m³ for PC.

The value of $E_{l.m.}$ can be determined by the construction of $E_p(\varphi_{cl})$ plotting, which is adduced in Figure 1.10. As one can see, this plot is approximately linear and its extrapolation to $\varphi_{cl} = 0$ gives the value $E_{l.m.}$. And at last, as it follows from the nanoclusters definition, one should accept $D_p \approx l_{st}$ for them and then the Eq. (1.8) accepts the following look [34]:

In Figure 1.11 the comparison of theoretical calculation according to the Eqs. (1.7) and (1.10) with experimental values of reinforcement degree $E_p/E_{l.m.}$ for PC is adduced. As one can see, both indicated equations give a good enough correspondence with the experiment: their average discrepancy makes up 5,6 percent in the Eq. (1.7) case and 9,6 percent for the Eq. (1.10). In other words, in both cases the average discrepancy does not exceed an experimental error for mechanical tests. This means, that both considered methods can be used for PC elasticity modulus prediction. Besides, it is necessary to note that the percolation relationship (1.7) qualitatively describes the dependence $E_p/E_{l.m.}(\varphi_{cl})$ better than the empirical Eq. (1.10).

$$\frac{E_n}{E_m} = 1 + 0{,}19\rho\phi_{cl}l_{st}^{1/2}. \tag{1.10}$$

The obtained results allowed to make another important conclusion. As it is known, the percolation relationship (1.7) assumes, that nanofiller is percolation system (polymer composite) solid-body component and in virtue of this circumstance defines this system elasticity modulus. However, for artificial polymer particulate-filled nanocomposites, consisting of polymer matrix and inorganic nanofiller, Eq. (1.7) in the cited form gives the understated values of reinforcement degree. The authors [9, 17] showed that for such nanocomposites the sum ($\varphi_n+\varphi_{if}$), where φ_{if} was interfacial regions relative fraction, was a solid-body component. The correspondence of experimental data and calculation according to the Eq. (1.7) demonstrates that amorphous polymer is the specific nanocomposite, in which interfacial regions are absent [24, 25]. This important circumstance is necessary to take into consideration at amorphous glassy polymers structure and properties description while simulating them as natural nanocomposites. Besides, one should note that unlike micromechanical models the Eqs. (1.7) and (1.10) do not take into account nanofiller elasticity modulus, which is substantially differed for PC nanoclusters and inorganic nanofillers [34].

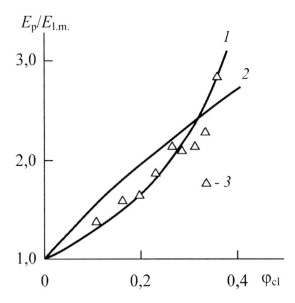

FIGURE 1.11 The dependences of reinforcement degree $E_p/E_{l.m}$ on nanoclusters relative fraction φ_{cl} for PC. 1—calculation according to the Eq. (1.7); 2—calculation according to the Eq. (1.10); 3—the experimental data [34].

ns application is micromechanical models application, developed for polymer composites mechanical behavior description [1, 37–39]. So, Takayanagi and Kerner models are often used for the description of reinforcement degree on composition for the indicated materials [38, 39]. The authors [40] used the mentioned models for theoretical treatment of natural nanocomposites reinforcement degree temperature dependence on the example of PC.

Takayanagi model belongs to a micromechanical composite models group, allowing empirical description of composite response upon mechanical influence on the basis of its constituent elements properties. One of the possible expressions within the frameworks of this model has the following look [38]:

$$\frac{G_c}{G_m} = \frac{\phi_m G_m + (\alpha + \phi_f)G_f}{(1+\alpha\phi_f)G_m + \alpha\phi_m G_f},\qquad(1.11)$$

where G_c, G_m and G_f are shear moduli of composite, polymer matrix and filler, accordingly, φ_m and φ_f are polymer matrix and filler relative fractions, respectively, α is a fitted parameter.

Kerner equation is identical to the formula (1.11), but for it the parameter α does not fit and has the following analytical expression [38]:

$$\alpha_m = \frac{2(4-5v_m)}{(7-5v_m)},\qquad(1.12)$$

where α_m and v_m are parameter α and Poisson's ratio for polymer matrix.

Let us consider the determination methods of the Eqs. (1.11) and (1.12) parameters, which are necessary for the indicated equations application in the case of natural nanocomposites, Firstly, it is obvious that in the last case one should accept $G_c = G_p$, $G_m = G_{l.m.}$, and $G_f = G_{cl}$, where G_p, $G_{l.m.}$ and G_{cl} are shear moduli of polymer, loosely packed matrix and nanoclusters, accordingly, and also $\varphi_f = \varphi_{cl}$, where φ_{cl} is determined according to the percolation relationship (in previous paper). Young's modulus for loosely packed matrix and nanoclusters can be received from the data of Figure 1.10 by the dependence $E_p(\varphi_{cl})$ extrapolation to $\varphi_{cl} = 1,0$, respectively. The corresponding shear moduli were calculated according to the general equation (in previous paper). The value of nanoclusters fractal dimension d_f^{cl} in virtue of their dense package is accepted equal to the greatest dimension for real solids (d_f^{cl} = 2,95 [40]) and loosely packed matrix fractal dimension $d_f^{l.m.}$ can be estimated.

However, the calculation according to the Eqs. (1.11) and (1.12) does not give a good correspondence to the experiment, especially for the temperature range of T = 373–413 K in PC case. As it is known [38], in empirical modifications of Kerner equation it is usually supposed, that nominal concentration scale differs from mechanically effective filler fraction ϕ_f^{ef}, which can be written accounting for the designations used above for natural nanocomposites as follows [41].

$$\phi_f^{ef} = \frac{(G_p - G_{l.m.})(G_{l.m.} + \alpha_{l.m.}G_{cl})}{(G_{cl} - G_{l.m.})(G_{l.m.} + \alpha_{l.m.}G_p)}, \qquad (1.13)$$

where $\alpha_{l.m.} = \alpha_m$. The value $\alpha_{l.m.}$ can be determined according to the Eq. (1.12), estimating Poisson's ratio of loosely packed matrix $\nu_{l.m.}$ by the known values $d_f^{l.m.}$ according to the equation (in previous paper).

Besides, one more empirical modification ϕ_f^{ef} exists, which can be written as follows [41]:

$$\phi_{cl_2}^{ef} = \phi_{cl} + c\left(\frac{\phi_{cl}}{2r_{cl}}\right)^{2/3}, \qquad (1.14)$$

where c is empirical coefficient of order one and r_{cl} is nanocluster radius, determined according to the equation (in previous paper).

At the value $\phi_{cl_2}^{ef}$ calculation according to the Eq. (1.14), magnitude c was accepted equal to 1,0 for the temperature range of $T = 293$–363 K and equal to 1,2—for the range of $T = 373$–413 K and $2r_{cl}$ is given in nm. In Figure 1.12 the comparison of values ϕ_{cl}^{ef}, calculated according to the Eqs. (1.13) and (1.14) ($\phi_{cl_1}^{ef}$ and $\phi_{cl_2}^{ef}$, accordingly), is adduced. As one can see, a good enough conformity of the values ϕ_{cl}^{ef}, estimated by both methods, is obtained (the average discrepancy of $\phi_{cl_1}^{ef}$ and $\phi_{cl_2}^{ef}$ makes up slightly larger than 20 percent). Let us note, that the effective value φ_{cl} exceeds essentially the nominal one, determined according to the relationship (in previous paper): within the range of $T = 293$–363K by about 70 percent and within the range of $T = 373$–413K—almost in three times.

In Figure 1.13 the comparison of experimental and calculated according to Kerner equation (using equations(1.11)),, (1.13) and (1.14) i in which t reinforcement degree is obtained by shear modulus $G_p/G_{l.m.}$ as a function of testing temperature T for PC.) As one can see in this case, at the usage of nanoclusters effective concentration scale (ϕ_{cl}^{ef} instead of φ_{cl}), the good conformity of theory and experiment is obtained (their average discrepancy makes up 6 percent).

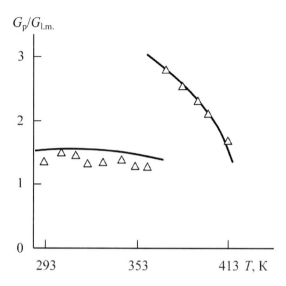

FIGURE 1.12 The comparison of nanoclusters effective concentration scale $\varphi_{cl_1}^{ef}$ and $\varphi_{cl_2}^{ef}$, calculated according to the Eqs. (1.13) and (1.14), respectively, for PC. A straight line shows the relation 1:1 [41].

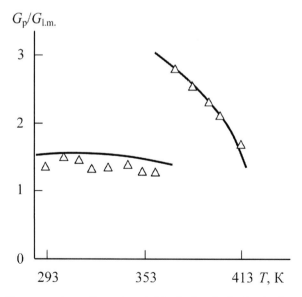

FIGURE 1.13 The comparison of experimental (*points*) and calculated values according to the Eqs. (1.11), (1.13), and (1.14) (*solid lines*)

Hence, the above stated results have shown the modified Kerner equation application correctness for natural nanocomposites elastic response description. Really this fact by itself confirms the possibility of amorphous glassy polymers treatment as nanocomposites. Microcomposite models usage gives the clear notion about factors influencing polymers' stiffness.

1.1.3 INTERCOMPONENT ADHESION IN NATURAL NANOCOPOSITES

Amorphous glassy polymers as natural nanocomposites puts forward to the foreground their study intercomponent interactions (i.e., interactions nanoclusters-loosely-packed matrix). This problem plays always one of the main roles at multiphase (multicomponent) systems consideration, since the indicated interactions or interfacial adhesion level defines to a great extent such systems properties [42]. Therefore the authors [43] studied the physical principles of intercomponent adhesion for natural nanocomposites on the example of PC.

The authors [44] considered three main cases of the dependence of reinforcement degree E_c/E_m on φ_f. In this work, the authors have shown that there are the following main types of the dependences $E_c/E_m(\varphi_f)$ exist:

1. The ideal adhesion between filler and polymer matrix, described by Kerner equation (perfect adhesion), which can be approximated by the following relationship:

$$\frac{E_c}{E_m} = 1 + 11,64\phi_f - 44,4\phi_f^2 + 96,3\phi_f^3 ; \qquad (1.15)$$

2. Zero adhesional strength at a large friction coefficient between filler and polymer matrix, which is described by the equation:

$$\frac{E_c}{E_m} = 1 + \phi_f ; \qquad (1.16)$$

3. The complete absence of interaction and ideal slippage between filler and polymer matrix, when composite elasticity modulus is defined practically by polymer cross-section and connected with the filling degree by the equation:

$$\frac{E_c}{E_m} = 1 - \phi_f^{2/3} . \qquad (1.17)$$

In Figure 1.14 the theoretical dependences $E_p/E_{l.m.}(\varphi_{cl})$ plotted according to the Eqs. (1.15) ÷ (1.17), as well as experimental data (points) for PC are shown. As it follows from the adduced in Figure 1.14 comparison at $T = 293 ÷ 363$ K the experimental data correspond well to Eq. (1.16) (i.e., in this case zero adhesional strength at a large friction coefficient is observed). At $T = 373 ÷ 413$ K the experimental data correspond to the Eq. (1.15) (i.e., the perfect adhesion between nanoclusters and loosely packed matrix is observed). Thus, the adduced in Figure 1.14 data demonstrated that depending on testing temperature, two types of interactions nanoclusters-loosely-packed matrix are observed: either perfect adhesion or large friction between them. For quantitative estimation of these interactions, it is necessary to determine their level, which can be made with the help of the parameter b_m, which is determined according to the equation [45]:

$$\sigma_f^c = \sigma_f^m K_s - b_m \phi_f \qquad (1.18)$$

where σ_f^c and σ_f^m are fracture stress of composite and polymer matrix, respectively, and K_s is stress concentration coefficient. It is obvious that since b_m increase results to σ_f^c reduction, then this means interfacial adhesion level decrease.

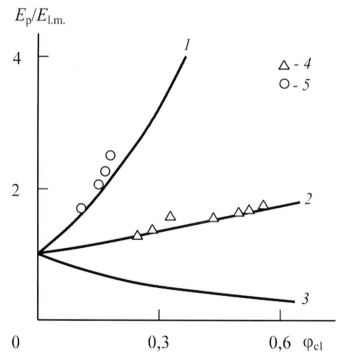

FIGURE 1.14 The dependences of reinforcement degree $E_p/E_{l.m.}$ on nanoclusters relative fraction ϕ_{cl}. 1–3—the theoretical dependences, corresponding to the Eqs. (1.15) (1.17), accordingly; 4, 5—the experimental data for PC within the temperature ranges: 293–363K(4) and 373–413K(5) [43].

The true fracture stress σ_f^{tr} for PC, taking into account sample cross-section change in a deformation process, was used as σ_f^c for natural nanocomposites, which can be determined according to the known formula:

$$\sigma_f^{tr} = \sigma_f^n (1+\varepsilon_f), \qquad (1.19)$$

where σ_f^n is nominal (engineering) fracture stress, ε_f is strain at fracture.

The value σ_f^m, which is accepted equal to loosely packed matrix strength $\sigma_f^{l.m.}$, was determined by graphic method, namely, by the dependence $\sigma_f^{tr}(\varphi_{cl})$ plotting, which proves to be linear, and by subsequent extrapolation of it to $\varphi_{cl} = 0$, that gives $\sigma_f^{l.m.} = 40$ MPa [43].

And at last, the value Ks can be determined with the help of the following Eq. [39]:

$$\sigma_f^{tr} = \sigma_f^{l.m.}\left(1 - \phi_{cl}^{2/3}\right) K_s. \qquad (1.20)$$

The parameter b_m calculation according to the above stated technique shows its decrease (intercomponent adhesion level enhancement) at testing temperature raising within the range of $b_m \approx 500 \div 130$.

For interactions nanoclusters-loosely-packed matrix estimation within the range of $T = 293 \div 373$K, the authors [48] used the model of Witten-Sander clusters friction, stated in paper [46]. This model application is due to the circumstance, that amorphous glassy polymer structure can be presented as an indicated clusters large number set [47]. According to this model, Witten-Sander clusters generalized friction coefficient t can be written as follows [46]:

$$f = \ln c + \beta \times \ln n_{cl}, \qquad (1.21)$$

where c is constant, β is coefficient, n_{cl} is statistical segments number per one nanocluster.

The coefficient β value is determined as follows [46]:

$$\beta = \left(d_f^{cl}\right)^{-1}, \qquad (1.22)$$

where d_f^{cl} is nanocluster structure fractal dimension, which is equal, as before, to 2,95 [40].

In Figure 1.15 the dependence $b_m(f)$ is adduced, which is broken down into two parts. On the first of them, corresponding to the range of $T = 293 \div 363$ K, the intercomponent interaction level is intensified at f decreasing (i.e., b_m reduction is observed and on the second one, corresponding to the range of $T = 373$–413 K, b_m = const independent on value f). These results correspond completely to the data of Figure 1.14, where in the first from the indicated temperature ranges the value

A Detailed Review on Characteristics, Application

$E_p/E_{l.m.}$ is defined by nanoclusters friction and in the second one by adhesion and, hence, it does not depend on friction coefficient.

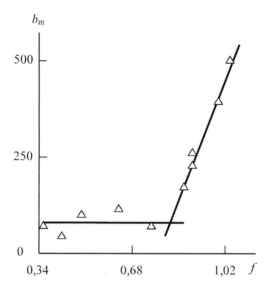

FIGURE 1.15 The dependence of parameter b_m on generalized friction coefficient f for PC [43].

As it has been shown in paper [48], the interfacial (or intercomponent) adhesion level depends on a number of accessible for the formation interfacial (intercomponent) bond sites (nodes) on the filler (nanocluster) particle surface N_u, which is determined as follows [49]:

$$N_u = L^{d_u}, \qquad (1.23)$$

where L is filler particle size, d_u is fractal dimension of accessible for contact ("nonscreened") indicated particle surface.

One should choose the nanocluster characteristic size as L for the natural nanocomposite which is equal to statistical segment l_{st}, determined according to the equation (in previous paper), and the dimension d_u is determined according to the following relationship [49]:

$$d_u = (d_{surf} - 1) + \left(\frac{d - d_{surf}}{d_w}\right), \qquad (1.24)$$

where d_{surf} is nanocluster surface fractal dimension and d_w is dimension of random walk on this surface, estimated according to Aarony-Stauffer rule [49]:

$$d_w = d_{surf} + 1. \quad (1.25)$$

The following technique was used for the dimension d_{surf} calculation. First the nanocluster diameter $D_{cl} = 2r_{cl}$ was determined according to the equation (in previous paper) and then its specific surface S_u was estimated [35]:

$$S_u = \frac{6}{\rho_{cl} D_{cl}}, \quad (1.26)$$

where ρ_{cl} is the nanocluster density, equal to 1,300 kg/m³ in the PC case.

And at last, the dimension d_{surf} was calculated with the help of the equation [20]:

$$S_u = 5,25 \times 10^3 \left(\frac{D_{cl}}{2} \right)^{d_{surf} - d} \quad (1.27)$$

In Figure 1.16 the dependence $b_m(N_u)$ for PC is adduced, which is broken down into two parts similarly to the dependence $b_m(f)$ (Figure 1.15). At $T = 293–363$ K, the value b_m is independent on N_u, since nanocluster-loosely-packed matrix interactions are defined by their friction coefficient. Within the range of $T = 373 \div 413$ K, intercomponent adhesion level enhancement (b_m reduction) at active sites number N_u growth is observed, as was to be expected. Thus, the data of both Figures 1.16 and 1.15 correspond to Figure 1.14 results.

With regard to the data of Figures 1.15 and 1.16, two remarks should be made. Firstly, the transition from one reinforcement mechanism to another corresponds to loosely packed matrix glass transition temperature, which is approximately equal to $T_g - 50$K [11]. Secondly, the extrapolation of Figure 1.16 plot to $b_m = 0$ gives the value $N_u \approx 71$, that corresponds approximately to polymer structure dimension $d_f = 2,86$.

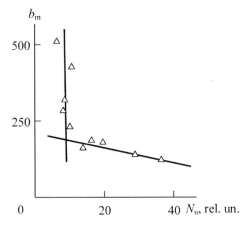

FIGURE 1.16 The dependence of parameter b_m on nanocluster surface active ("nonscreened") sites number N_u for PC [43].

A Detailed Review on Characteristics, Application

In this theme, completion of an interesting structural aspect of intercomponent adhesion in natural nanocomposites (polymers) should be noted. Despite the considered above different mechanisms of reinforcement and nanoclusters-loosely-packed matrix interaction realization, the common dependence $b_m(\varphi_{cl})$ is obtained for the entire studied temperature range of 293–413K, which is shown in Figure 1.17. This dependence is linear, that allows to determine the limiting values $b_m \approx 970$ at $\varphi_{cl} = 1,0$ and $b_m = 0$ at $\varphi_{cl} = 0$. Besides, let us note that the shown in Figures 1.14, 1.15, and 1.16, structural transition is realized at φ_{cl} 0,26 [43].

Hence, the above stated results have demonstrated that intercomponent adhesion level in natural nanocomposites (polymers) has structural origin and is defined by nanoclusters relative fraction. In two temperature ranges two different reinforcement mechanisms are realized, which are due to large friction between nanoclusters and loosely packed matrix and also perfect (by Kerner) adhesion between them. These mechanisms can be described successfully within the frameworks of fractal analysis.

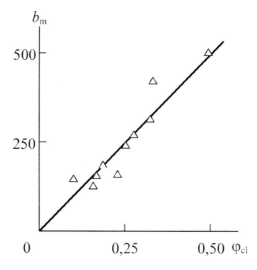

FIGURE 1.17 The dependence of parameter b_m on nanoclusters relative fraction φ_{cl} for PC [43].

The further study of intercomponent adhesion in natural nanocomposites was fulfilled in paper [50]. In Figure 1.18 the dependence $b_m(T)$ for PC is shown, from which b_m reduction or intercomponent adhesion level enhancement at testing temperature growth follows. In the same figure the maximum value b_m for nanocomposites polypropylene/Na[+]-montmorillonite [9] was shown by a horizontal shaded line. As one can see, b_m values for PC within the temperature range of $T = 373$–413 K by absolute value are close to the corresponding parameter for the indicated nanocomposite, that indicates high enough intercomponent adhesion level for PC within this temperature range.

Let us note an important structural aspect of the dependence $b_m(T)$, shown in Figure 1.18. According to the cluster model [4], the decay of instable nanoclusters occurs at temperature $T'_g \approx T_g - 50$ K, holding back loosely packed matrix in glassy state, owing to which this structural component is devitrificated within the temperature range of $T'_g - T_g$. Such effect results to rapid reduction of polymer mechanical properties within the indicated temperature range [51]. As it follows from the data of Figure 1.18, precisely in this temperature range the highest intercomponent adhesion level is observed and its value approaches to the corresponding characteristic for nanocomposites polypropylene/Na$^+$-montmorillonite.

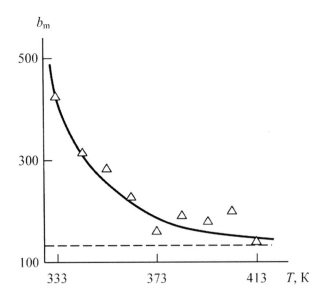

FIGURE 1.18 The dependence of parameter b_m on testing temperature T for PC. The horizontal shaded line shows the maximum value b_m for nanocomposites polypropylene/Na$^+$-montmorillonite [50].

A Detailed Review on Characteristics, Application

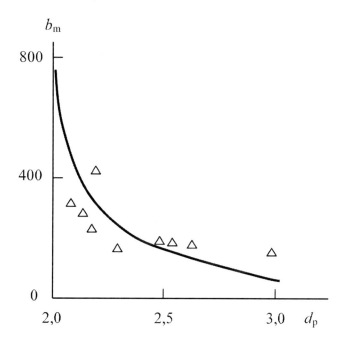

FIGURE 1.19 The dependence of parameter b_m on nanoclusters surface fractal dimension d_{surf} for PC [50].

It can be supposed with a high probability degree that adhesion level depends on the structure of nanoclusters surface, coming into contact with loosely packed matrix, which is characterized by the dimension d_{surf}. In Figure 1.19 the dependence $b_m(d_{surf})$ for PC is adduced, from which rapid reduction b_m (or intercomponent adhesion level enhancement) follows at d_{surf} growth or, roughly speaking, at nanoclusters surface roughness enhancement.

The authors [48] showed that the interfacial adhesion level for composites polyhydroxyether/graphite was raised at the decrease of polymer matrix and filler particles surface fractal dimensions difference. The similar approach was used by the authors of paper [50], who calculated nanoclusters d_f^{cl} and loosely packed matrix $d_f^{l.m.}$ fractal dimensions difference Δd_f:

$$\Delta d_f = d_f^{cl} - d_f^{l.m.}, \tag{1.28}$$

where d_f^{cl} is accepted equal to real solids maximum dimension ($d_f^{cl} = 2.95$ [40]) in virtue of their dense packing and the value $d_f^{l.m.}$ was calculated according to the mixtures rule (the equation from previous paper).

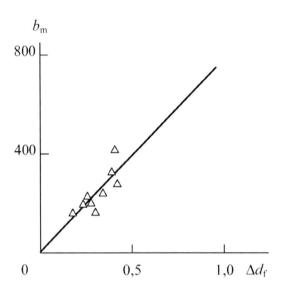

FIGURE 1.20 The dependence of parameter b_m on nanoclusters d_f^{cl} and loosely packed matrix $d_f^{l.m.}$ structures fractal dimensions difference Δd_f for PC [50].

In Figure 1.20 the dependence of b_m on the difference Δd_f is adduced, from which b_m decrease or intercomponent adhesion level enhancement at Δd_f reduction or values d_f^{cl} and $d_f^{l.m.}$ growing similarity follows. This dependence demonstrates, that the greatest intercomponent adhesion level, corresponding to $b_m = 0$, is reached at $\Delta d_f = 0,95$ and is equal to ~780.

The data of Figures 1.14 and 1.18 combination shows, that the value $b_m \approx 200$ corresponds to perfect adhesion by Kerner. In its turn, the Figures 1.16 and 1.17 plots data demonstrated, that the value $b_m \approx 200$ could be obtained either at $d_{surf} > 2,5$ or at $\Delta d_f < 0,3$, accordingly. The obtained earlier results showed [24], that the condition $d_{surf} > 2,5$ was reached at $r_{cl} < 7,5\text{Å}$ or $T > 373$ K, that again corresponded well to the above stated results. And at last, the $\Delta d_f \approx 0,3$ or $d_f^{l.m.} \approx 2,65$ according to the equation (in previous paper) was also obtained at $T \approx 373$K.

Hence, at the indicated above conditions, fulfillment within the temperature range of $T < T_g'$ for PC perfect intercomponent adhesion can be obtained, corresponding to Kerner equation, and then the value E_p estimation should be carried out according to the Eq. (1.15). At $T = 293$ K ($\varphi_{cl} = 0,56$, $E_m = 0,85$GPa) the value E_p will be equal to 8,9 GPa, that approximately in six times larger, than the value E_p for serial industrial PC brands at the indicated temperature.

Let us note the practically important feature of the above obtained results. As it was shown, the perfect intercomponent adhesion corresponds to $b_m \approx 200$, but not

$b_m = 0$. This means that the real adhesion in natural nanocomposites can be higher than the perfect one by Kerner, that was shown experimentally on the example of particulate-filled polymer nanocomposites [17, 52]. This effect was named as nano-adhesion and its realization gives large possibilities for elasticity modulus increase of both natural and artificial nanocomposites. So, the introduction in aromatic polyamide (phenylone) of 0,3 mas.% aerosil only at nano-adhesion availability gives the same nanocomposite elasticity modulus enhancement effect, as the introduction of 3 mas. % of organoclay, which at present is assumed as one of the most effective nanofillers [9]. This assumes, that the value $E_p = 8,9$ GPa for PC is not a limiting one, at any rate, theoretically. Let us note in addition that the indicated E_p values can be obtained at the natural nanocomposites nanofiller (nanoclusters) elasticity modulus magnitude $E_{cl} = 2,0$ GPa (i.e., at the condition $E_{cl} < E_p$). Such result possibility follows from the polymer composites structure fractal concept [53], namely, the model [44], in which the Eqs. (1.15), (1.16), and (1.17) do not contain nanofiller elasticity modulus, and reinforcement percolation model [35].

The condition $d_{surf} < 2,5$ (i.e., $r_{cl} < 7,5$ Å or $N_{cl} < 5$), in practice, can be realized by using the nanosystems mechanosynthesis principles, the grounds of which are stated in paper [54]. However, another more simple and, hence, more technological method of desirable structure attainment realization is possible, one which will be considered in subsequent section.

Hence, the above stated results demonstrated that the adhesion level between natural nanocomposite structural components depended on nanoclusters and loosely packed matrix structures closeness. This level change can result in polymer elasticity modulus significant increase. A number of this effect practical realization methods was considered [50].

The above mentioned dependence of intercomponent adhesion level on nanoclusters radius r_{cl} assumes more general dependence of parameter b_m on nanoclusters geometry. The authors [55] carried out calculation of accessible for contact sites of nanoclusters surface and loosely packed matrix number N_u according to the relationship (1.23) for two cases. the nanocluster is simulated as a cylinder with diameter D_{cl} and length l_{st}, where l_{st} is statistical segment length, therefore, in the first case its butt-end is contacting with loosely packed matrix nanocluster surface and then $L = D_{cl}$ and in the second case with its side (cylindrical) surface and then $L = l_{st}$. In Figure 1.21 the dependences of parameter b_m on value N_u, corresponding to the two considered above cases, are adduced. As one can see, in both cases, for the range of $T = 293-363$ K l_{st}, where interactions between the nanoclusters-loosely-packed matrix are characterized by powerful friction , the value b_m does not depend on N_u, as it was expected. For the range of $T = 373-413$ K, where between nanoclusters and loosely packed matrix perfect adhesion is observed, the linear dependences $b_m(N_u)$ are obtained. However, at using value D_{cl} as Lb_m reduction or intercomponent adhesion level enhancement at N_u decreasing is obtained and at $N_u = 0$ b_m value reaches its minimum magnitude $b_m = 0$. In other words, in this case the minimum level of intercomponent adhesion is reached at intercomponent bonds formation sites (nodes) absence that is physically incorrect [48]. And on the contrary at the condition $L = l_{st}b_m$ the reduction (intercomponent adhesion level enhancement) at the increase of contacts number N_u between nanoclusters and loosely packed matrix is observed, that is obvious from the physical point of view. Thus, the data of Figure 1.21 indicate unequivocally, that the intercomponent adhesion is realized over side (cylindrical) nanoclusters surface and butt-end surfaces in this effect formation do not participate.

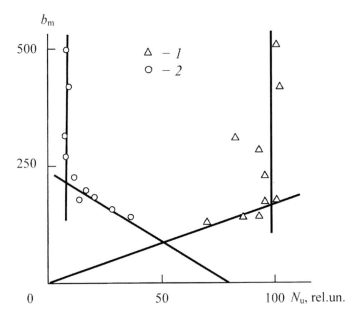

FIGURE 1.21 The dependences of parameter b_m on a number of accessible for intercomponent bonds formation sizes on nanocluster surface N_u at the condition $L = D_{cl}$ (1) and $L = l_{st}$ (2) for PC [55].

Let us consider geometrical aspects intercomponent interactions in natural nanocomposites. In Figure 1.22 the dependence of nanoclusters butt-end S_b and side (cylindrical) S_c surfaces areas on testing temperature T for PC are adduced. As one can see, the following criterion corresponds to the transition from strong friction to perfect adhesion at $T = 373$K [55]:

$$S_b \approx S_c. \tag{1.29}$$

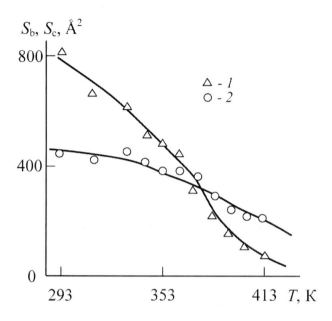

FIGURE 1.22 The dependences of nanoclusters butt-end S_b(1) and cylindrical S_c(2) surfaces areas on testing temperature T for PC [55].

Hence, the intercomponent interaction type transition from the large friction nanoclusters-loosely-packed matrix to the perfect adhesion between them is defined by nanoclusters geometry: at $S_b > S_c$, the interactions of the first type is realized and at $S_b < S_c$, the second one. Proceeding from this, it is expected that intercomponent interactions level is defined by the ratio S_b/S_c. Actually, the adduced in Figure 1.23 data demonstrate b_m reduction at the indicated ratio decrease, but at the criterion (29), realization of $S_b/S_c \approx 1$ S_b/S_c Sb/Sc decreasing does not result to b_m reduction and at $S_b/S_c < 1$ intercomponent adhesion level remains maximum high and constant [55].

Hence, the above stated results have demonstrated that interactions nanoclusters-loosely-packed matrix type (large friction or perfect adhesion) is defined by nanoclusters butt-end and side (cylindrical) surfaces areas ratio or their geometry that if the first from the mentioned areas is larger than the second one, then a large friction nanoclusters-loosely-packed matrix is realized; if the second one exceeds the first one, then between the indicated structural components perfect adhesion is realized. In the second from the indicated cases intercomponent adhesion level does not depend on the mentioned areas ratio and remains maximum high and constant. In other words, the adhesion nanoclusters-loosely-packed matrix is realized by nanoclusters cylindrical surface.

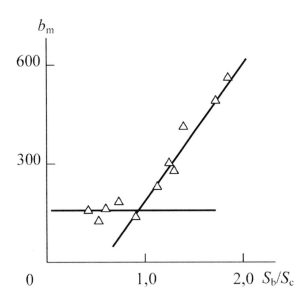

FIGURE 1.23 The dependence of parameter b_m on nanoclusters butt-end and cylindrical surfaces are ratio S_b/S_c value for PC [55].

The above stated results were experimentally confirmed by the EPR-spectroscopy method [56]. The Eqs. (1.1) and (1.6) comparison shows that dipole-dipole interaction energy ΔH_{dd} has structural origin, namely [56]:

$$\Delta H_{dd} \approx \left(\frac{v_{cl}}{n_{cl}}\right). \qquad (1.30)$$

As estimations according to the Eq. (1.30) showed, within the temperature range of $T = 293$ 413K for PC ΔH_{dd} increasing from 0,118 up to 0,328 Ersteds was observed.

Let us consider dipole-dipole interaction energy ΔH_{dd} intercommunication with nanoclusters geometry. In Figure 1.24 the dependence of ΔH_{dd} on the ratio S_c/S_b for PC is adduced. As one can see, the linear growth ΔH_{dd} at ratio S_c/S_b increasing is observed (i.e., either at S_c enhancement or at S_b reduction). Such character of the adduced in Figure 1.24 dependence indicates unequivocally that the contact nanoclusters-loosely-packed matrix is realized on nanocluster cylindrical surface. Such effect was to be expected, since emerging from the butt-end surface statistically distributed polymer chains complicated the indicated contact realization unlike relatively smooth cylindrical surfaces. It is natural to suppose that dipole-dipole interactions intensification or ΔH_{dd} increasing results in natural nanocomposites elasticity modulus E_p enhancement. The second is natural supposition at PC consideration as nanocomposite is the influence on the value E_p of nanoclusters (nanofiller) relative fraction φ_{cl}, which is determined according to the percolation relationship (in previous paper).

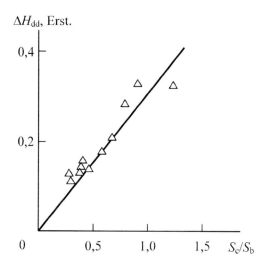

FIGURE 1.24 The dependence of dipole-dipole interaction energy ΔH_{dd} on nanoclusters cylindrical S_c and butt-end S_b surfaces areas ratio for PC [56].

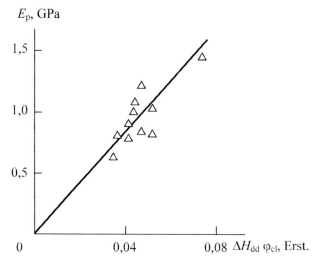

FIGURE 1.25 The dependence of elasticity modulus E_p on complex argument $(\Delta H_{dd}\varphi_{cl})$ for PC [56].

In Figure 1.25 the dependence of elasticity modulus E_p on complex argument $(\Delta H_{dd} \varphi_{cl})$ for PC is presented. As one can see, this dependence is a linear one, passes through coordinates origin, and is described analytically by the following empirical equation [56].

$$E_p = 21(\Delta H_{dd} \varphi_{cl}), \text{GPa}, \qquad (1.31)$$

which with the appreciation of the Eq. (1.30) can be rewritten as follows [56]:

$$E_p = 21 \times 10^{-26} \left(\frac{\phi_{cl} v_{cl}}{n_{cl}} \right), \text{GPa}. \qquad (1.32)$$

The Eq. (1.32) demonstrates clearly that the value E_p and, hence polymer reinforcement degree is a function of its structural characteristics, described within the frameworks of the cluster model [3–5]. Let us note that since parameters v_{cl} and φ_{cl} are a function of testing temperature, then the parameter n_{cl} is the most suitable factor for the value E_p regulation for practical purposes. In Figure 1.26 the dependence $E_p(n_{cl})$ for PC at $T = 293$ K is adduced, calculated according to the Eq. (1.32), where the values v_{cl} and φ_{cl} were calculated according to the equations (in previous paper). As one can see, at small n_{cl} (<10) the sharp growth E_p is observed and at the smallest possible value $n_{cl} = 2$ the magnitude $E_p \approx 13{,}5$ GPa. Since for PC $E_{l.m.} = 0{,}85$ GPa, then it gives the greatest reinforcement degree E_p/E_m 15,9. Let us note, that the greatest attainable reinforcement degree for artificial nanocomposites (polymers filled with inorganic nanofiller) cannot exceed 12 [9]. It is notable, that the shown in Figure 1.26 dependence $E_p(n_{cl})$ for PC is identical completely by dependence shape to the dependence of elasticity modulus of nanofiller particles diameter for elastomeric nanocomposites [57].

Hence, the above presented results have shown that elasticity modulus of amorphous glassy polycarbonate, considered as natural nanocomposite, are defined completely by its suprasegmental structure state. This state can be described quantitatively within the frameworks of the cluster model of polymers amorphous state structure and characterized by local order level. Natural nanocomposites reinforcement degree can essentially exceed analogous parameter for artificial nanocomposites [56].

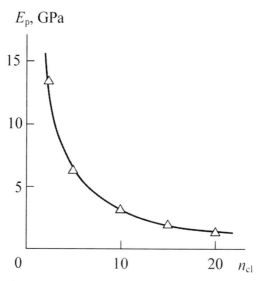

FIGURE 1.26 The dependence of elasticity modulus E_p on segments number n_{cl} per one nanocluster, calculated according to the Eq. (1.32) for PC at $T = 293K$ [56].

As it has been shown above (see the Eqs. (1.7) and (1.15)), the nanocluster relative fraction increasing results to polymers elasticity modulus enhancement, similar to nanofiller contents enhancement in artificial nanocomposites. Therefore the necessity of quantitative description and subsequent comparison of reinforcement degree for the two above indicated nanocomposites classes appears. The authors [58, 59] fulfilled the comparative analysis of reinforcement degree by nanoclusters and by layered silicate (organoclay) for polyarylate and nanocomposite epoxy polymer/ Na⁺-montmorillonite [60], accordingly.

In Figure 1.27 theoretical dependences of reinforcement degree E_n/E_m on nanofiller contents φ_n, calculated according to the Eqs. (1.15), (1.16), and (1.17), are adduced. Besides, in the same figure the experimental values (E_n/E_m) for nanocomposites epoxy polymer Na⁺-montmorillonite (EP/MMT) at $T < T_g$ and $T > T_g$ (where T and T_g are testing and glass transition temperatures, respectively) are indicated by points. As one can see, for glassy epoxy matrix the experimental data correspond to the Eq. (1.16){ (i.e., zero adhesional strength at a large friction coefficient and for devitrificated matrix)}—and for equation (1.15) (i.e., the perfect adhesion between nanofiller and polymer matrix), can be described by Kerner equation. Let us note that the authors [17] explained the distinction indicated above by a much larger length of epoxy polymer segment in the second case.

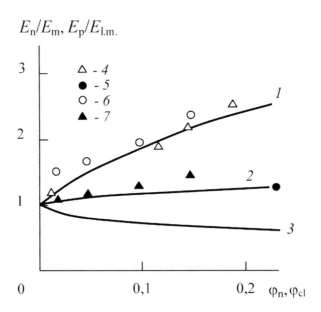

FIGURE 1.27 The dependences of reinforcement degree E_n/E_m and $E_p/E_{l.m.}$ on the contents of nanofiller φ_n and nanoclusters φ_{cl}, accordingly. 1–3—theoretical dependences (E_n/E_m) (φ_n), corresponding to the Eqs. (1.15), (1.16), and (1.17); 4,5—the experimental data $(E_p/E_{l.m.})$ for PAR at $T = T_g' - T_g$ (4) and $T < T_g'$ (5); 6, 7—the experimental data (E_n/E_m) (φ_n) for EP/MMT at $T > T_g$ (6) and $T < T_g$ (7) [59].

To obtain the similar comparison for natural nanocomposite (polymer) is impossible, since at $T \geq T_g$ nanoclusters are disintegrated and polymer ceases to be quasi-two-phase system [5]. However, within the frameworks of two-stage glass transition concept [11] it has been shown, that at temperature T_g', which is approximately equal to $T_g - 50$ K, instable (small) nanoclusters decay occurs, that results in loosely packed matrix devitrification at the indicated temperature [5]. Thus, within the range of temperature $T_g'-T_g$ natural nanocomposite (polymer) is an analog of nanocomposite with glassy matrix [58]. As one can see, for the temperatures within the range of $T = T_g' - T_g$ ($\varphi_{cl} = 0{,}06–0{,}19$), the value $E_p/E_{l.m.}$ corresponds to the Eq. (1.15), (i.e., perfect adhesion nanoclusters-loosely-packed matrix) and at $T < T_g'$ ($\varphi_{cl} > 0{,}24$), to the Eq. (1.16) (i.e., to zero adhesional strength at a large friction coefficient). Hence, the data of Figure 1.27 demonstrated clearly the complete similarity, both qualitative and quantitative, of natural (Par) and artificial (EP/MMT) nanocomposites reinforcement degree behavior. Another microcomposite model (e.g., accounting for the layered silicate particles strong anisotropy) application can change the picture quantitatively only. The data of Figure 1.27 qualitatively give the correspondence of reinforcement degree of nanocomposites indicated classes at the identical initial conditions.

Hence, the analogy in behavior of reinforcement degree of polyarylate by nanoclusters and nanocomposite epoxy polymer/Na⁺-montmorillonite by layered silicate gives another reason for the consideration of polymer as natural nanocomposite. Again strong influence of interfacial (intercomponent) adhesion level on nanocomposites of any class reinforcement degree is confirmed [17].

1.1.4 THE METHODS OF NATURAL NANOCOMPOSITES NANOSTRUCTURE REGULATION

As it has been noted above, at present it is generally acknowledged [2] that macromolecular formations and polymer systems are always natural nanostructural systems in virtue of their structure features. In this connection the question of using this feature for polymeric materials properties and operating characteristics improvement arises. It is obvious enough that for structure-properties relationships receiving the quantitative nanostructural model of the indicated materials is necessary. It is also obvious that if the dependence of specific property on material structure state is unequivocal, then there will be quite sufficient modes to achieve this state. The cluster model of such state [3–5] is the most suitable for polymers amorphous state structure description. It has been shown that this model basic structural element (cluster) is nanoparticles (nanocluster). The cluster model was used successfully for cross-linked polymers structure and properties description [61]. Therefore the authors [62] fulfilled nanostructures regulation modes and of the latter influence on rarely cross-linked epoxy polymer properties study within the frameworks of the indicated model.

In paper [62] the studied object was an epoxy polymer on the basis of resin UP5-181, cured by iso-methyltetrahydrophthalic anhydride in the ratio by mass 1:0,56. Testing specimens were obtained by the hydrostatic extrusion method. The indicated method choice is due to the fact that high hydrostatic pressure imposition in deformation process prevents the defects formation and growth, resulting in the material failure [63-64]. The extrusion strain ε_e was calculated and makes up 0,14, 0,25, 0,36, 0,43 and 0,52. The obtained by hydrostatic extrusion specimens were annealed at maximum temperature 353 K during 15 min.

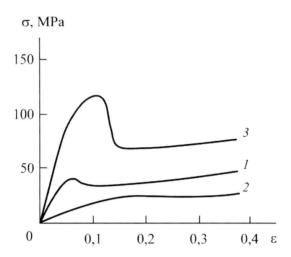

FIGURE 1.28 The stress—strain (σ – ε) diagrams for initial (1), extruded up to $\varepsilon_e = 0{,}52$ (2) and annealed (3) REP samples [62].

The hydrostatic extrusion and subsequent annealing of rarely cross-linked epoxy polymer (REP) result in very essential changes of its mechanical behavior and properties, in addition to unexpected ones also. The qualitative changes of REP mechanical behavior can be monitored according to the corresponding changes of the stress—strain (σ – ε) diagrams, shown in Figure 1.28. The initial REP shows the expected enough behavior and both its elasticity modulus E and yield stress σ_Y are typical for such polymers at testing temperature T being distant from glass transition temperature T_g on about 40 K [51]. The small (≈ 3 MPa) stress drop beyond yield stress is observed, that is also typical for amorphous polymers [61]. However, REP extrusion up to $\varepsilon_e = 0{,}52$ results to stress drop $\Delta\sigma_Y$ ("yield tooth") disappearance and to the essential E and σ_Y reduction. Besides, the diagram σ – ε itself is now more like the similar diagram for rubber, than for glassy polymer. This specimen annealing at maximum temperature $T_{an} = 353$ K gives no less strong, but diametrically opposite effect—yield stress and elasticity modulus increase sharply (the latter in about twice in comparison with the initial REP and more than one order in comparison with the extruded specimen). Besides, the strongly pronounced "yield tooth" appears. Let us note, that specimen shrinkage at annealing is small (≈10%), that makes up about 20 percent of ε_e [62].

The common picture of parameters E and σ_Y change as a function of ε_e is presented in Figures 1.29 and 1.30 accordingly. As one can see, both indicated parameters showed common tendencies at ε_e change: up to $\varepsilon_e \approx 0{,}36$ inclusive E and σ_Y weak increase at ε_e growth is observed, moreover their absolute values for extruded and annealed specimens are close, but at $\varepsilon_e > 0{,}36$ the strongly pronounced antibatness of these parameters for the indicated specimen types is displayed. The cluster model of polymers amorphous state structure and developed within its frameworks polymers yielding treatment allows to explain such behavior of the studied samples [35, 65].

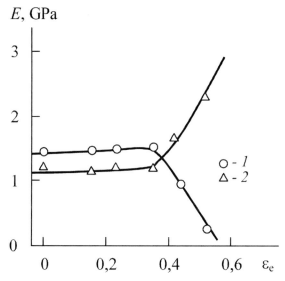

FIGURES 1.29 The dependences of elasticity modulus E_p on extrusion strain ε_e for extrudated (1.1) and annealed (1.2) REP [62].

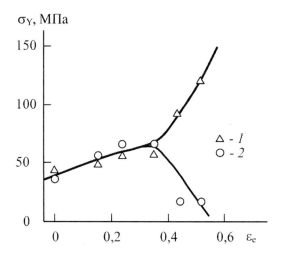

FIGURE 1.30 The dependences of yield stress σ_Y on extrusion strain ε_e for extrudated (1.1) and annealed (1.2) REP [62].

The cluster model supposes that polymers amorphous state structure represents the local order domains (nanoclusters), surrounded by loosely packed matrix. Nanoclusters consist of several collinear densely packed statistical segments of different macromolecules and in virtue of this they offer the analog of crystallite with stretched chains. There are two types of nanoclusters—stable, consisting of a relatively large segments number, and instable, consisting of a less number of such segments [65]. At temperature increase or mechanical stress application, the instable nanoclusters disintegrate in the first place that results in the two well-known effects. The first from them is known as two-stage glass transition process [11] and it supposes that at $T'_g = T_g - 50$ K disintegration of instable nanoclusters, restraining loosely packed matrix in glass state, occurs that defines devitrification of the latter [3, 5]. The well-known rapid polymers mechanical properties reduction at approaching to T_g [51] is the consequence of this. The second effect consists of instable nanoclusters decay at σ_Y under mechanical stress action, loosely packed matrix mechanical devitrification and, as consequence, glassy polymers rubber-like behavior on cold flow plateau [65]. The stress drop $\Delta\sigma_Y$ beyond yield stress is due to just instable nanoclusters decay and therefore $\Delta\sigma_Y$ value serves as characteristic of these nanoclusters fraction [5]. Proceeding from this brief description, the experimental results, adduced in Figures 1.28, 1.29 and 1.30, can be interpreted.

REP, on the basis of resin UP5-181, has low glass transition temperature T_g, which can be estimated according to shrinkage measurements data as equal ≈333K. This means, that the testing temperature $T = 293$ K and T'_g for it are close, that is confirmed by small $\Delta\sigma_Y$ value for the initial REP. It assumes nanocluster (nanostructures) small relative fraction φ_{cl} [3–5] and, since these nanoclusters have arbitrary orientation, ε_e increase results rapidly enough to their decay, that induces loosely packed matrix mechanical devitrification at $\varepsilon_e > 0{,}36$. Devitrificated loosely packed matrix gives insignificant contribution to E_p [66, 67], equal practically to zero, that results to sharp (discrete) elasticity modulus decrease. Besides, at $T > T'_g$ φ_{cl}, rapid decay is observed (i.e., segments number decrease in both stable and instable nanocluster) [5]. Since just these parameters (E and φ_{cl}) check σ_Y value, then their decrease defines yield stress' sharp lessening. Now extruded at $\varepsilon_e > 0{,}36$ REP presents as matter of fact rubber with high cross-linking degree, that is reflected by its diagram $\sigma - \varepsilon$ (Figure 1.28, curve 2).

The polymer oriented chains shrinkage occurs at the extruded REP annealing at temperature higher than T_g. Since this process is realized within a narrow temperature range and during a small time interval, then a large number of instable nanoclusters are formed. This effect is intensified by available molecular orientation (i.e., by preliminary favorable segments arrangement), and it is reflected by $\Delta\sigma_Y$ strong increase (Figures 1.28, curve 3).

The φ_{cl} enhancement results to E_p growth (Figures 1.29) and φ_{cl} and E_p combined increase – to σ_Y considerable growth (Figures 1.30).

The considered structural changes can be described quantitatively within the frameworks of the cluster model. The nanoclusters relative fraction φ_{cl} can be calculated according to the method, stated in paper [68].

The shown in Figures 1.31 dependences $\varphi_{cl}(\varepsilon_e)$ have the character expected from the adduced above description and are its quantitative conformation. The adduced in Figures 1.32 dependence of density ρ of REP extruded specimens on ε_e is similar to the dependence $\varphi_{cl}(\varepsilon_e)$, that was to be expected, since densely packed segments fraction decrease must be reflected in ρ reduction.

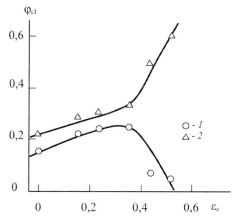

FIGURE 1.31 The dependences of nanoclusters relative fraction φ_{cl} on extrusion strain ε_e for extruded (1) and annealed (2) REP [62].

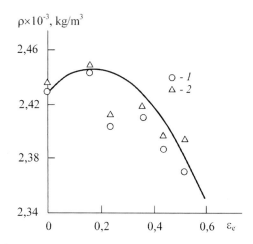

FIGURE 1.32 The dependence of specimens density ρ on extrusion strain ε_e for extruded (1) and annealed (2) REP [62].

In paper [69] the supposition was made that ρ change can be conditioned to microcracks network formation in specimen that results to ρ reduction at large ε_e (0,43 and 0,52), which are close to the limiting ones. The ρ relative change (Δρ) can be estimated according to the equation

$$\Delta\rho = \frac{\rho^{max} - \rho^{min}}{\rho^{max}}, \qquad (1.33)$$

where ρ^{max} and ρ^{min} are the greatest and the smallest density values. This estimation gives Δρ ≈ 0,01. This value can be reasonable for free volume increase, which is necessary for loosely matrix devitrification (accounting for closeness of T and T'_g), but it is obviously small if to assume as real microcracks formation. As the experiments have shown, REP extrusion at $\varepsilon_e > 0,52$ is impossible owing to specimen cracking during extrusion process. This allows to suppose that value $\varepsilon_e = 0,52$ is close to the critical one. Therefore the critical dilatation $\Delta\delta_{cr}$ value, which is necessary for microcracks cluster formation, can be estimated as follows [40]:

$$\Delta\delta_{cr} = \frac{2(1+\nu)(2-3\nu)}{11-19\nu}, \qquad (1.34)$$

where ν is Poisson's ratio.

Accepting the average value ν ≈ 0,35, we obtain $\Delta\delta_{cr} = 0,60$, that is essentially higher than the estimation Δρ made earlier. These calculations assume that ρ decrease at $\varepsilon_e = 0,43$ and 0,52 is due to instable nanoclusters decay and to corresponding REP structure loosening.

The stated above data give a clear example of large possibilities of polymer properties operation through its structure change. From the plots of Figure 1.29 it follows that annealing of REP extruded up to $\varepsilon_e = 0,52$ results to elasticity modulus increase in more than eight times and from the data of Figure 1.30 yield stress increase of six times follows. From the practical point of view the extrusion and subsequent annealing of REPs allow to obtain materials, which are just as good by stiffness and strength as densely cross-linked epoxy polymers, but exceeding the latter by plasticity degree. Let us note, that besides extrusion and annealing other modes of polymers nanostructure operation exist: plasticization [70], filling [26, 71], films obtaining from different solvents [72], and so on.

Hence, the stated above results demonstrated that neither cross-linking degree nor molecular orientation level defined cross-linked polymers final properties. The factor, controlling properties is a state of suprasegmental (nanocluster) structure, which, in its turn, can be goal-directly regulated by molecular orientation and thermal treatment application [62].

In the stated above treatment not only nanostructure integral characteristics (macromolecular entanglements cluster network density ν_{cl} or nanocluster relative fraction φ_{cl}), but also separate nanocluster parameters are important (see Section

A Detailed Review on Characteristics, Application

1.1). In this case of particulate-filled polymer nanocomposites (artificial nanocomposites), it is well-known that their elasticity modulus sharply increases at nanofiller particles size decrease [17]. The similar effect was noted above for REP, subjected to different kinds of processing (see Figure 1.28). Therefore the authors [73] carried out the study of the dependence of elasticity modulus E on nanoclusters size for REP.

It has been shown earlier on the example of PC, that the value E_p is defined completely by natural nanocomposite (polymer) structure according to the Eq. (1.32) (see Figure 1.26)

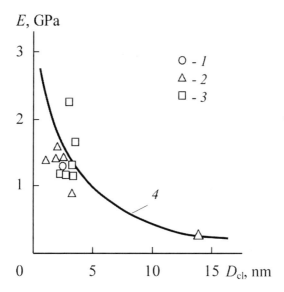

FIGURE 1.33 The dependence of elasticity modulus E_p on nanoclusters diameter D_{cl} for initial (1), extruded (2) and annealed (3) REP. 4—calculation according to the Eq. (1.32) [73].

In Figure 1.33 the dependence of E_p on nanoclusters diameter D_{cl}, determined according to the equation (in previous paper), for REP subjected to the indicated processing kinds at ε_e values within the range of 0,16–0,52 is adduced. As one can see, like in the case of artificial nanocomposites, for REP strong (approximately of order of magnitude) growth is observed at nanoclusters size decrease from 3 up to 0,9 nm. This fact confirms again, that REP elasticity modulus is defined by neither cross-linking degree nor molecular orientation level, but it depends only on epoxy polymer nanocluster structure state, simulated as natural nanocomposite [73].

Another method of the theoretical dependence $E_p(D_{cl})$ calculation for natural nanocomposites (polymers) is given in paper [74]. The authors [75] have shown that the elasticity modulus E value for fractal objects, which are polymers [4], is given by the following percolation relationship:

$$K_T, G \sim (p - p_c)^\eta, \qquad (1.35)$$

where K_T is bulk modulus, G is shear modulus, p is solid-state component volume fraction, p_c is percolation threshold, η is exponent.

The following equation for the exponent η was obtained at a fractal structure simulation as Serpinsky carpet [75]:

$$\frac{\eta}{v_p} = d - 1, \qquad (1.36)$$

where v_p is correlation length index in percolation theory, d is dimension of Euclidean space, in which a fractal is considered.

As it is known [4], the polymers nanocluster structure represents itself the percolation system, for which $p = \varphi_{cl}$, $p_c = 0{,}34$ [35] and further it can be written:

$$\frac{R_{cl}}{l_{st}} \sim (\phi_{cl} - 0{,}34)^{v_p}, \qquad (1.37)$$

where r_{cl} is the distance between nanoclusters, determined according to the equation (4.63), l_{st} is statistical segment length, v_p is correlation length index, accepted equal to 0,8 [76-77].

Since in the considered case the change E_p at n_{cl} variation is interesting first of all, then the authors [74] accepted v_{cl} = const = $2{,}5 \times 10^{27}$ m^{-3}, l_{st} = const = 0,434 nm. The value E_p calculation according to the Eqs. (1.35) and (1.37) allows to determine this parameter according to the formula [74]:

$$E_p = 28{,}9(\phi_{cl} - 0{,}34)^{(d-1)v_p}, \text{ GPa}. \qquad (1.38)$$

In Figure 1.34 the theoretical dependence (a solid line) of E_p on nanoclusters size (diameter) D_{cl}, calculated according to the Eq. (1.38) is adduced. As one can see, the strong growth E_p at D_{cl} decreasing is observed, which is identical to the shown one in Figure 1.33. The adduced in Figure 1.34 experimental data for REP, subjected to hydrostatic extrusion and subsequent annealing, correspond well enough to calculation according to the Eq. (1.38). The decrease D_{cl} from 3,2 up to 0,7 nm results again to E_p growth on order of magnitude [74].

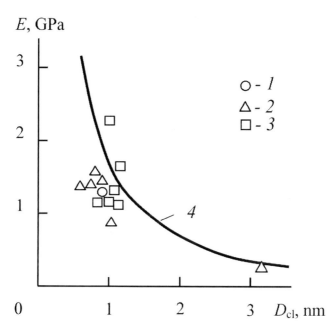

FIGURE 1.34 The dependence of elasticity modulus E_p on nanoclusters diameter D_{cl} for initial (1), extruded (2) and annealed (3) REP. 4—calculation according to the Eq. (1.38) [74].

The similar effect can be obtained for linear amorphous polycarbonate (PC) as well. Calculation according to the Eq. (1.38) shows n_{cl} reduction from 16 (the experimental value n_{cl} at $T = 293K$ for PC [5]) up to 2 results to F_p growth from 1,5 up to 5,8 GPa and making of structureless ($n_{cl} = 1$) PC will allow to obtain $E_p \approx 9,2$ GPa (i.e., comparable with obtained one for composites on the basis of PC).

Hence, the stated results in the present chapter give purely practical aspect of such theoretical concepts as the cluster model of polymers amorphous state stricture and fractal analysis application for the description of structure and properties of polymers, treated as natural nanocomposites. The necessary nanostructure goal-directed making will allow to obtain polymers, not yielding (and even exceeding) by their properties to the composites, produced on their basis. Structureless (defect-free) polymers are imagined the most perspective in this respect. Such polymers can be natural replacement for a large number of elaborated at present polymer nanocomposites. The application of structureless polymers as artificial nanocomposites polymer matrix can give much larger effect. Such approach allows to obtain polymeric materials, comparable by their characteristics with metals (e.g., with aluminum).

REFERENCES

1. Kardos, I. L.; and Raisoni I.; The potential mechanical response of macromolecular systems-F composite analogy. *Polymer Engng. Sci.*, **1975**, *15*(3), 183–189.
2. Ivanches, S. S.; and Ozerin, A. N.; A nanostructures. In Polymeric Systems.Vysokomolek Soed. B. (Eds), volume 48, 8, Springer, **2006**; pp 531–1544.
3. Kozlov, G. V.; and Novikov, V. U.; The cluster model of polymers amorphous state. *Uspekhi Fizicheskikh Nauk*, **2001**, *171*(7), 717–764.
4. Kozlov, G. V.; and Zaikov, G. E.; Structure of the Polymer Amorphous State. Brill Academic Publishers: Utrecht, Boston, **2004**; 465 p.
5. Kozlov, G. V.; Ovcharenko, E. N.; and Mikitaev, A. K.; Structure of the Polymer Amorphous State. Moscow, Publishers of the D. I. Mendeleev RKhTU, **2013**, 392 p.
6. Kozlov, G. V.; and Novikov, V. U.; Synergetics and Fractal Analysis of Cross-Linked Polymers. Moscow, Klassika, **2013**, 112 p.
7. Burya, A. I.; Kozlov G. V.; Novikov V. U.; and Ivanova, V. S.; Synergetics of Supersegmental Structure of Amorphous Glassy Polymers. Mater. of 3-rd Intern. In Conference Research and Development in Mechanical Industry-RaDMI-03," September 19–23, Herceg Novi, Serbia and Montenegro, **2003**; pp 645–647.
8. Bashorov, M. T.; Kozlov, G. V.; and Mikitaev, A. K.; Nanostructures and Properties of Amorphous Glassy Polymers. Moscow, Publishers of the D.I. Mendeleev RKhTU, **2010**; 269 p.
9. Malamatov, A.Kh.; Kozlov, G. V.; and Mikitaev, M. A.; Reinforcement Mechanisms of Polymer Nanocomposites. Moscow, Publishers of the D.I. Mendeleev RKhTU, **2006**; 240 p.
10. Kozlov, G. V.; Gazaev M. A., Novikov V. U., and Mikitaev, A. K.; Simulation of Amorphous Polymers Structure as Percolation Cluster. *Pis'ma v ZhTF*, **1996**, *22*(16), 31–38.
11. Belousov, V. N.; Kotsev, B. Kh., and Mikitaev, A. K.; Two-step of amorphous polymers glass transition Doklady ANSSSR, **1983**, 270(5), pp 1145–1147.
12. Ivanova, V. S.; Kuzee, I. R., and Zakirnichnaya, M. M.; Synergetics and Fractals. Universality of Metal Mechanical Behaviour. Ufa, Publishers of UGNTU, **1998**; 366p.
13. Berstein, V. A.; and Egorov, V. M.; Differential scanning calorimetry in Physics-Chemistry of the Polymers. Leningrad, Khimiya, Ozerin, **1990**; 256 p.
14. Bashorov, M. T.; Kozlov, G. V.; and Mikitaev, A. K.; A Nanoclusters Synergetics in Amorphous Glassy Polymers Structure. *Inzhenernaya Fizika*. **2009**, *4*, 39–42.
15. Bashorov, M. T.; Kozlov, G. V.; and Mikitaev, A. K.; A nanostructures in polymers: formation synergetics, regulation methods and influence on the properties. *Materialovedenie*. **2009**, *9*, 39–51.
16. Shevchenko, V.Ya.; and Bal'makov, M. D.; A particles-Centravs as nanoworld objects. *Fizika I Khimiya Stekla*. **2002**, *28*(6), 631–636.
17. Mikitaev, A. K.; Kozlov, G. V.; and Zaikov, G. E.; Polymer Nanocomposites: Variety of Structural Forms and Applications. Nova Science Publishers, Inc: New York, **2008**; 318 p.
18. Buchachenko, A. L.; The nanochemistry—Direct way to high technologies of new century. *Uspekhi Khimii*. **2003**, *72*(5), 419–437.
19. Formanis, G. E.; Self-assembly of nanoparticles is nanoworld special properties spite. In: Proceedings of Intern. Interdisciplinary Symposium "Fractals and Applied Synergetics," Moscow, Publishers of MGOU, **2003**, 303–308.
20. Bashorov, M. T.; Kozlov, G. V.; Shustov, G. B.; and Mikitaev, A. K.; The estimation of fractal dimension of nanoclusters surface in polymers. Izvestiya Vuzov, Severo-Kavkazsk. region, *estestv. Nauki*. **2009**, *6*, 44–46.
21. Magomedov, G. M.; and Kozlov, G. V.; Synthesis, structure and properties of cross-linked polymers and nanocomposites on its basis. Moscow, Publishers of Natural Sciences Academy, **2010**; 464 p.

22. Kozlov, G. V.; Polymers as natural nanocomposites: The missing opportunities. *Rec. Patents. Chem. Eng.* **2011**, *4*(1), 53–77.
23. Bovenko, V. N.; and Startsev, V. M.; The discretely-wave nature of amorphous poliimide supramolecular organization. *Vysokomolek. Soed. B.* **1994**, *36*(6), 1004–1008.
24. Bashorov, M. T.; Kozlov, G. V.; and Mikitaev, A. K.; Polymers as natural nanocomposites: an interfacial regions identification. In Proceedings of 12th Intern. Symposium "Order, Disorder and Oxides Properties." Rostov-na-Donu-Loo, September 17–22, **2009**; pp. 280–282.
25. Magomedov, G. M.; Kozlov, G. V.; and Amirshikhova, Z. M.; Cross-Linked Polymers as Natural Nanocomposites: An Interfacial Region Identification. Izvestiya DGPU, estestv. *I tochn. Nauki.* **2013**, *4*, 19–22.
26. Kozlov, G. V.; Yanovskii, Yu. G., and Zaikov, G. E.; Structure and Properties of Particulate-Filled Polymer Composites: The Fractal Analysis. Nova Science Publishers, Inc.: New York, **2010**; 282 p.
27. Bashorov, M. T.; Kozlov, G. V.; Shustov, G. B.; and Mikitaev, A. K.; Polymers as natural nanocomposites: The filling degree estimations. *Fundamental'nye Issledovaniya*, **2009**, *4*, 15–18.
28. Vasserman, A. M.; and Kovarskii, A. L.; A Spin Probes and Labels in Physics-Chemistry of Polymers. Moscow, Nauk, **2013**; 246 p.
29. Korst, N. N.; and Antsiferova, L. I.; A slow molecular motions study by stable radicals EPR method. *Uspekhi Fizicheskikh Nauk.* **1978**, *126*(1), 67–99.
30. Yech, G. S.; The general notions on amorphous polymers structure. local order and chain conformation degrees. *Vysokomolek. Soed. A.* **2013**, *21*(11), 2433–2446.
31. Perepechko, I. I.; Introduction in Physics of Polymers. Moscow, Khimiya, **1978**; 312 p.
32. Kozlov, G. V.; and Zaikov, G. E.; The generalized description of local order in polymers. In Fractals and Local Order in Polymeric Materials. Kozlov, G. V.; and Zaikov, G. E. Eds.; Nova Science Publishers: New York, Inc., **2001**; 55–63.
33. Tager, A. A.; Physics-Chemistry of Polymers. Moscow, Khimiya, 1978; 416 p.
34. Bashorov, M. T.; Kozlov, G. V.; Malamatov, A.Kh., and Mikitaev, A. K.; Amorphous Glassy Polymers Reinforcement Mechanisms by Nanostructures. Mater. of IV Intern. Sci.-Pract. Conf. "New Polymer Composite Materials." Nal'chik, KBSU, **2008**; pp 47–51.
35. Bobryshev, A. N.; Koromazov, V. N.; Babin, L. O.; and Solomatov, V. I.; Synergetics of Composite Materials. Lipetsk, NPO ORIUS, **1994**; 154p.
36. Aphashagova, Z.Kh., Kozlov, G. V.; Burya, A. T.; and Mikitaev, A. K.; The Prediction of particulate-Filled Polymer Nanocomposites Reinforcement Degree. *Materialovedenie*, **2007**, *9*, 10–13.
37. Sheng, N.; Boyce, M. C.; Parks, D. M.; Rutledge, G. C.; and Ales, J. I.; Cohen R. E.; Multiscale micromechanical modeling of Polymer/Clay Nanocomposites and the Effective Clay Particle. *Polymer.* **2004**, *45*, 2, 487–506.
38. Dickie, R. A.; The mechanical properties (Small Strains) of multiphase polymer blends. In: *Polymer Blends*. Paul D. R., Newman S. (Ed) Academic Press: New York/San Francisko/London, **1980**, 1; pp. 397–437.
39. Ahmed, S.; and Jones, F. R.; A review of particulate reinforcement theories for polymer composites. *J. Mater. Sci.* 1990, *25*(12), 4933–4942.
40. Balankin, A. S.; Synergetics of Deformable Body. Publishers of Ministry Defence SSSR: Moscow, **2013**; 404 p.
41. Bashorov, M. T.; Kozlov, G. V.; and Mikitaev, A. K.; Polymers as natural nanocomposites: description of elasticity modulus within the frameworks of micromechanical models. *Plast. Massy.* **2010**, *11*, 41–43.
42. Lipatov, Yu. S.; Interfacial phenomena in polymers. Kiev, Naukova Dumka, **1980**; 260p.
43. Yanovskii, Yu. G.; Bashorov, M. T.; Kozlov, G. V.; and Karnet, Yu. N.; Polymeric Mediums as Natural Nanocomposites: Intercomponont Interactions Geometry. In Proceedings of All-

Russian Conf. "Mechanics and Nanomechanics of Structurally-Complex and Heterogeneous Mediums Achievements, Problems, Perspectives." Moscow, IPROM, **2012**, 110–117.
44. Tugov, I. I.; and Shaulov, A. Yu.; A particulate-filled composites elasticity modulus. *Vysokomolek. Soed. B.* **1990**, *32*, (7), 527–529.
45. Piggott, M. R.; and Leidner, Y.; Microconceptions about filled polymers. *Y. Appl. Polymer Sci.* **1974**, *18*(7), 1619–1623.
46. Chen, Z.-Y.; Deutch, Y. M.; and Meakin, P.; Translational friction coefficient of diffusion limited aggregates. *Y. Chem. Phys.* **1984**, *80*(6), 2982–2983.
47. Kozlov, G. V.; Beloshenko, V. A.; and Varyukhin V. N.; Simulation of cross-linked polymers structure as diffusion-limited aggregate. *Ukrainskii Fizicheskii Zhurnal*, **1998**, *43*(3), 322–323.
48. Novikov, V. U.; Kozlov, G. V.; and Burlyan, O. Y.; The fractal approach to interfacial layer in filled polymers. *Mekhanika Kompozitnykh Materialov*, **2013**, *36*(1) 3–32.
49. Stanley, E. H.; A fractal surfaces and "Termite" model for two-component random materials. In: *Fractals in Physics*. Pietronero, L., Tosatti, E. Eds. Amsterdam, Oxford, New York, Tokyo, North-Holland, **1986**, 463–477.
50. Bashorov, M. T.; Kozlov, G. V.; Zaikov, G. E.; and Mikitaev, A. K.; Polymers as natural nanocomposites: Adhesion between structural components. *Khimicheskaya Fizika i Mezoskopiya.* **2013**, *11*(2), 196–203.
51. Dibenedetto, A. T.; and Trachte, K. L.; The brittle fracture of amorphous thermoplastic polymers. *Y. Appl. Polym. Sci.* **1970**, *14*(11), 2249–2262.
52. Burya, A. I.; Lipatov, Yu. S.; Arlamova, N. T.; and Kozlov, G. V.; Patent by useful model N27 199. Polymer composition. It is registered in Ukraine Patents State Resister October 25, **2007**.
53. Novikov, V. U.; and Kozlov, G. V.; Fractal parametrization of filled polymers structure. *Mekhanika Kompozitnykh Materialov*, **1999**, *35*(3), 269–290.
54. Potapov, A. A.; A nanosystems design principles. *Nano- i Mikrosistemnaya Tekhnika.* **2008**, *3*(4) 277–280.
55. Bashorov, M. T.; Kozlov, G. V.; Zaikov, G. E.; and Mikitaev, A. K.; Polymers as natural nanocomposites. 3. The geometry of intercomponent interactions. *Chem. Chem. Technol.* **2009**, *3*(4), 277–280.
56. Bashorov, M. T.; Kozlov, G. V.; Zaikov, G. E.; and Mikitaev, A. K.; Polymers as Natural Nanocomposites. 1. The reinforcement structural model. *Chem. Chem. Technol.* **2009**, *3*(2), 107–110.
57. Edwards, D. C.; Polymer-filler interactions in rubber reinforcement. I. *Mater. Sci.*, **1990**, *25*(12), 4175–4185.
58. Bashorov, M. T.; Kozlov, G. V.; and Mikitaev, A. K.; Polymers as natural nanocomposites: The comparative analysis of reinforcement mechanism. *Nanotekhnika.* **2009**, *4*, 43–45.
59. Bashorov, M. T.; Kozlov, G. V.; Zaikov, G. E.; and Mikitaev, A. K.; Polymers as natural nanocomposites. 2. The comparative analysis of reinforcement mechanism. *Chem. Chem. Technol.* **2009**, *3*(3), 183–185.
60. Chen, Y.-S.; Poliks, M. D.; Ober, C. K.; Zhang, Y.; Wiesner, U.; and Giannelis, E. Study of the interlayer expansion mechanism and thermal-mechanical properties of surface-initiated Epoxy nanocomposites. *Polymer.* **2002**, *43*(17), 4895–4904.
61. Kozlov, G. V.; Beloshenko, V. A.; Varyukhin, V. N.; and Lipatov, Yu. S.; Application of cluster model for the description of Epoxy polymers structure and properties. *Polymer.* **1999**, *40*(4), 1045–1051.
62. Bashorov, M. T.; Kozlov, G. V.; and Mikitaev, A. K.; Nanostructures in cross-linked epoxy polymers and their influence on mechanical properties. *Fizika I Khimiya Obrabotki Materialov.* **2013**, *2*, 76–80.

63. Beloshenko, V. A.; Shustov, G. B.; Slobodina, V. G.; Kozlov, G. V.; Varyukhin, V. N.; Temiraev, K. B.; and Gazaev, M. A.; Patent on Invention "The Method of Rod-Like Articles Manufacture from Polymers." Clain for Invention Rights N95109832. Patent N2105670. Priority: 13 June 1995. It is Registered in Inventions State Register of Russian Federation February 27 **1998**.
64] Aloev, V. Z.; and Kozlov, G. V.; Physics of Orientational Phenomena in Polymeric Materials. Nalchik, Polygraph-service and T, **2002**; 288 p.
65. Kozlov, G. V.; Beloshenko, V. A.; Garaev, M. A.; and Novikov, V. U.; Mechanisms of yielding and forced high-elasticity of cross-linked polymers. *Mekhanika. Kompozitnykh. Materialov.* **2013**, *32*(2), 270–278.
66. Shogenov, V. N.; Belousov, V. N.; Potapov, V. V.; Kozlov, G. V.; and Prut, E. V.; The glassy polyarylatesurfone curves stress-strain description within the frameworks of high-elasticity concepts. *Vysokomolek.Soed.F.* **1991**, *33*(1), 155–160.
67. Kozlov, G. V.; Beloshenko, V. A.; and Shogenov, V. N.; The Amorphous Polymers Structural Relaxation Description within the Frameworks of the Cluster Model. *Fiziko-Khimicheskaya Mekhanika Materialov*, **2013**, *35*(5), 105–108.
68. Kozlov, G. V.; Burya, A. I.; and Shustov, G. B.; The Influence of Rotating Electromagnetic Field on Glass Transition and Structure of Carbon Plastics on the Basis of lhenylone. *Fizika I Khimiya Obrabotki Materialov*. **2005**, *5*, 81–84.
69. Pakter, M. K.; Beloshenko, V. A.; Beresnev, B. I.; Zaika, T. R.; Abdrakhmanova, L. A.; and Berai, N. I.; Influence of hydrostatic processing on densely cross-linked Epoxy polymers structural organization formation. *Vysokomolek.Soed. F*, **2013**, *32*(10), 2039–2046.
70. Kozlov, G. V.; Sanditov, D. S.; Lipatov, and Yu, S.; Structural and mechanical properties of amorphous polymers in yielding region. In: Fractals and Local Order in Polymeric Materials. Kozlov G. V.; and Zaikov G. E. Eds. Nova Science Publishers Inc.: New York, **2001**; 65–82.
71. Kozlov, G. V.; Yanovskii, Yu, G., and Zaikov, G. E.; Synergetics and Fractal Analysis of Polymer Composites Filled with Short Fibers. Nova Science Publishers, Inc.: New York, **2011**; 223p.
72. Shogenov, V. N.; and Kozlov, G. V.; Fractal clusters in physics-chemistry of polymers. *Nal'chik, Polygraphservice and T*, **2002**; 270 p.
73. Kozlov, G. V.; and Mikitaev, A. K.; Polymers as Natural Nanocomposites: Unrealized Potential. Saarbrücken, Lambert Academic Publishing, **2010**; 323 p.
74. Magomedov, G. M.; Kozlov, G. V.; and Zaikov, G. E.; Structure and Properties of Cross-Linked Polymers. Shawbury, A Smithers Group Company; **2011**; 492 p.
75. Bergman, D. Y.; and Kantor, Y.; Critical Properties of an Elastic Fractal. *Phys. Rev. Lett.*, **1984**, *53*(6), 511–514.
76. Malamatov, A. Kh.; and Kozlov, G. V.; The Fractal Model of Polymer-Polymeric Nanocomposites Elasticity. In Proceedings of Fourth Intern. Interdisciplinary Symposium "Fractals and Applied Synergetics FaAS-05." Moscow, Interkontakt Nauka, **2005**; pp 119–122.
77. Sokolov, I. M.; Dimensions and other geometrical critical exponents in percolation theory. *Uspekhi Fizicheskikh Nauk*, **2013**, *151*(2), 221–248.

CHAPTER 2

STRUCTURE OF GRAPHITIC CARBONS: A COMPREHENSIVE REVIEW

HEINRICH BADENHORST

2.1 INTRODUCTION

Graphite in its various forms is a very important industrial material it is utilized in a wide variety of specialized applications. These include high temperature uses where the oxidative reactivity of graphite is very important, such as electric arc furnaces and nuclear reactors. Graphite intercalation compounds are utilized in lithium ion batteries or as fire retardant additives. These may also be exfoliated and pressed into foils for a variety of uses including fluid seals and heat management.

Graphite and related carbon materials have been the subject of scientific investigation for longer than a century. Despite this fact there is still a fundamental issue that remains, namely the supramolecular constitution of the various carbon materials [1, 2]. In particular it is unclear how individual crystallites of varying sizes are arranged and interlinked to form the complex microstructures and defects found in different bulk graphite materials.

Natural graphite flakes are formed under high pressure and temperature conditions during the creation of metamorphosed siliceous or calcareous sediments [3]. Synthetic graphite on the other hand is produced via a multistep, reimpregnation process resulting in very complex microstructures and porosity [4]. For both of these highly graphitic materials the layered structure of the ideal graphite crystal is well established [5]. However, in order to compare materials for a specific application the number of exposed, reactive edge sites are of great importance.

This active surface area (ASA) is critical for quantifying properties like oxidative reactivity and intercalation capacity. The ASA is directly linked to the manner in which crystalline regions within the material are arranged and interconnected. The concept of ASA has been around for a long time [6–11]; however, due to the nature of these sites and the very low values of the ASA for macrocrystalline graphite, it is difficult to directly measure this parameter accurately and easily. Hence it is been difficult to implement in practice and an alternative method must be employed to assess the microstructures found in graphite materials.

New developments in the field of scanning electron microscopy (SEM) allow very high resolution imaging with excellent surface definition [12, 13]. The use of high-brightness field-emission guns and in-lens detectors allow the use of very low (~1kV) acceleration voltages. This limits electron penetration into the sample and significantly enhancing the surface detail which can be resolved. Due to the beneficiation and processing of the material, graphite exhibits regions of structural imperfection which conceal the underlying microstructure. Since oxygen only gasifies graphite at exposed edges or defects, these regions may be largely removed by oxidation, leaving behind only the underlying core flake structure. This oxidative treatment will also reveal crystalline defects such as screw dislocations.

Furthermore, the oxidative reactivity of graphite is very sensitive to the presence of very low levels of impurities that are catalytically active. The catalytic activity is directly dependent on the composition of the impurity not just the individual components. Hence a pure metal will behave differently from a metal oxide, carbide, or carbonate [14–16], a distinction which is impossible to ascertain from elemental impurity analysis. The very low levels required to significantly affect oxidative properties are also close to the detection limits of most techniques, as such the only way to irrefutably verify the absence of catalytic activity is through visual inspection of the oxidized microstructure.

Thus, in combination these two techniques are an ideal tool for examining the morphology of graphite materials. Only once a comprehensive study of the microstructure of different materials has been conducted can their ASA related properties be sensibly compared.

2.1 MATERIALS AND METHODS

Four powdered graphite samples will be compared. The first two are proprietary nuclear grade graphite samples, one from a natural source (NNG) and one synthetically produced material (NSG). Both samples were intended for use in the nuclear industry and were subjected to high levels of purification including halogen treatment. The ash contents of these samples were very low, with the carbon content being >99.9 mass percent. The exact histories of both materials are not known. The third graphite (RFL) was obtained from a commercial source (Graphit Kropfmühl AG Germany). This is a large flake, natural graphite powder and was purified by the supplier with an acid treatment and a high temperature soda ash burn up to a purity of 99.91 mass percent. A fourth sample was produced for comparative purposes by heating the RFL sample to 2,700°C for 6 h in a TTI furnace (Model: 1000–2560–

FP20). This sample was designated as pure RFL of (PRFL) since the treatment was expected to further purify the material. All thermal oxidation was conducted in a TA Instruments SDT Q600 thermogravimetric analyzer (TGA) in pure oxygen. The samples were all oxidized to a burn-off of around 30 percent, at which point the oxidizing atmosphere was rapidly changed to inert. SEM images were obtained using an ultrahigh resolution field-emission microscope (Zeiss Ultra Plus 55 FEGSEM) equipped with an in-lens detection system.

2.3 MICROSTRUCTURE OF GRAPHITE

2.3.1 FEGSEM RESOLUTION

Initially the RFL sample was only purified up to a temperature of 2,400°C. When this sample was subsequently oxidized, the purification was found to have been only partially effective. It was possible to detect the effects of trace levels of catalytic impurities, as can be seen from Figure 2.1.

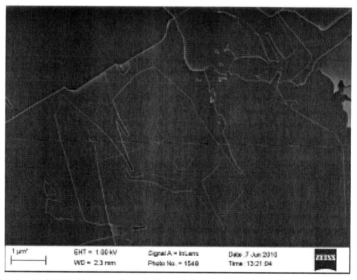

FIGURE 2.1 FEGSEM image of partially purified RFL (30k x magnification).

Since the catalyst particles tend to trace channels into the graphite, as seen in Figure 2.2, the consequences of their presence can be easily detected. As a result it is possible to detect a single, minute catalyst particle which is active on a large graphite flake. This effectively results in the ability to detect impurities that are present at extremely low levels.

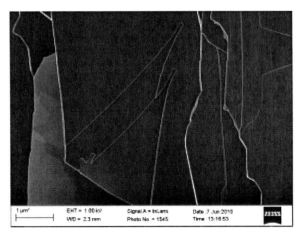

FIGURE 2.2 Channeling catalyst particles (40k x magnification).

When the tips of these channels are examined, the ability of the high resolution FEGSEM, operating at low voltages to resolve surface detail and the presence of catalytic particles, is further substantiated. As can be seen from Figure 2.3, the microscope is capable of resolving the catalyst particle responsible for the channeling.

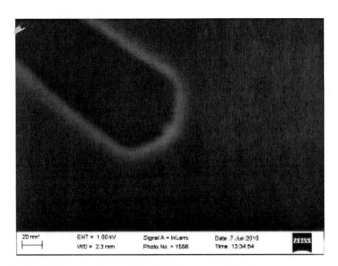

FIGURE 2.3 Individual catalyst particle (1000k x magnification).

In this case the particle in the image has a diameter of around ten nanometers. This demonstrates the powerful capability of the instrument and demonstrates its ability to detect the presence of trace impurities.

Structure of Graphitic Carbons: A Comprehensive Review

2.3.2 AS-RECEIVED MATERIAL

When the as-received natural graphite flakes are examined in Figure 2.4, their high aspect ratio and flat basal surfaces are immediately evident.

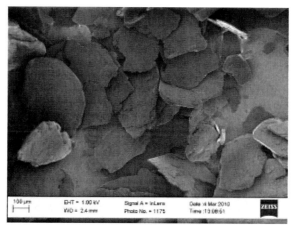

FIGURE 2.4 Natural graphite flakes (175 x magnification).

When some particles are examined more closely, they were found to be highly agglomerated, as shown in Figure 2.5.

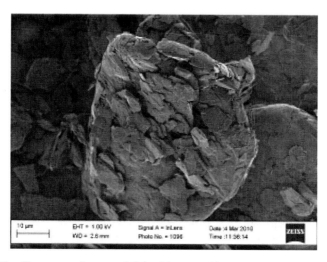

FIGURE 2.5 Close-up agglomerated flake (3k x magnification).

All samples were subsequently wet-sieved in ethanol to break up the agglomerates. When the sieved flakes are examined they are free of extraneous flakes but still appear to be composite in nature with uniform edges, as can be seen in Figure 2.6.

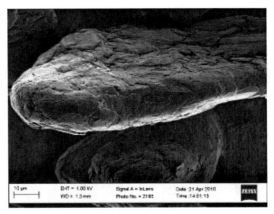

FIGURE 2.6   Close-up of sieved flake (3k x magnification).

This surface deformation is due to the beneficiation process, during which the edges tend to become smooth and rounded. As such the expected layered structure is largely obscured.

2.3.3 PURIFIED NATURAL GRAPHITE

However, when the oxidized natural graphite flakes are examined, their layered character is immediately apparent as seen in Figure 2.7.

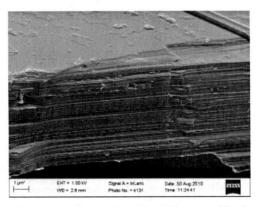

FIGURE 2.7   Layered structure of natural graphite (20k x magnification).

Structure of Graphitic Carbons: A Comprehensive Review 57

When the edges are examined from above, the crisp 120° angles expected for the hexagonal crystal lattice of graphite are evident as in Figure 2.8.

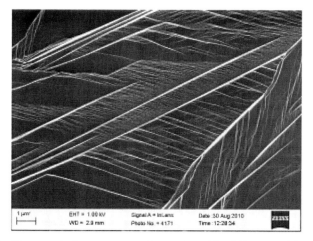

FIGURE 2.8 Hexagonal edge structures of natural graphite (20k x magnification).

The flat, linear morphology expected for a pristine graphite crystal is now more visible in Figure 2.9.

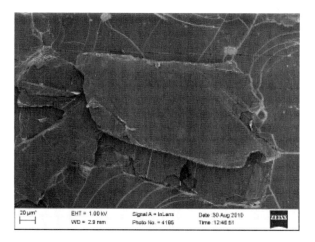

FIGURE 2.9 Oxidized natural graphite flake (800 x magnification).

When the basal surface is examined more closely as in Figure 2.10, the surface is smooth and flat across several tens of micrometers.

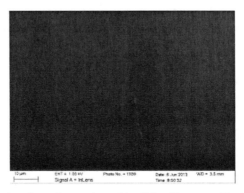

FIGURE 2.10 Basal surface of natural graphite flake (3k x magnification).

Since the graphite atoms are bound in-plane by strong covalent bonding, the basal surface is expected to be comparatively inert. This surface shows no signs of direct oxidative attack, with only some minor surface steps visible. Thus the highly crystalline nature of the material is readily evident but further investigation does reveal some defects are present. Four possible defect structures are generally found in graphite [16, 17], namely:

(i) Basal dislocations
(ii) Nonbasal edge dislocations
(iii) Prismatic screw dislocations
(iv) Prismatic edge dislocations

Due to the fact that the breaking of carbon-carbon bonds is required for nonbasal dislocations, the existence of type (ii) defects is highly unlikely [17, 18]. Given the very weak van der Waals bonding between adjacent layers, however, basal dislocations of type (i) are very likely and a multitude have been documented [5, 18]. However, such defects will not be visible in the oxidized microstructure.

The next possible defect is type (iii), prismatic screw dislocations. These dislocations are easily distinguishable by the large pits that form during oxidation with a characteristic corkscrew shape [19], as visible in Figure 2.11.

Structure of Graphitic Carbons: A Comprehensive Review 59

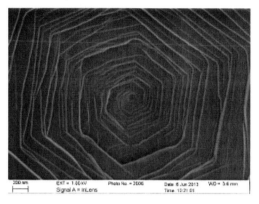

FIGURE 2.11 Prismatic screw dislocation (90k x magnification).

In general, no more than a few discrete occurrences of these defects were found in any given flake. A more prevalent defect is twinning, which is derived by a rotation of the basal plane along the armchair direction of the graphite crystal. These defects usually occur in pairs, forming the characteristic twinning band visible in Figure 2.12.

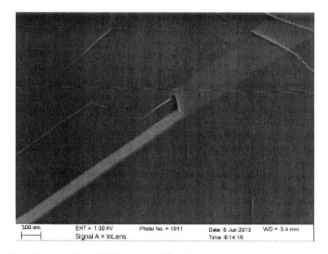

FIGURE 2.12 Twinning band (60k x magnification).

The angled nature of these defects is more evident when they are examined edge-on as in Figure 2.13.

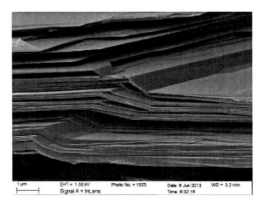

FIGURE 2.13 Twinning band edge (26k x magnification).

These folds are usually caused by deformation but may also be the result of the formation process whereby impurities became trapped within the macro flake structure and were subsequently removed by purification. As is visible in the lower left hand corner of Figure 2.13, a single sheet can undergo multiple, successive rotations and as can be seen in Figure 2.14, the rotation angle is variable.

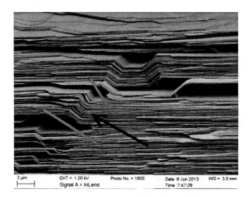

FIGURE 2.14 Rotation of twinning angle (11k x magnification).

The final defect to be considered is type (iv), prismatic edge dislocations. These involve the presence of an exposed edge within the flake body, an example is shown in Figure 2.15.

Structure of Graphitic Carbons: A Comprehensive Review 61

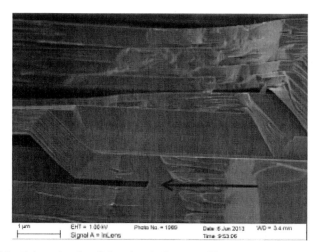

FIGURE 2.15 Prismatic edge dislocation (40k x magnification).

If the edge is small enough the structural order may be progressively restored, leading to the creation of a slit shaped pored that gradually tapers away until it disappears. An example of this behavior is demonstrated in Figure 2.16.

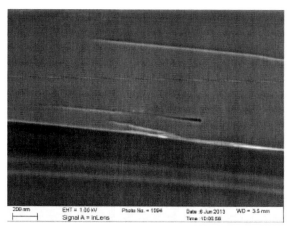

FIGURE 2.16 Gradual disappearance of small edge dislocation (125k x magnification).

If the edge dislocation is larger, when the stack collapses it will lead to a surface step, analogous to a twinning band. This can result in the formation of some very complex structures, such as the one shown in Figure 2.17.

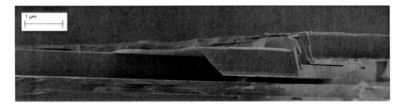

FIGURE 2.17 Complex surface structures (50k x magnification).

Thus despite being highly crystalline with an apparently straightforward geometry, complex microstructures can still be found in these natural graphite flakes.

2.3.4 CONTAMINATED NATURAL GRAPHITE

A very different microstructure is evident when the same natural graphite flakes are examined which have not been purified. Since the flakes are formed under geological processes involving high temperatures and pressures, the heat treatment step is not expected to have modified the flake microstructure. As expected, the high aspect ratio and general flat shape of the flakes are still visible in Figure 2.18.

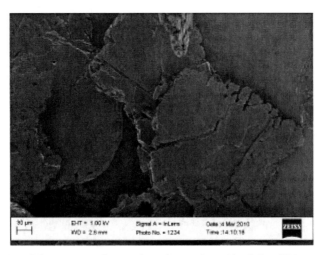

FIGURE 2.18 Flake structure of contaminated natural graphite flakes (500 x magnification).

However, when the edges of these particles are examined more closely as in Figure 2.19, highly erratic, irregular edge features are observed.

Structure of Graphitic Carbons: A Comprehensive Review 63

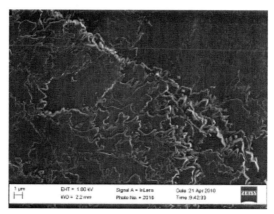

FIGURE 2.19 Erratic edge of contaminated natural graphite flakes (10k x magnification).

When the edges are scrutinized more closely, as in Figure 2.20, the reason for these edge formations becomes clear. They are caused by minute impurities which randomly trace channels into the graphite.

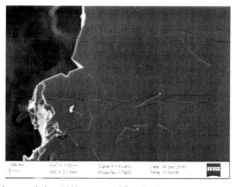

FIGURE 2.20 Catalyst activity (65k x magnification).

In certain cases, the activity is very difficult to detect, requiring the use of excessive contrast before they become noticeable as shown in Figure 2.21a, b.

64 Materials Science of Polymers

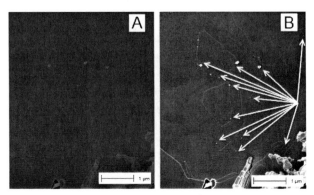

FIGURE 2.21 Contrast detection of catalyst activity (38k x magnification).

A very wide variety of catalytic behaviors were found. Broadly, these could be arranged into three categories. The first, show in Figure 2.22, are small, roughly spherical catalyst particles. Channels resulting from these particles are in most cases triangular in nature. In general it was found that these particles tend to follow preferred channeling directions, frequently executing turns at precise, repeatable angles, as demonstrated in Figure 2.22b. However, exceptions to these observed behaviors were also found, as illustrated in Figure 2.22c.

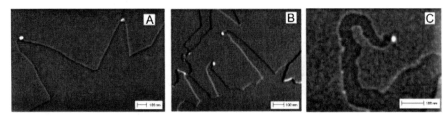

FIGURE 2.22 Small, spherical catalyst particles.

The second group contained larger, erratically shaped particles, some examples of which are shown in Figure 2.23.

FIGURE 2.23 Small, spherical catalyst particles.

Structure of Graphitic Carbons: A Comprehensive Review 65

These particles exhibited random, erratic channeling. Where it is likely that the previous group may have been in the liquid phase during oxidation, this is not true for this group, since the particles are clearly capable of catalyzing channels on two distinct levels simultaneously, as can be seen in Figure 23b, c. The final group contains behaviors which could not be easily placed into the previous two categories, of which examples are shown in Figure 2.24.

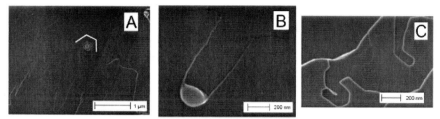

FIGURE 2.24 Small, spherical catalyst particles.

The fairly large particle in Figure 2.24a cannot be clearly distinguished as having been in the liquid phase during oxidation, yet the tip of the channel is clearly faceted with 120° angles. The particle in Figure 2.24b was clearly molten during oxidation as it has deposited material on the channel walls. It is interesting to note that since the channel walls have expanded a negligible amount compared to the channel depth, the activity of the catalyst deposited on the wall is significantly less than that of the original particle. Finally a peculiar behavior was found in the partially purified material, where a small catalyst particle is found at the tip of a straight channel, ending in a 120 ° tip (clearly noticeable in Figure 2.3). The width of the channel is roughly an order of magnitude larger than the particle itself, as seen in Figure 2.24c. In this case channeling was always found to proceed along preferred crystallographic directions.

Such a wide variety of catalytic behaviors are not unexpected for the natural graphite samples under consideration. Despite being purified, the purification treatments are unlikely to penetrate the graphite particles completely. As such inclusions which may have been trapped within the structure during formation will not be removed and will be subsequently exposed by the oxidation. These impurities can have virtually any composition and hence lead to the diversity of observed behaviors.

In addition to the irregular channels, erratically shaped pits are also found in the natural graphite sample, as shown in Figure 2.25.

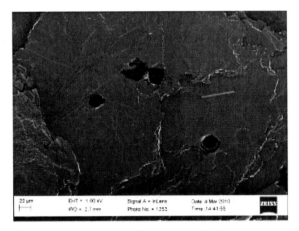

FIGURE 2.25 Pitting in natural graphite (650 x magnification).

Underdeveloped pits are often associated with erratically shaped impurities, as shown in Figure 2.26a, b.

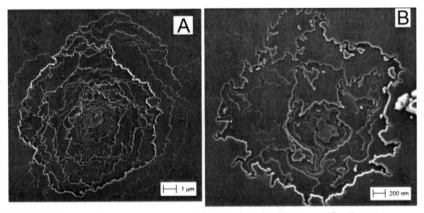

FIGURE 2.26 Impurity particles associated with pitting (35k x magnification).

The myriad of different catalytic behaviors found in this high purity natural graphite sample coupled with the enormous impact catalyst activity has on reactivity, demonstrates the danger of simply checking the impurity levels or ash content as a basis for reactivity comparison. A final morphological characteristic of this material is the presence of spiked or sawtooth-like edge formations, as can be seen in Figure 2.27.

Structure of Graphitic Carbons: A Comprehensive Review

FIGURE 2.27 Sawtooth edge formations (15k x magnification).

Closer inspection reveals that invariably the pinnacle of these structures is capped by a particle, as seen in Figure 2.28.

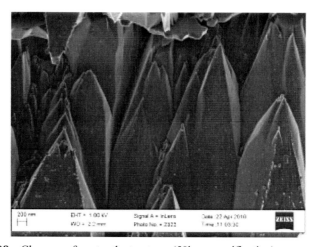

FIGURE 2.28 Close-up of sawtooth structures (50k x magnification).

Thus these formations are caused by inactive particles which shield the underlying graphite from attack. These layers protect subsequent layers leading to the formation of pyramid like structures crowned with a single particle. In some cases as on the left hand side of Figure 2.29, these start off as individual structures, but then as oxidation proceeds around them, the particles are progressively forced closer together to form inhibition ridges, as can be observed on the right hand side of Figure 2.29.

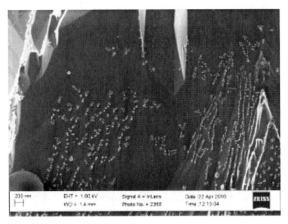

FIGURE 2.29 Inhibiting particles stacked along ridges (50k x magnification).

In extreme cases these particles may remain atop a structure until it is virtually completely reacted away, for example resulting in the nanopyramid shown in Figure 2.30.

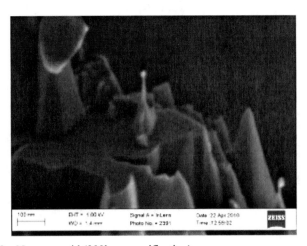

FIGURE 2.30 Nanopyramid (300k x magnification).

In some cases particles are found which appear to neither catalyze nor inhibit the reaction, such as the spherical particles seen in Figure 2.31.

Structure of Graphitic Carbons: A Comprehensive Review 69

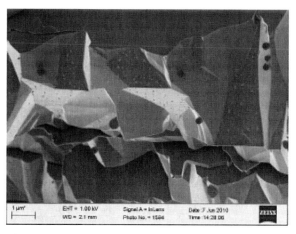

FIGURE 2.31 Spherical edge particles (25k x magnification).

These may be catalyst particles which agglomerate and deactivate due to their size. The graphite is oxidized away around them, until they are left at an edge, as seen in Figure 2.32.

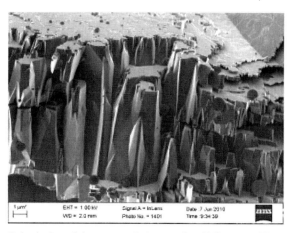

FIGURE 2.32 Spherical particles accumulating at edge (15k x magnification).

The accumulation of inhibiting particles at the graphite edge will inevitably lead to a reduction in oxidation rate as the area covered by these particles begins to constitutes a significant proportion of the total surface area. This may appreciably affect the shape of the observed conversion function, especially at high conversions.

2.3.5 NUCLEAR GRADE NATURAL GRAPHITE

The as-received nuclear grade natural graphite (NNG) exhibits a different morphology from that found in the commercial flake natural graphite. In this case the particles appear rounded and almost spherical, as shown in Figure 2.33.

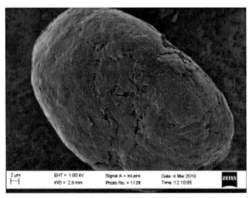

FIGURE 2.33 Rounded nuclear graphite particle (5k x magnification).

When the oxidized NNG microstructures are examined in Figure 2.34, fairly complex and irregular structures are found.

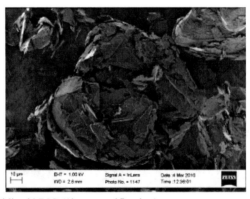

FIGURE 2.34 Oxidized NNG (1k x magnification).

The particles are extensively damaged and crumpled; however, the fact that they remain intact indicates that this is one continuous fragment. As the outer roughness is removed by oxidation, the multifaceted features of the particle interior are revealed. It may be concluded that these particles are in fact an extreme case of the damaged structure shown in Figure 2.6. This material has been extensively jet-milled to create so-called "potato-shaped" graphite. Initially the particles may have resembled the commercial natural graphite flakes; however, the malleability of graphite coupled with the impact deformation of jet-milling has caused them to buckle and collapse into a structure similar to a sheet of paper crumpled into a ball. Despite the high levels of purification, this material still exhibits extensive catalytic activity, similar to the flake natural graphite, as shown in Figure 2.35.

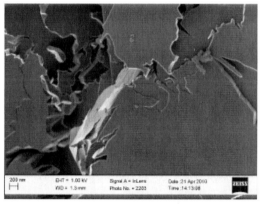

FIGURE 2.35 NNG catalytic activity (50k x magnification).

In spite of the catalytic activity and structural damage, in some regions the basal surface is still fairly smooth and flat across several micrometer, as can be seen in Figure 2.36, indicating that the material still has good underlying crystallinity.

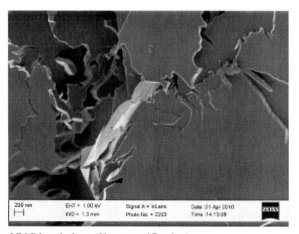

FIGURE 2.36 NNG basal plane (9k x magnification).

Thus this despite being naturally derived and evidently highly crystalline, the microstructure of the NNG material is very complex due to the extensive particle deformation during processing.

2.3.6 NUCLEAR GRADE SYNTHETIC GRAPHITE

The as-received nuclear grade synthetic graphite (NSG) exhibits a remarkably different behavior from the natural graphite samples. At first glance it is possible to distinguish between two distinct particle morphologies in Figure 2.37.

FIGURE 2.37 Oxidized NSG (700 x magnification).

Firstly, long, thin particles are noticeable with a high aspect ratio. During the fabrication of synthetic graphite a filler material known as needle coke is utilized. These particles are most likely derived from the needle coke with its characteristic elongated, needlelike shape. This filler is mixed with a binder which can be either coal tar or petroleum derived pitch. The pitch is in a molten state when added and the mixture is then either extruded or molded. The resulting artifact can then be reimpregnated with pitch if a high density product is required. The second group of particles have a complex, very intricate microstructure and are most likely derived from this molten pitch. They are highly disordered with a characteristic mosaic texture probably derived from the flow phenomena during impregnation. When examined edge-on, the layered structure of the needle coke derived particles is still readily evident, as seen in Figure 2.38.

Structure of Graphitic Carbons: A Comprehensive Review

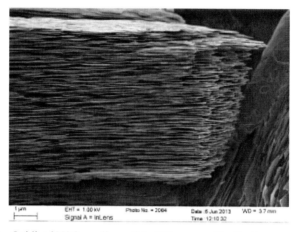

FIGURE 2.38 Oxidized NSG needle particle (20k x magnification).

The needle particles bear some resemblance to the natural graphite flakes, with the basal plane still readily identifiable in Figure 2.39.

FIGURE 2.39 Oxidized NSG needle particle (4k x magnification).

However, when the basal plane is examined more closely in Figure 2.40 there is a stark contrast with the natural graphite basal plane. The basal surface is severely degraded, with attack possible virtually anywhere.

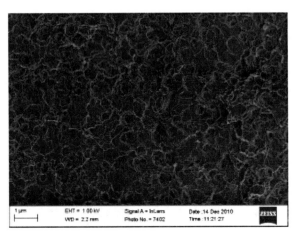

FIGURE 2.40 Oxidized NSG needle particle basal plane (25k x magnification).

The cavities were extensively investigated and no traces of impurities were found to be present. Instead the oxidation hollow has the characteristic corkscrew like shape of a screw dislocation as can be seen from Figure 2.41. In addition, the pits tend to have a vaguely hexagonal shape.

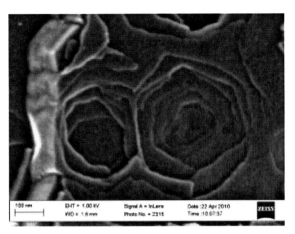

FIGURE 2.41 NSG screw dislocation (320k x magnification).

It is also important to notice that in some regions the defect density is not as high as in others, as can be seen for the different horizontal bands in Figure 2.42a and also the different regions visible in Figure 2.42b. This may imply different levels of crystalline perfection in these regions.

Structure of Graphitic Carbons: A Comprehensive Review

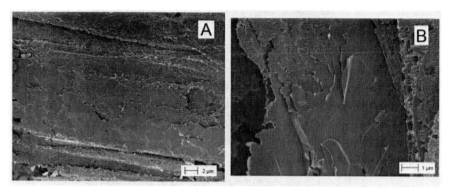

FIGURE 2.42 NSG crystallinity differences (16k x magnification).

When examined edge-on as in Figure 2.43, it can be seen that the needle particles retained their original structure; however, any gaps or fissures in the folds have grown in size. This implies the development of complex slit-like porosity, probably initiated by "Mrozowski" cracks, which would not have occurred to the same extent if direct basal attack was not possible to a large degree.

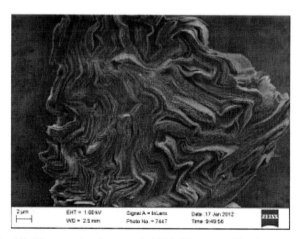

FIGURE 2.43 Slit-like pore development in NSG (8k x magnification).

When the particles edges are examined more closely in Figure 2.44, the low level of crystalline perfection is further evident. The maximum, continuous edge widths are no more than a few hundred nanometers, far less than the several micron observable in the natural samples, such as Figure 2.7.

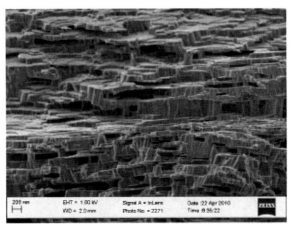

FIGURE 2.44 Degraded edge structure of NSG (50k x magnification).

The complex microstructural development characteristic of this sample is even more pronounced in the pitch particles, as can be seen from Figure 2.45.

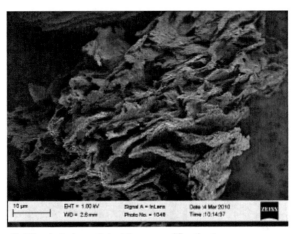

FIGURE 2.45 Oxidized pitch particle (4k x magnification).

These particles lack any long-range order; however, when their limited basal-like surfaces are examined more closely as in Figure 2.46, a texture very similar to the basal plane of needle particles is found, indicating possibly similar levels of crystalline perfection.

Structure of Graphitic Carbons: A Comprehensive Review

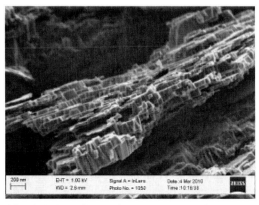

FIGURE 2.46 Oxidized pitch particle surface (88k x magnification).

On the whole, the synthetic material has the most intricate microstructural arrangement and despite the layered nature of the needle coke derived particles being readily evident, the basal surface is severely degraded indicating a high defect density. Thus this graphite can be expected to have the highest inherent ASA of all the samples considered.

2.3.6 REACTIVITY

As a comparative indication of reactivity the samples were subjected to oxidation in pure oxygen under a temperature program of 4°C/min in the TGA. The measured reaction rate as a function temperature is shown in Figure 2.47.

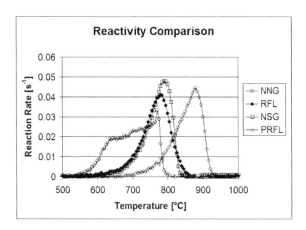

FIGURE 2.47 Reactivity comparison.

As a semiquantitative indication of relative reactivity the onset temperatures were calculated and are shown in Table 2.1.

TABLE 2.1 Onset temperatures

	Temp (°C)
NNG	572
RFL	696
NSG	704
PRFL	760

It is clear from Table 2.1 and Figure 2.46 that the NNG sample has the highest reactivity and PRFL the lowest. The NSG and RFL samples have similar intermediate reactivity, although the NSG sample does exhibit a higher peak reactivity. Given the microstructure and impurities found in the respective samples this result is not unexpected. The NNG and NSG samples have comparably complex microstructures which would both have relatively high surface areas. Despite the higher crystalline perfection of the NNG sample the presence of impurities increases its reactivity significantly above that of the NSG. The sample with the lowest reactivity is the purified PRFL sample, which is not surprising since it exhibited no catalytic activity coupled with a highly crystalline structure and a flake geometry with a large aspect ratio.

The RFL material also has excellent crystallinity and a disk structure with low edge surface area. However, despite its high purity (>99.9%) the RFL sample still contains considerable amounts of catalytically active impurities, thus increasing its reactivity. It is remarkable that despite the comparatively high amount of defects and consequently high ASA, the NSG sample achieves a reactivity comparable to the RFL materials. This indicates the dramatic effect even very low concentrations of catalytically active impurities can have on the oxidation rate of graphite. As can be seen from Figure 2.2, these minute impurities rapidly create vast amounts of additional surface area through their channeling action. This raises the ASA of the idealized, flat natural flakes to a level comparable to the synthetic material.

2.4 CONCLUSIONS

The ASA of graphite is important for a wide variety of applications. Through the use of oxidation to expose the underlying microstructure and high resolution surface imaging it is possible to discern between graphite materials from different origins, irrespective of their treatment histories. This establishes a direct link between the

ASA based characteristics, like the oxidative reactivity, of disparate samples and their observed microstructures. This enables like-for-like comparison of materials for selection based on the specific application.

Despite a highly crystalline structure, the oxidative behavior of natural graphite can be dramatically altered through the presence of trace catalytically active impurities and structural damage induced by processing. These differences would be difficult to detect using analytical techniques such as X-ray diffraction, Raman spectroscopy, or X-ray fluorescence, due to the similarity of the materials. Synthetic graphite has a much higher defect density than the natural graphite but a similar reactivity to these materials can be achieved if the material is free of catalytic impurities.

In addition, this technique enables insights regarding the extent to which the properties of different materials can be enhanced by further treatments. For example, on the basis of this investigation, it is clear that the oxidative reactivity of the NNG sample may be improved by purification, but due to the damaged structure it cannot achieve the stability observed for the PRFL material, despite both being natural graphite samples. Furthermore, despite having similar reactivities, the NSG and RFL materials have vastly different microstructures and therefore would not be equally suitable for applications where, for example, inherent surface area is very important. In conclusion, given the complexity found in different graphite materials, it is critical that the microstructure should be considered in conjunction with kinetic and other ASA related parameters to afford a comprehensive understanding of the material properties.

This is by no means an exhaustive study of all possible morphologies found in natural and synthetic graphite materials but it does demonstrate some of the intricate structures that are possible.

REFERENCES

1. Radovich L. R.; Physicochemical properties of carbon materials: a brief overview. In: Serp P, Figueiredo JL, eds. Carbon Materials for Catalysis, Hoboken: Wiley; **2009**, pp 1–34.
2. Harris P. J. F.; New perspectives on the structure of graphitic carbons. *Crit. Rev. Solid. State. Mater. Sci.* **2005**, *30*, 235–253.
3. Luque F. J.; Pasteris, J. D.; Wopenka, B.; Rodas, M.; and Barranechea, J. F.; Natural fluid-deposited graphite: mineralogical characteristics and mechanisms of formation. *Am. J. Sci.* **1998**, *298*, 471–498.
4. Pierson, H. O.; Handbook of Carbon, Graphite, Diamond and Fullerenes. Properties, Processing and Applications. New Jersey, USA: Noyes Publications; **1993**.
5. Reynolds, W. N.; Physical Properties of Graphite. Amsterdam: Elsevier, **1968**.
6. Laine, N. R., Vastola, F. J., and Walker, P. L.; Importance of active surface area in the carbon-oxygen reaction. *J. Phys. Chem.* **1963**; *67*, 2030–2034.
7. Thomas, J. M.; Topographical studies of oxidized graphite surfaces: a summary of the present position. *Carbon.* **1969**, *7*, 350–364.
8. Bansal, R. C.; Vastola, F. J.; and Walker, P. L.; Studies on ultra-clean carbon surfaces – III. Kinetics od chemisorption of hydrogen on graphon. *Carbon.* **1971**, *9*, 185–192.

9. Radovic, L. R., and Walker, P. L. R.G. J.; Importance of carbon active sites in the gasification of coal chars. *Fuel.* **1983**, *62*, 849–856.
10. Walker, P. L. R. L. J, J. M. T. An update on the carbon-oxygen reaction. *Carbon.* **1991**, *29*, 411–421.
11. Arenillas, A.; Rubiera, F.; Pevida, C.; Ania, C. O.; and Pis, J. J.; Relationship between structure and reactivity of carbonaceous materials. *J. Ther. Anal. Calorimetry.* **2004**, 76, 593–602.
12. Cazaux, J.; From the physics of secondary electron emission to image contrasts in scanning electron microscopy. *J. Elect. Micros* (Tokyo). **2012**, *61*(5), 261–284.
13. Lui, J.; The Versatile FEG-SEM: From Ultra-High Resolution To Ultra-High Surface Sensitivity. *Microscop. Microanal*, **2008**, *9*,144–145.
14. Baker, R. T. K.; Factors controlling the mode by which a catalyst operates in the graphite-oxygen reaction. *Carbon.* **1986**, *24*, 715–717.
15. Yang, R. T.; Wong, C.; Catalysis of carbon oxidation by transition metal carbides and oxides. *J. Catal.* **1984**, *85*, 154–168.
16. McKee, D. W.; and Chatterji, D.; The catalytic behaviour of alkali metal carbonates and oxides in graphite oxidation reactions. *Carbon.* **1975**, *13*, 381–390.
17. Fujita, F. E., and Izui, K.; Observation of lattice defects in graphite by electron microscopy, Part 1. *J. Phys. Soc. Japan.* **1961**, *16*(2), 214–217.
18. Suarez-Martinez, I.; Savini, G.; Haffenden, G.; Campanera, J. M.; and Heggie, M. I. Dislocations of Burger's Vector c/2 in graphite. *Phys. Status. Solidi C* **2007**, *4*(8), 2958–2962.
19. Rakovan, J., and Jaszczak, J. A.; Multiple length scale growth spirals on metamorphic graphite {001} surfaces studied by atomic force microscopy. *Am. Mineral.* **2002**, *87*, 17–24.

CHAPTER 3

RADIATION CROSS-LINKING OF ACRYLONITRILE-BUTADIENE RUBBER

KATARZYNA BANDZIERZ, DARIUSZ M. BIELINSKI,
ADRIAN KORYCKI, and GRAZYNA PRZYBYTNIAK

3.1 INTRODUCTION

Radiation modification of polymer materials has been gaining increasing popularity, not only in academic research, but also in industrial applications [1]. Among numerous advantages of radiation modification method, the noteworthy issue is the simplicity to control the ionizing radiation dose, which is absorbed by the modified material, dose rate, and energy of ionizing radiation. The resulting properties can be therefore "tailored" and the whole process is highly controllable and repeatable.

Radiation cross-linking is an interesting alternative for thermal cross-linking [2–4] or its complement [5–7]. One of the extensively studied polymers in respect to its radiation cross-linking is acrylonitrile-butadiene rubber (NBR) [8–12], which belongs to group of polymers that effectively cross-link on irradiation with ionizng radiation.

As a result of high-energetic irradiation, radicals are generated directly on polymer chains. By recombination, they form carbon-carbon (C-C) cross-links between the chains. Due to the fact that radiation cross-linking leads to formation of C-C cross-links and the mechanism is radical, it is often compared to peroxide cross-linking [3, 13]. It is noteworthy to enhance that the processes induced by ionizing radiation are very complicated and therefore not thoroughly understood [14]. C-C cross-links provide good elastic properties and are resistant to thermal aging, but they are short, stiff, and do not provide satisfactory properties for dynamic loadings.

According to Dogadkin's theory, cross-links of various structure provide better mechanical properties, owing to different lengths of the bridges between polymer chains, which do not break at the same time [15]. To provide optimal properties required for the end-use product, the researchers endeavor to design materials with hybrid (mixed) cross-links. To obtain hybrid type network with both C-C and longer sulfide cross-links, studies on thermal simultaneous cross-linking with two types of curatives, such as organic peroxide and sulfur cross-linking system, were carried out. The results generally showed lowered efficiency of cross-linking due to competing reactions, in which sulfur and cross-linking accelerator are involved in reactions with peroxide radicals [16–19].

In our previous research [20] concerning radiation cross-linking of NBR rubber with sulfur cross-linking system in composition, hybrid cross-link structure was proved to be formed upon irradiation. Inhibiting effect of sulfur cross-linking system on total cross-link density, formed in the irradiation process, was observed. To investigate in detail the contribution coming from particular components of the cross-linking system, such as rhombic sulfur and cross-linking accelerator DM, a set of samples with various ratios of sulfur and accelerator was prepared. The effect of these two components on radiation cross-linking process was studied by determination of basic mechanical properties, total cross-link density and analysis of cross-link structure.

3.2 EXPERIMENTAL

3.2.1 MATERIALS

Acrylonitrile—butadiene rubber Europrene N3325 (bound ACN content 33%) was supplied by Polimeri Europa (Italy). Precipitated silica Ultrasil VN3 was obtained from Evonik Industries (Germany). Vinyltrimethoxysilane U-611 was obtained from Unisil (Poland). Rhombic sulfur was provided by Siarkopol Tarnobrzeg (Poland) and zinc oxide, stearic acid and dibenzothiazole disulfide (DM), by Lanxess (Germany).

3.2.2 SAMPLES PREPARATION

Rubber mixes were prepared in two-stage procedure. In the first stage, rubber premixes of NBR, filled with 40 phr of precipitated silica and vinyltrimethoxysilane in amount of 10 wt. % of silica in the composite mix, were prepared with the use of Brabender Plasticorder internal micromixer (Germany) at temperature of mixing chamber of 120°C, with rotors speed of 20 RPM during components incorporation and 60 RPM during 25 min lasting homogenization process. In the second stage, components of cross-linking system, such as zinc oxide, stearic acid, rhombic sulfur and DM, were incorporated into the premix with David Bridge two-roll open mixing mill (UK) at 40°C and homogenized for 10 min. The samples composition is given in Table 3.1.

One mm rubber mixes sheets were compression molded in an electrically heated press at temperature of 110°C under pressure of 150 bar for 4 min.

TABLE 3.1 Composition of rubber mixes. The samples are designated as x/y. Here, x indicates the amount of sulfur and y amount of dibenzothiazole disulfide, respectively

Rubber mixes x/y Component, wt. %	0/0	0/1.5	2/0	2/1.5
NBR, Europrene N3325	100	100	100	100
Silica, Ultrasil VN3	40	40	40	40
Silane, U611	4	4	4	4
Zinc oxide, ZnO	5	5	5	5
Stearic acid	1	1	1	1
Rhombic sulfur, S_8	0	0	2	2
Dibenzothiazole disulfide, DM	0	1.5	0	1.5

3.2.3 SAMPLES IRRADIATION

The molded rubber sheets were subjected to electron beam (EB) irradiation at Elektronika 10/10 linear electron accelerator (Russia), located at the Institute of Nuclear Chemistry and Technology (Poland). The absorbed doses were 50, 122, and 198 kGy. Irradiation process was carried out in air atmosphere at room temperature. The rubber sheets were placed horizontally in the front of pulsed, scanned beam. The total doses were obtained by multipass exposure (approx. 25 kGy per pass).

3.2.4 SAMPLES CHARACTERIZATION

3.2.4.1 CROSS-LINK DENSITY DETERMINATION

Total cross-link density of the irradiated samples was determined taking advantage of equilibrium swelling in toluene and calculated on the basis of Flory-Rehner equation [21]. The Flory-Huggins interaction parameter used in the calculations for toluene—NBR rubber was 0.435 [22].

3.2.4.2 CROSS-LINK STRUCTURE DETERMINATION

The cross-link structure was analyzed and quantified by thiol—amine analysis, which is based on treatment of the crosslinked material with a set of thiol—amine chemical probes, specifically cleaving particular cross-links types [23]. Polysulphide cross-links are cleaved by treatment of crosslinked rubber samples with 2-propanethiol (0.4 M) and piperidine (0.4 M) in toluene for 2 h under inert gas atmosphere (argon) at room temperature, while polysulfide and disulfide cross-links can be cleaved by treatment under the same conditions with 1-dodecanethiol (1 M) in piperidine for 72 h.

3.2.4.3 MECHANICAL PROPERTIES

Mechanical tests were carried out with the use of "Zwick 1435" universal mechanical testing machine (Germany), according to ISO 37. The crosshead speed was 500 mm/min and the temperature was 23 ± 2°C. Five dumbbell specimens were tested for each sample and the average is reported here.

3.3 RESULTS AND DISCUSSION

3.3.1 CROSS-LINKDENSITY

The cross-link densities of samples irradiated with doses of 50, 122, and 198 kGy, calculated from equilibrium swelling in toluene, are presented in Figure 3.1. For all samples studied, cross-link densities formed during EB irradiation process are increasing linear function of dose.

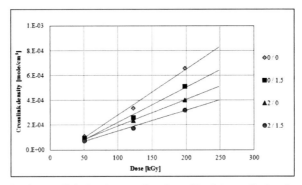

FIGURE 3.1 Total cross-link density as a function of ionizing radiation dose.

The inhibiting effect of DM and sulfur on the radiation cross-linking process was observed. According to experimental work, the inhibiting effect of sulfur (sample 2/0) and DM (sample 0/1.5) is not additive, comparing to corresponding inhibition coming from the same amount of sulfur and DM combined in one sample (2/1.5). The "experimental inhibition" (sample 2/1.5) is lower that the "theoretical inhibition" (summed up inhibition of samples 2/0 and 0/1.5), as shown in Figure 3.2. The probable explanation of the fact can be sulfur—accelerator complex formed during sheets molding process at 110°C. The complex formed facilitate formation of sulfide cross-links and possibly makes the reactions more effective—sulfur is used rather for formation of bridges between the polymer chains, than for formation of cyclic structures modifying the chains.

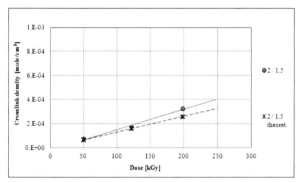

FIGURE 3.2 "Experimental inhibition" (solid line) and "theoretical inhibition" (dashed line) of radiation cross-linking process.

Inhibiting effect of DM arises from the presence of aromatic rings in its structure. The aromatic compounds are known to influence the radiation-induced modification by effect of resonance energy dissipation [24–25]. In the structural formula of DM, heteroatoms, such as sulfur and nitrogen are also present. The sulfur moieties are known to inhibit the effect of ionizing radiation action on matter [26–27], due to the fact that sulfur groups act as sinks of the radiation energy [28].

Rhombic sulfur itself also causes large inhibiting effect of subjected to ionizing radiation polymer. It has to be enhanced, that sulfur undoubtlessly is strong radiation-protecting agent, but probably the cross-linking efficiency is reduced because of intramolecular reactions, which result in modification of polymer chains by sulfur cyclic structures. The observed inhibiting effect of sulfur has therefore twofold contribution.

3.3.2 CROSS-LINK STRUCTURE

For samples with rhombic sulfur in composition, the cross-link structure investigation was carried out (Figure 3.3). Presence of both C-C and polysulfide cross-links was proved. The C-C cross-links are "regular" effect of polymer irradiation. The sulfide cross-links were formed as a result of breakage of S-S bonds in highly puckered ring structure of rhombic sulfur by the action of accelerated electrons. The S-S bond energy is low—approx. 240 kJ/mole, what makes it susceptible to break and generate sulfur radicals [6], what consequently leads to suphide cross-links formation.

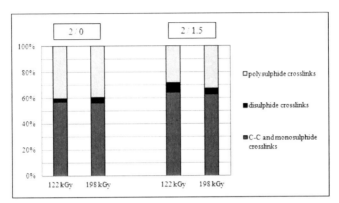

FIGURE 3.3 Cross-link structure of samples 2/0 and 2/1.5, irradiated with 122 and 198 kGy. Network density formed in the samples irradiated with a dose of 50 kGy was very low and the results on cross-link structure obtained from the thiol—amine analysis was not reliable.

The cross-link structure study showed that during irradiation of the sample containing sulfur, but without cross-linking accelerator (sample 2/0), the participation of polysuphide cross-links in the total cross-link density is approx. 40 percent. The difference between the number of polysulfide cross-links formed upon irradiation with 122 and 198 kGy is very little. In sample 2/1.5 in which both sulfur and cross-linking accelerator are present, the number of polysulfide cross-links is lower than in sample 2/0, and it slightly increases with irradiation dose (from 28 percent for 122 kGy up to 32 percent for 198 kGy). The presence of complex of cross-linking accelerator with sulfur promoted thereby formation of shorter cross-links.

3.3.3 MECHANICAL PROPERTIES

The mechanical properties of all samples studied are presented in Table 3.2.

TABLE 2.2 Mechanical properties (SE_{100}, SE_{200}, SE_{300}, TS, Eb) of samples irradiated with 50, 122, and 198 kGy.

Sample	Dose [kGy]	Cross-linkDensity [mol/cm³]	SE_{100} [MPa]	SE_{200} [MPa]	SE_{300} [MPa]	TS [MPa]	E_b [MPa]
0/0	50	$1.1 \cdot 10^{-4}$	2.7	4.0	5.6	10.6	599
	122	$3.4 \cdot 10^{-4}$	5.1	10.2	17.2	25.1	397
	198	$6.6 \cdot 10^{-4}$	9.8	21.1	–	23.6	219

0/1.5	50	$9.3 \cdot 10^{-5}$	2.0	2.9	4.0	7.6	670
	122	$2.6 \cdot 10^{-4}$	3.9	7.6	12.5	20.8	441
	198	$5.1 \cdot 10^{-4}$	7.5	15.8	–	23.7	277
2/0	50	$8.3 \cdot 10^{-5}$	1.9	2.6	3.5	7.6	756
	122	$2.4 \cdot 10^{-4}$	4.2	7.8	12.5	25.2	503
	198	$4.0 \cdot 10^{-4}$	6.1	12.6	21.6	30.9	388
2/1.5	50	$7.3 \cdot 10^{-5}$	2.1	2.7	3.5	6.5	756
	122	$1.7 \cdot 10^{-4}$	3.0	5.3	8.2	17.4	549
	198	$3.2 \cdot 10^{-4}$	4.8	9.7	16.3	26.7	429

In sample 2/0, the generated sulfur radicals inserted into polymer chains, forming long, polysulfide cross-links, which have significant participation in the total cross-link density. The presence of polysulfide cross-links is evident in mechanical properties—high tensile strength is provided by these long, labile bridges, which effectively dissipate the energy. Due to this effect, sample 2/0 showed the highest tensile strength among all analized samples (Figure 3.4). The lowest value of tensile strength exhibited the sample 0/1.5. The presence of the DM not only inhibited the formation of cross-links, but also considerably deteriorated the resulting mechanical properties of the rubber sample. Modification of the polymer chain by the products of DM transformation upon irradiation is probable.

Tensile strength curve of sample containing both sulfur and DM, is located between the corresponding curves of samples containing solely DM or sulfur. Its tensile strength is higher than of sample without sulfur nor DM (0/0), due to presence of mixed, diversified cross-links in sample 2/1.5, and exclusively uniform C-C cross-links in sample 0/0.

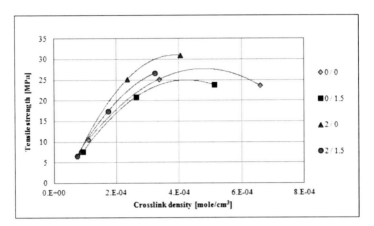

FIGURE 3.4 Tensile strength of samples as a function of cross-link density.

3.4 CONCLUSIONS

In our study, the influence of particular components of sulfur cross-linking system, such as rhombic sulfur and cross-linking accelerator DM, on the process of radiation cross-linking of NBR was investigated.

- Inhibition of radiation cross-linking by both sulfur and DM was proved. Due to complex nature of the investigated system and complicated processes induced by high energy radiation, it is difficult to unambiguously identify the mechanisms responsible for the inhibiting effect.
- Irradiation of samples with sulfur in composition leads to formation of hybrid network type, characterizing itself with both short C-C cross-links and longer sulfide ones. The presence of diversified cross-links guarantee high tensile strength of the rubber samples.

The reactions induced in polymer matrix with sulfur cross-linking system in composition are probably multistage. To comprehend mechanisms of the reactions initiated by ionizing radiation, further investigation within this area is needed.

ACKNOWLEDGMENTS

The work was performed in the frame of Young Scientists' Fund at the Faculty of Chemistry, Lodz University of Technology, Grant W-3/FMN/6G/2013.

KEYWORDS

- Cross-linkdensity and structure
- Mechanical properties
- Nitrile rubber
- Radiation cross-linking

REFERENCES

1. Clough, R. L.; High-energy radiation and polymers: A review of commercial processes and emerging applications. *Nucl. Instrum. Meth. B.* **2001**, *185*(1–4), 8–33.
2. Bhowmick, A. K.; and Vijayabaskar, V.; Electron beam curing of elastomers. *Rubber. Chem. Technol.* **2006**, *79*(3), 402–428.
3. Manaila, E.; Stelescu, M. D.; and Craciun, G.; Advanced elastomers—technology, properties and applications. Aspects Regarding Radiation Crosslinking of Elastomers. In Tech, **2012**, 3–34.
4. Bik, J.; Gluszewski, W.; Rzymski, W. M.; and Zagorski Z. P.; EB radiation crosslinking of elastomers. *Radiat. Phys. Chem.* **2003**, *67*(3), 421–423.
5. Stepkowska, A.; Bielinski, D. M.; and Przybytniak, G.; Application of electron beam radiation to modify crosslink structure in rubber vulcanizates and its tribological consequences, *Acta Phys. Pol. A.* **2011**, *120*(1), 53–55.
6. Vijayabaskar, V.; Costa, F. R.; and Bhowmick, A. K.; Influence of electron beam irradiation as one of the mixed crosslinking systems on the structure and properties of nitrile rubber. *Rubber Chem. Technol.* **2004**, *77*(4), 624–645.
7. Vijayabaskar, V.; and Bhowmick, A. K.; Dynamic mechanical analysis of electron beam irradiated sulphur vulcanized nitrile rubber network—some unique features. *J. Mater. Sci.* **2005**, *40*(11), 2823–2831.
8. Yasin, T.; Ahmed, S.; Yoshii, F.; and Makuuchi, K., Radiation vulcanization of acrylonitrile-butadiene rubber with polyfunctional monomers. *React. Funct. Polym.* **2002**, *53*(2–3), 173–181.
9. Bik, J. M.; Rzymski, W. M.; Gluszewski, W.; and Zagorski, Z. P.; Electron beam crosslinking of hydrogenated acrylonitrile-butadiene rubber. *Kaut. Gummi Kunstst.* **2004**, *57*(12), 651–655.
10. Stephan, M.; Vijayabaskar, V.; Kalaivani, S.; Volke, S.; Heinrich, G.; Dorschner, H.; Wagenknecht, U.; and Bhowmick, A. K.; Crosslinking of nitrile rubber by electron beam irradiation at elevated temperatures. *Kaut. Gummi Kunstst.* **2007**, *60*(10), 542–547.
11. Vijayabaskar, V.; Tikku, V. K.; and Bhowmick, A. K.; Electron beam modification and crosslinking: Influence of nitrile and carboxyl contents and level of unsaturation on structure and properties of nitrile rubber. *Radiat. Phys. Chem.* **2006**, *75*(7), 779–792.
12. Hill, D. J. T.; O'Donnell, J. H.; Perera, M. C. S.; and Pomery, P. J.; An investigation of radiation-induced structural changes in nitrile rubber. *J. Polym. Sci. Pol. Chem.* **1996**, *34*(12), 2439–2454.
13. Loan, L. D.; Peroxide crosslinking reactions of polymers. *Pure Appl. Chem.* **1972**, *30*(1–2), 173–180.
14. Zagorski, Z. P.; Modification, degradation and stabilization of polymers in view of the classification of radiation spurs. *Radiat. Phys. Chem.* **2002**, *63*(1), 9–19.

15. Dogadkin, B. A.; Tarasova, Z. N.; Golberg I. I.; and Kuanyshev K. G.; Effect of vulcanization structures on the strength of vulcanizates. *Kolloid. Zh.* **1962**, *24*, 141–151.
16. Manik, S. P.; and Banerjee, S.; Studies on dicumylperoxide vulcanization of natural rubber in presence of sulfur and accelerators. *Rubber Chem. Technol.* **1969**, *42*(3), 744–758.
17. Manik, S. P.; and Banerjee, S.; Sulfenamide accelerated sulfur vulcanization of natural rubber in presence and absence of dicumyl peroxide. *Rubber Chem. Technol.* **1970**, *43*(6), 1311–1326.
18. Bakule, R.; and Havránek, A.; The dependence of dielectric properties on crosslinking density of rubbers. *J. Polym. Sci. Polym. Symp.* **1975**, *53*(1), 347–356.
19. Bakule, B.; Honskus, J.; Nedbal, J.; and Zinburg, P.; Vulcanization of natural rubber by dicumyl peroxide in the presence of sulphur. *Collect. Czech. Chem. Commun.* **1973**, *38*(2), 408–416.
20. Bandzierz, K.; and Bielinski, D. M.; Radiation methods of polymers modification: hybrid crosslinking of butadiene—acrylonitrile rubber. **2013**, 244–247. Abstracts Collection on New Challenges in the European Area: Young Scientists, Baku, Azerbaijan.
21. Flory, P. J.; and Rehner, J.; Statistical mechanics of crosslinked polymer networks II. Swelling. *J. Chem. Phys.* **1943**, *11*(11), 521–526.
22. Hwang, W.-G.; Wei, K.-H.; and Wu, C.-M. Mechanical, thermal, and barrier properties of NBR/Organosilicate nanocomposites. *Polym. Eng. Sci.* **2004**, *44*(11), 2117–2124.
23. Saville, B.; and Watson, A. A.; Structural characterization of sulfur-vulcanized rubber networks. *Rubber Chem. Technol.* **1967**, *40*(1), 100–148.
24. Głuszewski, W.; and Zagórski, Z. P. Radiation effects in polypropylene/polystyrene blends as the model of aromatic protection effects. *Nukleonika.* **2008**, *53*(1), 21–24.
25. Seguchi, T.; Tamura, K.; Shimada, A.; Sugimoto, M.; and Kudoh, H. Mechanism of antioxidant interaction on polymer oxidation by thermal and radiation ageing. *Radiat. Phys. Chem.* **2012**, *81*(11), 1747–1751.
26. Charlesby, A.; Garratt, P. G.; and Kopp, P. M.; Radiation protection with sulphur and some sulphur-containing compounds. *Nature.* **1962**, *194*, 782.
27. Charlesby, A.; Garratt, P. G.; and Kopp, P. M. The use of sulphur as a protecting agent against ionizing radiations. *Int. J. Radiat. Biol.* **1962**, *5*(5), 439–446.
28. Nagata, C.; and Yamaguchi, T. Electronic structure of sulfur compounds and their protecting action against ionizing radiation. *Radiat. Res.* **1978**, *73*(3), 430–439.

CHAPTER 4

RUBBER VULCANIZATES CONTAINIG PLASMOCHEMICALLY MODIFIED FILLERS

DARIUSZ M. BIELIŃSKI, MARIUSZ SICIŃSKI, JACEK GRAMS, and MICHAŁ WIATROWSKI

4.1 INTRODUCTION

Powders are commonly used as fillers for rubber mixes. The most popular are carbon black, silica, kaolin, or more modern like graphene, fullerenes, and carbon nanotubes. The nature of their surface is the main attribute of fillers, as surface energy and specific area determine the compatibility of filler with rubber matrix and the affinity to other c ingredients. One of the major problems is the tendency of fillers to agglomeration—formation of bigger secondary structures, associated with lower level of filler dispersion, what is reflected by the decrease of mechanical properties of rubber vulcanizates [1]. Surface modification of powder can improve interaction between rubber matrix and filler. Application of low-temperature plasma treatment for this purpose has been drawing increasing attention recently [2, 3].

Silica is one of the most popular mineral filler used in rubber technology. Three types of silica can be distinguished, namely precipitated, fumed, and surface-modified silica. As an amorphous material with randomly placed functional silanol groups (Figure 4.1), it readily generates hydrogen bonds with surrounding molecules [4].

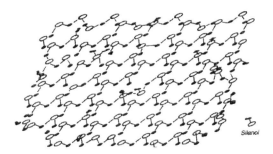

FIGURE 4.1 Surface chemistry of silica [4].

Polar character and big specific surface area enable various modifications of silica surface. Modifying by coupling compounds is the most popular one [4]. In subject literature [5, 6] and patent declarations [7], many references on the modification processes, their kinetics, current opportunities, and proposals of further development can be found. All chemical methods have a significant disadvantage: emission of large amounts of chemical waste, usually in the form of harmful solvents.

Taking into account the necessity of their utilization, application of "clean" plasma modification has to be considered as a cost effective possibility for significant reduction of environmental hazard.

Low-temperature plasma can be generated with a discharge between electrodes in a vacuum chamber. The process used to be carried out in the presence of gas (i.e., Ar_2, O_2, N_2, methane, or acetylene). Depending on the medium applied, surface of modified material can be purified, chemically activated, or grafted with various functional groups.

This paper presents the results of low-temperature, oxygen plasma activation of silica, kaolin, and wollastonite. Fillers were modified in a tumbler reactor, enabling rotation of powders in order to modify their entire volume effectively. Based on our previous work [8], the process was carried out with 100W discharge power. The time of modification varied from 8 to 64 min. Additionally, for the most favorable (in terms of changes to surface free energy) time of modification for kaolin, the process was repeated and ended with a flushing of the reactor chamber with hydrogen, in order to reduce of carboxyl groups content, generated on filler surface. Rubber mixes, filled with the modified powders, based on SBR or NBR were prepared and vulcanized. Mechanical properties of the vulcanizates were determined and explained from the point of view rubber—filler interactions and filler, estimated from micromorphology of the materials.

4.2 EXPERIMENTAL

4.2.1 MATERIALS

4.2.1.1 RUBBER VULCANIZATES

Three fillers were the objects of study: micro silica Arsil (Z. Ch. Rudniki S.A., Poland), kaolin KOM (Surmin-Kaolin S.A., Poland), and wollastonite Casiflux (Sibelco Specialty Minerals Europe, The Netherlands). Rubber mixes, prepared with their application, were based on: styrene-butadiene rubber (SBR), KER 1500 (Synthos S.A., Poland), and acrylonitrile-butadiene rubber (NBR) NT 1845 (Lanxess, Germany).

Rubber mixes were prepared with a Brabender Plasticorder laboratory micromixer (Germany), operated with 45 rpm, during 30 min. Their composition is presented in Table 4.1. The only one variable was the type of modified mineral filler (see Section 4.2.1).

TABLE 4.1 Composition of the rubber mixes studied

Components	Content [phr]	
SBR KER1500	100	0
NBR NT1845	0	100
ZnO	5	5
Stearine	1	1
CBS	2	2
Sulphur	2	2
Arsil Silica	20	20
Modified filler	20	20

Samples were vulcanized in 160 °C, time of vulcanization: 6 min. (for NBR vulcanizates) and 15 min. (for SBR vulcanizates).
Symbols of the prepared vulcanizate samples:
  NBR-X—composites based on NBR rubber, X—modified filler;
  SBR-X—composites based on SBR rubber, X—modified filler.

4.2.1.2 PLASMOCHEMICAL MODIFICATION OF FILLERS

Fillers studied were modified with a Diener tumbler plasma reactor (Germany). The reactor operated with the frequency of 40 kHz and the maximum discharge power of 100 W. Scheme of the reactor is shown in Figure 4.2.

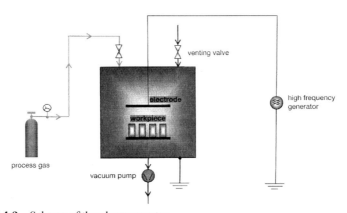

FIGURE 4.2 Scheme of the plasma reactor.

Mineral fillers were subjected to the oxygen plasma treatment during various points of time. Efficiency of process gas flow was 20 cm³/min, and the pressure in the reactor chamber was maintained at 30 Pa. Symbols of the modified fillers are as follows:

A-REF—silica Arsil, virgin reference filler;
A-XX—silica Arsil, modified during time of XX min (XX = 16; 32; 48; 64);
K-REF—kaolin, virgin reference filler;
K-XX—kaolin, modified during time of XX min (XX = 8; 16; 32);
K-16H—kaolin, modified during 16 min, the process terminated with hydrogen;
W-REF—ollastonite, virgin reference filler;
W-XX—wollastonite, modified during time of XX min (XX = 16; 32; 48; 64);

4.2.2 TECHNIQUES

4.2.2.1 SURFACE FREE ENERGY OF FILLERS

Effectiveness of plasmochemical modification of the fillers is represented by changes to their surface free energy (SFE) and its components—polar and dispersion one. SFE was examined with a K100 MKII tensiometer (KRÜSS GmbH, Germany). Contact angle was determined using polar (water, methanol, ethanol) and nonpolar (n-hexane, n-heptane) liquids. SFE and its components were calculated by the method proposed by Owens-Wendt-Rabel-Kaeble [9].

4.2.2.2 MICROMORPHOLOGY OF RUBBER

Micromorphology of rubber vulcanizates was studied with an AURIGA (Zeiss, Germany) scanning electron microscope (SEM). Secondary electron signal (SE) was used for surface imaging. Accelerating voltage of the electron beam was set to 10 keV. Samples were fractured by breaking after dipping in liquid nitrogen.

4.2.2.3 MECHANICAL PROPERTIES OF RUBBER VULCANIZATES

Mechanical properties of the vulcanizates studied were determined with a Zwick 1435 universal mechanical testing machine (Germany). Tests were carried out on "dumbbell" shape, 1.5 mm thick and 4 mm width specimens, according to PN-ISO

37:1998 standard. The following properties of the materials were determined: elongation at break (Eb), stress at elongation of 100% (SE100), 200% (SE200), 300% (SE300), and tensile strength (TS).

4.3 RESULTS AND DISCUSSION

Our previous studies revealed that low-temperature plasma causes changes to SFE and its component of carbon nanotubes [10]. Plasma modification is a good method of CNT purification as an amorphous carbon is eliminated from their surface during process [11]. Purifying changes the properties of CNT, and affects its dispersion in rubber matrix. It encouraged us to try plasma modification to silica, kaolin, and wollastonite. The objective of the study was to characterize changes to filler surface and its susceptibility to oxygen activation, being expected to be an intermediate step in surface functionalization with various chemical groups/compounds.

4.3.1 SURFACE FREE ENERGY (SFE)

Reference silica powder represents relatively low value of surface energy and its polar component (Figure 4.3a)—probably because of physically adsorbed water present on filler surface. After modification—regardless time of the process—SFE remains constant, however the dispersive component decreases in favor of the polar component increasing. Generally, silica remains resistant to plasma modification under the experimental conditions. However, it seems likely to change under higher discharge power.

Wollastonite behaves in a different way (Figure 4.3b). After 48 min of plasma treatment value of its SFE reaches a maximum. Its polar component becomes almost doubled—probably because grafting of oxygen groups on filler surface. After 48 min of the treatment polar component of SFE is decreasing, probably because the process balance moves toward the surface cleaning.

Plasma treatment of kaolin (Figure 4.3c) during 16 min results in an increase of SFE value and its polar component. After this time further changes are not observed. Hydrogen termination of the process (lasting 2 min) results in almost doubled the polar component of SFE, probably being the effect of surface present carbonyl groups reduction to the more stable carboxyl ones.

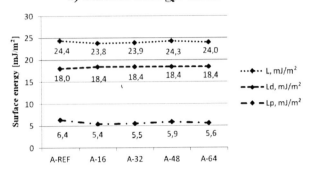
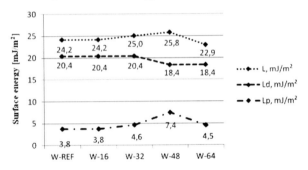
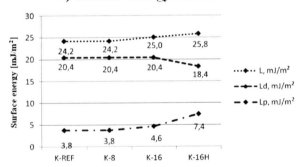

FIGURE 4.3 Results of the analysis of total surface free energy of fillers: (a) Silica, (b) Wollastonite, (c) Kaolin (L—total surface free energy, L_d—dispersive part, L_p—polar part).

4.3.2 MORPHOLOGY OF RUBBER VULCANIZATES

In order to determine the influence of rubber matrix polarity on filler dispersion and rubber-filler interaction, two kinds of rubber, NBR and SBR, were chosen. SEM pictures of the rubber vulcanizates, filled with reference and 48 min plasma-treated wollastonite, are presented in Figure 4.4. Morphology of SBR/wollastonite samples does not reveal any changes, explaining strengthening of the material (see the next section).

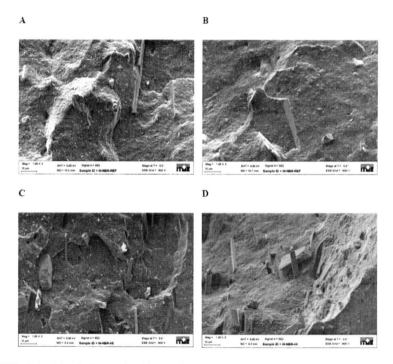

FIGURE 4.4 Morphology of rubber vulcanizates studied: (a), (b)—NBR-W-REF; (c), (d)—NBR-W-48; magnification 5.000x.

Pictures of NBR-W-REF samples (Figure 4.4a, b) present broken needles of wollastonite in the area of fracture, whereas in the case of NBR-W-48 sample (Figure 4.4c, d) needles of wollastonite are nonbroken but "pulled out" from rubber matrix. This change to morphology, reflected by lower rubber-filler interactions, responsible for worse mechanical properties of rubber vulcanizates (see the next section), is undoubtedly the result of an increase of SFE polar component of filler after plasma treatment. The SEM pictures of the vulcanizates, no matte, containing virgin or modifies wollastonite particles, do not reveal any filler agglomeration.

Morphology of SBR-K-REF and SBR-K-16H samples are presented in Figure 4.5. Agglomerates of kaolin can be seen in vulcanizate containing reference filler (Figure 4.6a, b). Modified kaolin does not exhibit tendency to agglomeration (Figure 4.5c, d). Better filler dispersion suggests on higher mechanical properties of the SBR vulcanizates filled with plasma-treated kaolin.

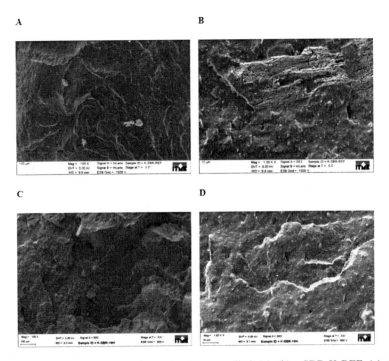

FIGURE 4.5 Morphology of rubber vulcanizates studied: (a), (b)—SBR-K-REF; (c), (d)—SBR-K-16H; magnification 100x (A,C) and 1000x (B,D)

4.3.3 MECHANICAL PROPERTIES OF RUBBER VULCANIZATES

Mechanical properties of the vulcanizates studied, containing virgin and plasma modified fillers, are presented in Figure 4.6a, b, c, d, e, f.

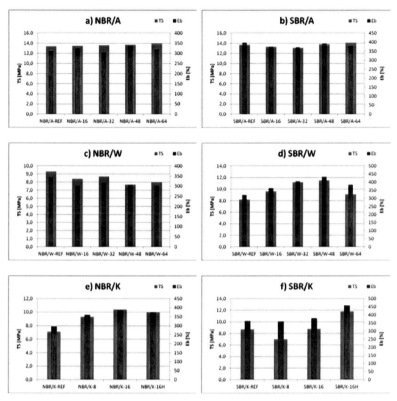

FIGURE 4.6 Mechnical properties of elastomer composites based on NBR and SBR filled with: (a) (b) silica, (c) (d) wollastonite, (e) (f) kaolin; TS—tensile strength, Eb—elongation at break.

Plasma modification does not cause any changes to SFE of silica. This is clearly reflected by the mechanical properties of silica filled rubber vulcanizates (Figure 4.6a, b). Changes to the values of material stress at elongation 100, 200, and 300 percent (SE100, SE200 and SE300), its TS and Eb, being the result of plasma treatment of the filler, are negligible.

Mechanical properties of wollastonite filled NBR vulcanizates decrease due to plasma modification of the filler (Figure 4.6c), whereas in case of vulcanizates based on SBR an increase of TS and Eb is observed (Figure 4.6d)—especially for the most effective 48 min treatment. SEM pictures of the vulcaniates confirm on adequate changes to their morphology.

For the rubber vulcanizates filled with kaolin (Figure 4.6e, f), despite the biggest changes to SFE and its components (observed for 16 min plasma treatment followed by hydrogen termination), determined changes to mechanical properties are different in comparison to the wollastonite filled vulcanizates. The biggest increase of TS and Eb is observed for SBR/K-16H sample - about 30 percent as compared to the reference sample (containing virgin filler). Reinforcement of rubber seems to be dependent on overlapping effects originated from rubber-filler interactions and dispersion of filler in rubber matrix.

4.4 CONCLUSIONS

1. Oxygen plasma treatment can activate surface of mineral fillers, by grafting oxygen groups on the surface.
2. The efficiency of the treatment depends on the filler. Changes to surface free energy and its components are observed for kaolin and wollastonite, whereas practically no energetic effect is present in the case of silica.
3. Any changes to filler particles SFE and its components effect on mechanical properties of rubber vulcanizates filled with the modified filler. Improvement of mechanical properties of the materials originates increased rubber-filler interaction and better dispersion of filler particles in rubber matrix.

KEYWORDS

- Mineral fillers
- Low-temperature plasma
- Surface modification
- Rubber vulcanizates

ACKNOWLEDGEMENT

The project was funded by the National Science Centre Poland (NCN) conferred on the basis of the decision number DEC-2012/05/B/ST8/02922

REFERENCES

1. Wolff, S.; and Wang, J.; Filler—Elastomer Interactions. Part IV. The effect of the surface energies of fillers on elastomer reinforcement. *Rubber. Chem. Technol.* **1992**, *65*, 329–342.
2. Dierkes, W. K.; Guo, R.; Mathew, T.; Tiwari, M.; Datta, R. N.; Talma, A. G.; Noordemeer, J. W. M.; and van Ooij, W. J.; A key to enhancement of compatibility and dispersion in elastomer blends. *Kautschuk Gummi Kunststoffe.* **2011**, *64*, 28–35.
3. Chityala, A.; and van Ooij, W. J.; Plasma deposition of polymer films on pmma powders using vacuum fluidisation techniques. *Surf. Eng.* **2000**, *16*, 299–302.
4. Wang, M.-J.; Effect of polymer-filler and filler-filler interactions on dynamic properties of filled vulcanizates. *Rubb. Chem. Technol.* **1998**, *71*, 520–589.

5. Hair, M. L.; and Hertl, W.; Reaction of chlorosilanes with silica. *J. Phys. Chem.* **1971**, *14*, 2181–2185.
6. Blume, A.; Kinetics of the Silica-Silane Reaction. *Kautschuk und Gummi Kunststoffe.* **2011**, *4*, 38–43,
7. Revis, A.; Chlorosilane blends for treating silica. *US Patent* **2003**, 6613139B1.
8. Bieliński, D.; Parys, G.; and Szymanowski, H.; Plazmochemiczna modyfikacja powierzchni sadzy jako napełniacza mieszanek gumowych. *Przemysł Chemiczny.* **2012**, *91*, 1508–1512.
9. Owens, D. K.; and Wendt, R. C.; Estimation of the surface free energy of polymers. *J. Appl. Polym. Sci.* **1969**, *13*, 1741–1747.
10. Siciński, M.; Bieliński, D.; Gozdek, T.; Piątkowska, A.; Kleczewska, J.; and Kwiatos K.; Kompozyty elastomerowe z dodatkiem grafenu lub MWCNT modyfikowanych plazmochemicznie. *Inżynieria Materiałowa.* **2013**, *6*,
11. Xu, T.; Yang, J.; Liu, J.; and Fu, Q.; Surface modification of multi-walled carbon nanotubes by O_2 plasma. *Appl. Surf. Sci.* **2007**, *253*, 8945–8951.

CHAPTER 5

MODIFICATION OF THE INDIAN RUBBER IN THE FORM OF LATEX WITH OZONE

L. A. VLASOVA, P. T. POLUEKTOV, S. S. NIKULIN, and V. M. MISIN

5.1 INTRODUCTION

Various functional groups involved in the macromolecules of elastomers impart all kind of rubbers a new complex of service properties as it was shown in practice. This fact considerably expands the market for rubbers as well as the areas of rubbers application.

According to a number of Russian and foreign scientists some distinctive features of the Indian rubber unlike of its nearest synthetic analog (cis-polyisoprene of different grades) are first of all connected with the absence of the functional groups in the macromolecules of synthetic isoprene rubber. The presence of such functional groups in the Indian rubber is provided by existence of nonrubber components inside, mainly of the protein type. Moreover, the difference is connected with a content and structure of gel, and besides, molecular-mass distribution of the Indian rubber macromolecules [1–4].

This viewpoint is supported by increase of the cohesion strength of the rubber compounds due to the modification of rubbers with polar compounds, for example, grafting of maleic anhydride to lithium polyisoprene [5] or by the treatment of SKI-3 rubber with n-nitrosodiphenylamine [6]. Modification of the Indian rubber is made similar to synthetic rubbers by polymer-like reactions, for example, by epoxidation [6], maleinization, hydroxylation, and using some other techniques. All enumerated reactions impart a rubber some additional properties that are characteristic for the chemical properties of either functional group. The most extensive possibilities for proceeding of the chemical reactions are inherent for epoxy, as well as carbonyl groups in aldehydes and ketones because the double bond in the carbonyl groups of these compounds is strongly polarized. For example, in order to modify

Indian rubber by latex epoxidation peroxyformic acid was applied (system in-situ: formic acid—hydrogen peroxide). As a result, a modification of the Indian rubber was performed in the industrial environment that was intended for special purposes, particularly, for producing of shock-absorbing rubber soles at the railways [6].

In spite of a wide application of ozone in the laboratory routine, as well as numerous investigations of the mechanism of ozonization reaction and the study of the structure of a lot of polymer materials and chemical compounds, ozone was begun to be applied in the industrial technology only recently when high-duty ozonization units have appeared in the market [6]. Now ozonolysis of unsaturated rubbers in a solution on an industrial scale is used for production of oligomers with the terminal functional groups that can be applied as the binders for a solid propellant, as the components of rubber products, tires, film-forming composites [1]. Oligoisobutylenes with ketone and carboxyl groups [2] became the first oligomers obtained in industry by ozonolysis of butyl-rubber and they were intended for the use as thickener additives to the automotive oils.

The works enumerated above are related to performing of polymer ozonolysis in the organic solvents. As a rule, the usual disadvantage of this process is a high viscosity of polymer solutions and hence, the corresponding difficulties while stirring of polymer solution with the gas mixture containing ozone. Moreover, while using ozone-air mixtures in contact with a solvent, a usual release of hydrocarbon solvent from the bulk of reaction mass is observed. This can lead not only to the loss of solvent but also to the formation of flammable and explosive air-gas mixtures.

These disadvantages are completely excluded under ozonization of polymers at the stage of latex [1-2]. The essence of elaborated technique is in the fact that latex of butadiene-styrene copolymer is treated with air-ozone mixture with a simultaneous regulating of pH value of latex within 9.5–10.5 range. In this case, practically instantaneous and complete absorption of ozone by latex polymer takes place. Ozonization degree is regulated by the volume of supplied ozone-air mixture within the required mass ratio value: dry latexsubstance-absorbed ozone. Information on the ozonization of the Indian rubber in the form of latex is not known in scientific literature, as well as the data on the properties of the ozonizated latex and Indian rubber.

5.2 EXPERIMENTAL PART

The study of ozonization process of the Indian rubber in latex was performed with the use of the sample of nonconcentrated latex of the Indian rubber received from Vietnam. Basing on the results of analyses made by the authors, latex was characterized by the following factors:

Factor	Value
Dry substance concentration, mass %	33.0
Rubber content, mass %	31.0
Content of nonrubber share, mass %	2.9
pH value	10.6
Ammonium content calculated for NH_3, mass %	0.97
Surface tension, mN/m	46.2
Solubility of rubber in toluene, mass %	93.3
Gel content in a rubber, evolved from latex, mass %	6.1
Averaged-viscous molecular mass of rubber, a.u.	1188000

Rubber for making the investigations was filled with nonstaining anti-oxidant of phenolic type "Agidol—2" at the rating of 0.5 mass percent, after that it was released from the Indian rubber latex sample by acidifying with acetic acid with the following thorough washing of polymer with distilled water and drying it up to residual moisture content of 0.2 mass percent.

The process of latex ozonization was performed with the use of pilot laboratory plant comprising of the unit applied for electrosynthesis of ozone and a reactor where interaction of the Indian rubber latex with air-ozone mixture proceeded. This pilot plant permitted to vary ozonolysis modes: to control and support pH value within the required limits, to control the process temperature, the rate of air ozone mixture and latex supply, to control ozone concentration in the air flow in front of and after reactor, as well as the time of contact of the air-ozone mixture with latex. The unit of ozone electrosynthesis was composed of the system for air preparation and just of ozone generator operating at 220 V 50 Hz AC and working supply voltage at the electrodes up to 6000 V. The control for ozone concentration in the air-ozone mixture was made with the use of iodometric titration. Round-bottomed flask supplied with mechanical stirrer and a bubbler for supplying of the air-ozone mixture was used as a reactor. Reactor was attached to the unit for a continuous control of pH value of the ozonizated latex and also the dosing device specifying the required value of 2 percent aqueous solution of potassium hydroxide for supporting of pH value in the latex within 9.5–10.5.

5.3 RESULTS AND DISCUSSION

Latex of the Indian rubber is known to have a number of specific features in its colloid-chemical properties as compared with those ones of synthetic rubber latexes. The main differences are much more sizes of the latex particles that attain 200.0–350.0 nm as well as a specific property of their protective adsorption layer consisting of a set of the natural high-molecular fatty acids, alcohols, resinous acids and protein-like compounds [3]. In this case, content of the dry substances soluble in water is of about 3.0–3.5 mass percent in a virgin latex, according to [3]. This is approximately in accordance with the result of analysis of the investigated latex (2.9 mass %). Fatty acids in latex are presented by oleic, linoleic ones as well as by other carboxyl-containing compounds that have concentration of 1.05–2.05 mass percent in the acetone extract for Indian rubber latex. Content of the protein substances in the freshly gathered latex was up to 4.0 mass percent, while free amino acids were found in the concentration of 0.2 mass percent.

This rather complicated composition of compounds participating in the formation of adsorption layer on the globules of Indian rubber latex provides an ability of its existence in a dependence of pH value in anion (pH>8.0) or in cation (pH<5.0) form keeping to some extent its aggregative stability. As a rule, Indian rubber latex is filled with ammonia in order to stabilize it and is characterized by pH = 10.0–10.5. Acidifying of latex during its storage or under its targeted filling with acids, for example, acetic acid, results in latex coagulation. This specific feature was taken into account while refining ozonization conditions since under latex ozonization they are somewhat acidified due to the formation of carboxylic groups in polymer. In this case, pH is reduced up to the neutral or even low-acid value that is capable to cause an untimely latex coagulation.

Alkaline reaction leads to the formation of the corresponding fatty-acid salts (anion-active surface-active substances), adsorbed on the surface of polymer particles. Surface tension of the investigated latex determined by tensometric method was of 46.2 mN/m thus defining an extra aggregative stability of Indian rubber latex during air-ozone bubbling.

Experimentally determined reaction rate constant of ozone interaction with –C = C- bonds in the polymer of Indian rubber latex are of the order of $2-6 \cdot 10^4$ l/mole·s. Therefore reaction rate of the unsaturated polymer of isoprene is mainly limited by supply rate of ozone into reactor.

Process of Indian rubber latex ozonization was made in the following way. Latex in the amount necessary for making the experiment was supplied to the round-bottom flask with a mechanical stirrer and after its activation aqueous emulsion of antifoam agent was added. Next, air-ozone mixture with the concentration of 16–18 mg/dm^3 was also supplied through the bubbler. The temperature of the process was supported within 18–20°C, while pH value was within 9.5–10.5. Quantitative absorption of ozone with latex was controlled by iodometric analysis of the outgoing air evolving from the reactor flask through absorbing 2 percent solution of potassium iodide.

In order to reveal the changes taking place with Indian rubber latex and the rubber in the process of ozonization we made the investigations of ozone absorption

TABLE 5.1 The change of colloid-chemical properties of latex and Indian rubber in the process of ozonization

No	Amount of the Bound Ozone Relative to the Rubber, Mass %	Solid Residue, Mass %	pH Value After Ozonization	Surface Tension, mN/m	Content of Carbonyl Groups in Rubber, Mass %	Content of Carboxyl Groups in Rubber, Mass %	Solubility of Rubber in Toluene, Mass %	Gel Content, Mass %	Averaged-Viscous Molecular Mass (MM) of Rubbers Released from Rubber Samples
1	2	3	4	5	6	7	8	9	10
2	0	33.00	10.70	46.20	Not determined	Not determined	93.90	6.10	1188000
3	0.30	32.80	10.50	42.50	0.11	0.10	93.00	6.90	1082100
4	0.70	32.90	10.40	41.40	0.30	0.32	93.60	6.40	997274
5	0.80	32.80	10.40	40.50	0.40	0.40	92.30	7.70	990120
6	1.00	33.20	10.30	38.90	0.50	0.41	91.70	8.30	993000
7	1.50	31.90	10.10	36.00	0.71	0.70	89.50	10.50	825037
8	4.00	33.00	9.90	35.00	2.00	1.80	85.00	15.00	394094
9	9.00	32.50	8.70	35.00	4.50	4.10	83.70	16.30	138943

Note: for the Indian rubber, separated from the given latex sample, stabilized with anti-oxidant Agidol-2 (0,5 mass %) after plasticization with rolling mills at the temperature of 25–30°C and the gap of 2 mm for 10 min Mooney viscosity was reduced up to 50,0–55,0 units.

with latex basing on the estimation of 0.3; 0.7; 0.8; 1.0; 1.5; 4.0; 8.0 mass percent of ozone absorption relative to the rubber. An increase of ozonization degree did not result in any complications in the process connected with latex aggregative stability.

Results of analysis of the main colloid-chemical properties for the modified and original latexes and Indian rubber are presented in Table 5.1.

As it is seen from the obtained data, the ozonization process of the Indian rubber latex with an increase of the amount of the bound ozone is entailed by the regular decrease of the surface tension value from 46.0 mN/m for the original latex to 35.0 mN/m for the ozonizated latexes. Consequently, the degree of adsorption saturation for the surface of latex globules increased from 55–60 percent to 100 percent.

This phenomenon is connected with the formation of intermediate ozonide cycles under ozonolysis of the binary bonds and dissociation of these cycles up to the fragments of macromolecules with carboxyl and carbonyl terminal groups. Then carboxyl groups in the ionized form (pH—9.5–10.5) can additionally participate in the formation of protective adsorption layer on the surface of latex particles. Due to increase of the adsorption saturation, Indian rubber latex gained the enhanced aggregative stability that was observed after extraction of rubber samples from ozonizated latexes by coagulation method. It should be noted that latex concentration in the process of its treatment with ozone was not actually changed and it was not accompanied by formation of coagulum; this fact confirmed technological ability of the considered way to modify Indian rubber.

While investigating the properties of ozonizated rubbers, a continuous decrease of the averaged-viscosity MM was observed with an increase of the ozone amount attached to the rubber. At the same time the content of gel fraction remained almost the same and slightly increased with the rise of ozonization up to 4–8 mass percent and, consequently, with the increase of the number of functional groups in the polymer chain.

It was found that the value of mean molecular weight of original Indian rubber present in the fraction soluble in toluene was equal approximately to 1 188 000 units. Ozonization of the Indian rubber in latex with the attachment of 0.8–1.0 mass percent of ozone reduced mean molecular weight up to 990 000–993 000 units (i.e., to the value that approximately corresponded to the working viscosity of mechanically plasticized rubber with Mooney viscosity equal to 50–55 units). Dependence of the change of MM for the ozonizated rubber on the amount of the bound ozone is presented in Table 5.1. In all of the samples of modified polymers extracted from the ozonizated Indian rubber latex content of the carbonyl and carboxyl functional groups regularly increased with an increase of amount of the attached ozone. The presence of carbonyl groups related to aldehyde or ketone groups in the ozonizated rubbers was confirmed by the results of IR spectroscopy in accordance with absorption band in the range of 1,710–1,740 cm^{-1} and by UV spectroscopy by the presence of absorption at 270–280 nm. The amount of carboxyl groups was determined by acid-base titration of rubber solution in toluene with 0.1 N alcohol solution of potassium hydroxide. It should be noted that according to the known scheme of ozonization reaction functional groups incorporated into the Indian rubber can be arranged at the ends of molecular chain. It is connected with the fact that Indian rubber has cis-1,4—structure and actually does not involve the links of 1,2—or 3,4—attachments. Therefore with a release of internal binary bonds under the effect of ozone Indian rubber can form only terminal functional groups. This feature in the arrangement of functional groups in the ozonizated polymers can play an essential positive role since their participation in the chemical reactions with the components of rubber compounds facilitates elongation of molecular chains and it will not lead to scorching. Similar disadvantage is characteristic of synthetic carboxylate rubbers which have statistical distribution of the carboxylic groups along the copolymer chain.

To estimate the properties of ozonizated Indian rubber in the rubber compound a sample of 0.4 kg mass was obtained with ozonization degree of 0.8 mass percent (sample № 6 in Table 5.1). This sample was used for the preparation of the rubber compound according to the standard formulation approved in tire production. Rubber compound prepared on the basis of the rubber extracted from the sample of original latex that was preliminarily plasticized with the use of cold rolling mills with the gap of 2 mm for 3 min was applied as the reference sample. The choice of plasticization time was connected with the necessity of decrease of the rubber MM up to the value of that one obtained from the ozonizated latex.

Rubber compounds were vulcanized at 133°C for 30 min. Results of the tests are presented in Table 5.2.

TABLE 5.2 Properties of the rubber compounds on the basis of initial and ozonizated Indian rubbers

N	Name of the Factors	Initial Rubber	Ozonizated Rubber
1	2	3	4
1	Mooney viscosity, MB 1+4 (100°C) rubber compound	18.00	22.00
2	Cohesion strength of rubber compound, kg-force/cm^2	3.90	7.21
3	Strength of bond with metal according to GOST 209-62, kg-force/cm^2	15.50	15.20
4	Tension under 300 % elongation, MPa	7.00	8.40
5	Conditional strength under rupture, MPa	31.50	33.40
6	Relative elongation under rupture, %	670.00	625.00
7	Relative residual deformation after rupture, %	28.00	26.00

Note: factors of NN 3-7 are presented for vulcanizates

5.4 CONCLUSIONS

Analyzing the obtained experimental data one can conclude that rubber ozonization did not make worse the basic physicomechanical quality factors of the rubbers made on its basis. At the same time cohesion strength of the raw rubber considerably increased and the strength of the vulcanized rubber somewhat increased as well. It should be especially noted that the tested sample of ozonizated Indian rubber does not require placticization with the use of roll mills, that is rather power- and labor-consuming operation accepted in the technology of tire production.

KEYWORDS

- **Indian rubber**
- **Latex**
- **Ozonization**
- **Physicomechanical properties**

REFERENCES

1. Garmonov, I. V.; Caoutchouc and rubber, **1973**, № 5, pp. 6–15.
2. Poddubnyj, I. Ya.; Grechanovskii V.A.; and Ivanova L. S.; Molecular structure and microscopic properties of synthetic cis-polyisoprene. Report at the international symposium on isoprene rubber. Moscow, 20-24/IX, 1972. M. : TsNIITEneftekhim, **1972**, p. 19.
3. Briston, J.M.; Canin, J. I.; and Mullins, L.; Comparison of the properties and performance specifications of Indian rubber and synthetic cis-polyisoprene. Report at the international symposium on isoprene rubber. Moscow, 20-24/IX, 1972. M. : TsNIITEneftekhim, **1972**, p. 40.
4. Greg, E. S.; and McKey, J. H.; Differences of technological properties of synthetic polyisoprene and Indian rubber. Influence of non-rubber components. Report at the international symposium on isoprene rubber. Moscow, 20-24/IX, 1972. M.: TsNIITEneftekhim, **1972**, p. 35.
5. Sobolev, V. M.; and Borodina, I. V;. Industrial synthetic rubbers. M.: Khimia, **1977**, p. 256.
6. Lykin, A. S.; et al. Investigations of n—nitrosodiphenylamine effect on cohesion strength and stability of the coating rubbers made from SKI-3. Report at the international symposium on isoprene rubber. Moscow, 20-24/IX, 1972. M.: TsNIITEneftekhim, **1972**, p. 18.

CHAPTER 6

INFLUENCE OF THE STRUCTURE OF POLYMER MATERIAL ON MODIFICATION OF THE SURFACE LAYER OF IRON COUNTERFACE IN TRIBOLOGICAL CONTACT

DARIUSZ M. BIELIŃSKI, MARIUSZ SICIŃSKI, JACEK GRAMS, and MICHAŁ WIATROWSKI

6.1 INTRODUCTION

Interest toward chemical reactions accompanying friction has been growing since last years. This is reflected by significant progress in very important area of tribology, called tribochemistry [1]. One of its priorities are studies on chemical reactions taking place in the surface layer of materials constituting the friction couple and their exploitation consequences, e.g. concerning creation of protective layers, lowering wear, etc.

An increase of temperature in tribological contact during friction is well known. It facilitates the phenomenon of selective transfer of polymer components, followed by their chemical reaction with the surface layer of metal counterface, in the case of rubber–metal friction couple. The modification can not only effect composition and structure of the surface layer of polymer but metal as well [2]. Our previous studies confirmed on the possibility to modify the surface layer of iron counterface by sliding friction against sulfur vulcanizates of styrene–butadiene rubber [3, 4]. Extend of modification is related to the kind of dominated sulfur cross-links. An increase of temperature accompanying friction facilities breaking of cross-links present in vulcanizate, especially polysulfide ones [5]. The highest degree of modification was detected for Armco iron specimen working in tribological contact with rubber cross-linked by an effective sulfur system of short: mono- and di- to long polysulfide cross-links ratio equal to 0.55. Polysulphide cross-links characterize themselves by the lowest energy from the range of cross-links created during conventional sulfur vulcanization (-C-C-, -C-S-C-, -C-S_2-C-, -C-S_n-C-; where n≥3) [6]. So, their breaking as first is the most probable. As a result, the release of sulfur ions, representing high chemical reactivity to iron, takes place. FeS layer of 100–150 nm thickness was detected on iron specimen subjected to friction against sulfur vulcanizates of SBR [3]. It lubricates efficiently the surface of metal, reducing the coefficient of friction [7]. As a compound of low shearing resistance, FeS is easily spreaded in the friction zone, adheres to metal counterface, penetrating its microroughness. Even very thin layer of FeS showed to be effective due to high adhesion to iron. Metal oxides (mainly Fe_3O_4) being created simultaneously on the metal surface, act synergistically with FeS, making it wear resistance significantly increased [8]. This paper is to compare other polymer materials containing sulfur to conventional rubber vulcanizates in terms of their ability to the surface modification of iron. The polymers studied vary from sulfur vulcanizates either according to cross-link density (ebonite) or the kind of sulfur incorporation in macromolecules (polysulphone—rigid material and polysulfide rubber—elastomer).

6.2 EXPERIMENTAL

6.2.1 MATERIALS

Surface polished specimen made of Armco iron were subjected to extensive friction against:

polysulphone PSU 1000 (Quadrant PP, Belgium),
cross-linked polysulfide rubber LP-23 (Toray, Japan),
ebonite based on natural rubber [9], or
carbon black filled sulfur vulcanizate of styrene-butadiene rubber Ker 1500 (Z. Chem. Dwory, Poland).

Composition of the materials studied is given in Table 6.1.

TABLE 6.1 Composition of the polymer materials studied

Material Components	Polysulphide rubber	Ebonite	Conventional Rubber (SBR)	Polysulphone
Styrene-butadiene rubber, Ker 1500			100	
Natural rubber, RSS II		100		
Polysulphide rubber, LP-23	100			
Poly(sulphone), PSU 1000				100
Stearic acid			1	
Zinc oxide, ZnO			3	

Influence of the Structure of Polymer Material   113

HAF carbon black, Corax N 326			50
Ebonite powder		50	
Linseed oil		2	
Isostearic acid	0.20		
Manganese dioxide, MnO$_2$	10		
Tetramethylthiuram disulfide, TMTD	0.50		
N-third buthyl-di-benzothiazolilosulphenamide, TBBS			2.20
Zinc dithiocarbamate, Vulkacit		1	
Sulphur, S$_8$		42	0.80

Rubber mixes were prepared with a David Bridge (UK) roller mixer. Specimen for further examinations were vulcanized in a steel mold at 160°C, during time $\tau_{0.9}$, determined rheometrically with a WG 05 instrument (Metalchem, Poland), according to ISO 3417. Liquid polysulfide rubber was cured at room temperature by means of chemical initiator, activated by MnO$_2$. Polysulphone specimen were prepared by cutting off from a rod.

Modification of Armco iron counterface was performed by rubbing of polymer materials studied against metal specimen. The process was realized with a T-05 tribometer (IteE-PIB, Poland).

6.2.2 TECHNIQUES

6.2.2.1 TOF-SIMS

Studies were carried out by means of an ION-TOF SIMS IV instrument (Germany), operating with a pulse ^{69}Ga ion gun of beam energy 25 kV. Primary ion dose was any time kept below 3 10^{11}cm^{-2} (static mode). Negative and positive ion spectra of iron specimen were collected in the range of m/z 1–800, before and after friction against polymer materials. Analysis was narrowed to the range of m/z<35, which showed to be the most relevant in terms of sulfur modification. The most informative signals can be subscribed to: H$^-$ (m/z = 1), C$^-$ (m/z = 12), CH$^-$ (m/z = 13), O$^-$ (m/z = 16), OH$^-$ (m/z = 17), C$_2^-$ (m/z = 24), S$^-$ (m/z = 32), and SH$^-$ (m/z = 33). Any time counts of S$^-$ and SH$^-$ were normalized to the total number of counts present in the spectrum.

6.2.2.2 RAMAN SPECTROSCOPY

Studies were carried out by means of a Jobin—Yvon T64000 (France) instrument, operating with a laser of 514.5 nm line and power of 50–100 W. The spectrometer was coupled with a BX40 Olympus confocal microscope, operating with Olympus LMPlanFI 50 (NA = 0.50) or LMPlan 50 (NA = 0.75) objectives. The surface of iron specimen was examined with acquisition time of 240–540 s, at least in two distant places, before and after friction. The internet RASMIN database [10] was used for material identification. Characteristic absorption bands for FeS are present at wavelengths of 270 and 520 cm^{-1}—Figure 6.1.

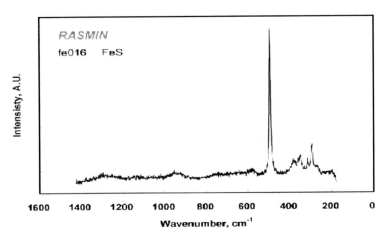

FIGURE 6.1 FT-Raman spectrum of iron sulfide [10].

6.2.2.3 TRIBOLOGICAL CHARACTERISTICS

Tribological characteristics of the materials studied were determined with a T-05 (ITeE – PIB, Poland) tribometer, operating with a block-on-ring friction couple. Ring made of polymer material was rotating over a still block made of Armco iron. The instrument worked together with a SPIDER 8 Hottinger Messtechnik (Germany) electronic system for data aquisition. The way for data analysis has been described in our previous work [11]. Polymer rings of diameter 35 mm, rotating with a speed of 60 rpm were loaded within the range of 5–100 N, during 60–120 min.

6.3 RESULTS AND DISCUSSION

In order to explain the influence of the way sulfur is bonding in polymer materials on modification of the surface layer of iron, the metal counterface was subjected to sliding friction against:

- polysulphone (load of 20 N/ time 2 hrs),
- ebonite (load either 20 or 100 N/ time 2 hrs),
- SBR vulcanizate (load 20 N/ time 2 hrs), and
- polysulfide rubber (load 5 N/ time 1 hr).

The load and time of friction in the last case have to be decreased due to low mechanical strength of polysulfide rubber.

From the specific spectra of secondary ions (Figure 6.2) and comparison between normalized counts for particular cases (Figure 6.3) it follows, that the highest amount of sulfur, in a form of SH⁻ ions, was transferred to the surface layer of iron counterface by ebonite. In the case of polysulphone, due to strong sulfur bonding to macromolecular backbone (Figure 6.4) and different from other polymers studied mechanisms of mechanodegradation, the expected effect of sulfur transfer is practically absent. The amount of iron sulfide, created in the surface layer of iron counterface depends on reactivity of sulfur containing polymer fragment being released during friction and their concentration in the friction zone. From possible substrates, involved in the creation of FeS, the highest affinity to iron exhibit polysulfide cross-links and ionic products of their destruction, released from some polymer materials subjected to intensive friction against the metal counterface. They can be produced only in the case of SBR vulcanizate and ebonite, what can be explained by their chemical structure. One should pay attention to different load being applied for the polymer materials studied. In the case of unfilled polysulfide rubber, the time of friction has additionally to be limited due to low mechanical strength of the material. However, an example of ebonite demonstrates that an increase of loading not necessarily has to lead to higher extent of modification of iron counterface during friction—Figures 6.2 and 6.3.

It can be the result of prevailing, under extreme friction conditions, radical degradation of macromolecules, accompanied by intensive oxidation of polymer. Distribution of all ions in the surface layer of metal is uniform. Apart from the sulfur ones released from polymer materials, oxides also are formed, which facilitates antifriction properties of metal [12]. Iron sulfide present can further be oxidized during friction to sulphones, which are even better lubricating agents. However, under too high loading conditions ionic mechanism of cross-link breaking is not able to show up, losing to macromolecular degradation, what explains lower efficiency of the modification of iron with sulfur. In order to confirm TOF–SIMS data on FeS presence in the surface layer of Armco iron subjected to friction against various polymer materials, complementary studies with Raman spectroscopy were carried out. Comparing collected spectra (Figure 6.5) to the standard FeS spectrum from the database (Figure 6.1), only the spectra of iron surface after friction against SBR vulcanizate and polysulfide rubber indicate on possible sulfur modification. Analysis of the surface of Armco iron specimen subjected to friction against polysulphone or ebonite did not bring unique results. FT–Raman spectra contain signals the most likely coming from degraded fragments of macromolecules or unidentified compounds containing carbon, oxygen and hydrogen. For the above two cases any absorption peak in the region characteristic for the FeS could not be assigned.

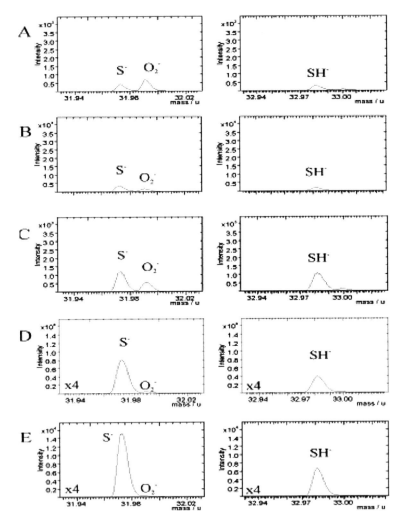

FIGURE 6.2 Specific TOF-SIMS spectra collected from the surface layer of iron Armco specimen, subjected to friction against various polymer materials studied: A—virgin, B—polysulphone, C—SBR vulcanizate, D—ebonite/100 N, E—ebonite/20 N.

Influence of the Structure of Polymer Material 117

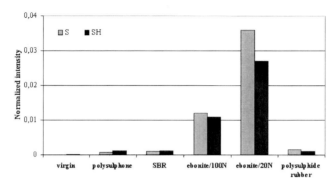

FIGURE 6.3 Normalized TOF-SIMS spectra counts of S⁻ and SH⁻ ions for the surface layer of Armco iron specimen subjected to friction against various polymer materials studied.

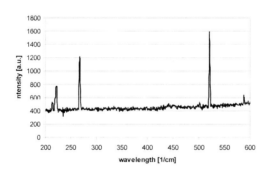

FIGURE 6.4 Chemical structure of polysulphone.

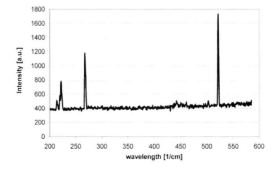

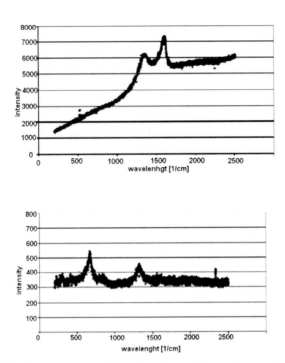

FIGURE 6.5 FT—Raman spectra of the surface layer of Armco iron specimen subjected to friction against various polymer materials studied (**a**) polysulfide rubber (**b**) SBR vulcanizate (**c**) ebonite/100 N (**d**) D − polysulphone.

Tribological characteristics of polymer materials sliding against Armco iron are demonstrated in Figure 6.6. Friction force and energy curves vary from material to material due to their different chemical structure and related mechanical properties. From tribological point of view, the most efficient modification has to be subscribed to ebonite—iron friction couple. In this case median of the friction force and discrete levels of energy exhibit the most stable courses among all the friction couples studied. The second run, repeated for new ebonite roll after 2 hrs of previous frictional modification of Armco iron, resulted in above 10 percent reduction of the coefficient of friction (Figure 6.6b). The first run for SBR vulcanizate (Figure 6.7c) exhibits "classical" course with characteristic maximum of the coefficient of friction, which appears already after some minutes from the start of experiment. For the first 100 min value of the friction force gradually decreased from 38 down to 26 N, eventually stabilizing at this level. In the second run, the friction force come back to the initial value, but right after beginning immediately goes down to the final value after the first run. It means, that in the case of SBR vulcanizates, the modification of metal counterface is the most important for the beginning of friction. Tribological characteristics determined for polysulfide rubber (Figure 6.6e, f) are not so stable

Influence of the Structure of Polymer Material 119

as for ebonite or SBR vulcanizates, probably because of poor mechanical properties of polysulfide rubber. Nevertheless, the modification of the surface layer of Armco iron, confirmed by TOF—SIMS and Raman spectroscopy, is also reflected by tribological data. In the first cycle, the friction force is maintained at the level of 17–18 N during the first 25 min, and suddenly goes down to 10 N, which level is kept constant till the end of experiment. The drop is reflected by significant increase of the energy component responsible for high energy vibrations (200–600 Hz). The vibrations of such energy are not present in tribological characteristics of elastomers [11]. Similar to SBR vulcanizate, the second run for polysulfide rubber starts from higher value of the friction force, which quickly goes down and stabilizes itself at the final level of the first run. In the case of polysulphone (Figure 6.6g, h), any tribological effects able to be subscribed to the surface modification of iron counterface have not been observed. During the first run, value of the friction force is increasing, eventually reaching stabilization at the level of 9 N, shortly before the end of experiment. The second run starts from the friction force value of 5 N, which gradually increases up to the final level of the first run. At this moment, which requires about 40 min from start, its course becomes very unstable, probably because of intensive wear of iron counterface influencing experimental data.

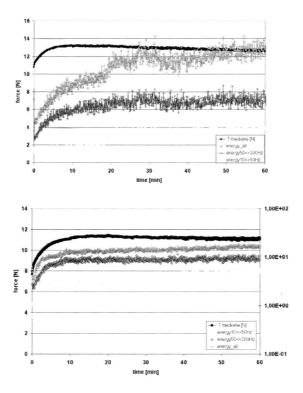

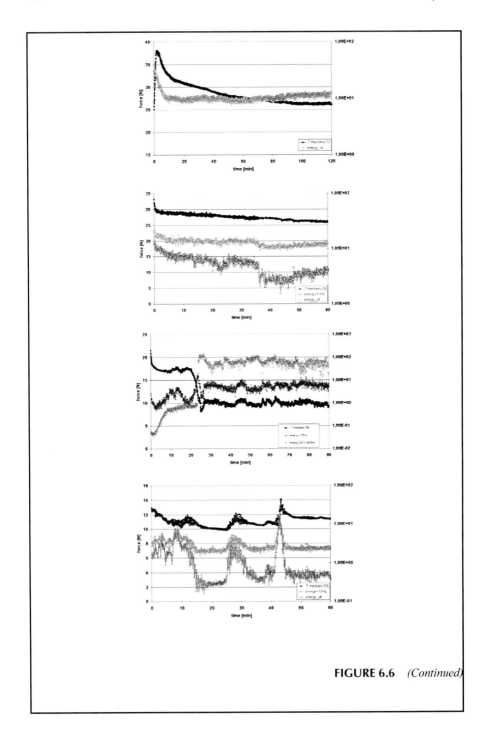

FIGURE 6.6 *(Continued)*

Influence of the Structure of Polymer Material 121

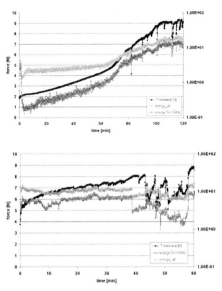

FIGURE 6.6 Tribological characteristics of Armco iron - polymer material friction couple. (**a**) ebonite/20 N—1st cycle, (**b**) ebonite/20 N—2nd cycle (**c**) SBR vulcanizate—1st cycle (**d**) SBR vulcanizate—2nd cycle (**e**) polysulfide rubber—1st cycle (**f**) polysulfide rubber—2nd cycle (**g**) polysulphone—1st cycle (**h**) polysulphone—2nd cycle.

6.4 CONCLUSIONS

1. Extensive friction against polymer materials containing sulfur can result in surface modification of Armco iron counterface with FeS. Extent of the modification depends on the way of sulfur bonding to macromolecules.
2. The results obtained point of higher efficiency of modification, when sulfur is present in ionic cross-links of polymer material—SBR vulcanizate and ebonite, contrary to constituing its macromolecules—polysulphone. Degradation of the latter is of radical character and its products immediately react with atmospherics oxygen. Only ionic species, produced by breaking of sulphidic cross-links, are able to react with iron.
3. The extent of modification is straightly related to the amount of sulfur being present in cross-links. The highest amount of sulfur in the case of ebonite results in the highest degree of modification of the surface layer of Armco iron after tribological contact. Application of cured polysulfide rubber is less effective due to the lack of sulfide cross-links in structure of the material. Additionally, its poor mechanical properties are responsible for a transfer of low molecular weight products of wear onto the metal counterface, whereas "strong" polysulphone makes Armco iron sample worn off.

4. TOF-SIMS data for Armco iron point on the highest degree of sulfur modification being the result of friction against ebonite, SBR vulcanizate and cured polysulfide rubber. The spectra represent the highest amount of species containing sulfur.
5. Tribological characteristics confirm the influence of sulfur modification of metal counterface on lowering friction for the metal–polymer couples studied. In the case of ebonite, the coefficient of friction reduced significantly for the whole experimental run, whereas application of SBR vulcanizate or polysulfide rubber was effective only during the first period of experimental cycles. Any improvement of tribological characteristics was not assigned for polysulphone. The polymer was observed to worn the surface of iron counterface, what resulted in increase of the coefficient of friction in this case.

KEYWORDS

- Iron
- Polymers
- Sliding friction
- Surface modification

REFERENCES

1. Płaza, S.; Physics and Chemistry of Tribological Processes, University of Łódź Press: Łódź; **1997**.
2. Rymuza, Z.; Tribology of Sliding Polymers, WNT, Warsaw; **1986**.
3. Bieliński, D. M.; Grams, J.; Paryjczak, T.; and Wiatrowski, M.; Tribological modification of metal counterface by rubber. *Tribol. Lett.* **2006**, *24*, 115–118.
4. Bieliński, D. M.; Siciński, M.; Grams, J.; and Wiatrowski, M.; Influence of the crosslink structure in rubber on the degree of modification of the surface layer of iron in elastomer-metal friction pair. *Tribologia.* **2007**, *212*, 55–64.
5. Boochathum, P.; and Prajudtake, W.; Vulcanization of cis- and trans-polyisoprene and their blends: cure characteristics and crosslink distribution. *Eur. Polym.* **2001**, J. *37*, 417–427.
6. Morrison, N. J.; and Porter, M.; Temperature effects on structure and properties during vulcanization and service of sulphur-crosslinked rubbers. *Plast. Rubber. Proc. Appl.* **1983**, *3*, 295–304.
7. Grossiord, C.; Martin, J. M.; Le Mogne, Th.; and Palermo, Th.; In situ MoS formation and selective transfer from MoDPT films. *Surf. Coat. Technol.* **1998**, *108–109*, 352–359.
8. Wang, H.; Xu, B.; Liu, J.; and Zhuang, D.; Investigation on friction and wear behaviors of FeS films on L6 steel surface. *Appl. Surf. Sci.* **2005**, *252*, 1084–1091.
9. Gaczyński, R. ed.; Rubber. Handbook for Engineers and Technicians, WNT, Warsaw **1981**, Tab. IV-35, p. 323.
10. http:\\ www.aist.go.jp/RIOBD/rasmin/E-index.htm

11. Głąb, P.; Bieliński, D. M.; and Maciejewska, K.; An attempt to analysis of stick-slip phenomenon for elastomers, Tribologia **2004**, *197*, 43–50.
12. Wang, H.; Xu, B.; Liu, J.; and Zhuang, D.; Characterization and tribological properties of plasma sprayed FeS solid lubrication coatings. *Mater. Charact.* **2005**, *55*, 43–49.

CHAPTER 7

BORON OXIDE AS A FLUXING AGENT FOR SILICONE RUBBER-BASED CERAMIZABLE COMPOSITES

R. ANYSZKA, D. M. BIELIŃSKI, and Z. PĘDZICH

7.1 INTRODUCTION

Ceramizable (ceramifiable) silicone rubber-based composites are fire-resistant materials developed especially for cable covers application. In case of fire, electrical installations are endangered of short circuit effect which can deactivate lots of important devices, like fire sprinklers, elevators, fire alarms or lamps indicating route to emergency exits. Ceramizable composites are able to sustain functioning of electric circuit on fire and high temperature up to 120 min by producing ceramic, porous layer protecting copper wire inside a cable.

Ceramizable silicone-based materials are dispersion type of composites, in which mineral particles (refractory fillers and, in some compositions, fluxing agent particles) are dispersed in continuous phase of silicone rubber [1–18]. Mechanism of protecting ceramic shield creation on the border between fire and material includes.

1. Production of amorphous silica during thermal degradation of polysiloxane matrix under oxidizing atmosphere which results in creation of mineral bridges between refractory filler particles (Figure 7.1).
2. Sintering of mineral filler particles due to condensation of hydroxyl groups present on their surface (Figure 7.2) [19].
3. Production of new mineral phases as a result of reaction between primary filler particles and silica produced as a result of thermal degradation of silicone matrix (Figure 7.3).
4. Creation of strong glassy bridges between particles of high thermal resistive minerals by low temperature softening amorphous glass particles (Figure 7.4). This type of ceramization made possible to use hydrocarbon polymers as a continuous phase for ceramizable composites. In the subject literature it is possible to find papers describing ceramizable composites based on polyethylene [20], poly(vinyl acetate) [21], poly(vinyl chloride), and ethylene-propylene-diene rubber (EPDM) [22].

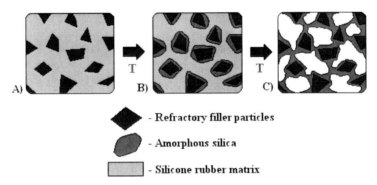

FIGURE 7.1 Scheme of ceramization process based on creation of amorphous silica microbridges between refractory filler particles during thermo oxidizing degradation of silicone rubber matrix.

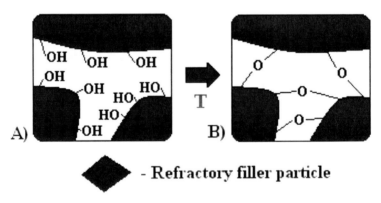

FIGURE 7.2 Scheme of sintering process of refractory mineral particles involving hydroxyl group's condensation.

$$PDMS \longrightarrow SiO_2 + cyclosiloxanes \quad (>500\ °C)$$
$$CaCO_3 \longrightarrow CaO + CO_2 \quad (>600\ °C)$$
$$CaO + SiO_2 \longrightarrow CaSiO_3 (Wollastonite)\ (800\ °C)$$
$$2CaO + SiO_2 \longrightarrow Ca_2SiO_4 (Larnite) \quad (800\ °C)$$

FIGURE 7.3 Possible chemical reactions between components of ceramizable composites, taking place at temperature.

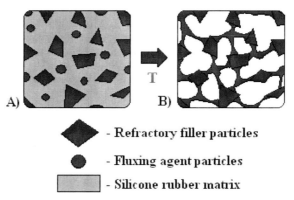

FIGURE 7.4 Scheme of the ceramization process based on creation of microbridges between refractory filler particles by melted particles of fluxing agent.

In this study particles of boron oxide were used as a fluxing agent in order to enhance mechanical strength of ceramic layer after high temperature treatment of silicone rubber-based composites. Due to relatively low meting point temperature of crystalline boron oxide (T_m = 450°C), which is comparable, or even lower than the temperature of softening point of glass oxide frits, very effective ceramization process, resulting in desirable thermal and mechanical properties of the composites, is expected.

7.2 EXPERIMENTAL

7.2.1 MATERIALS

Silicone rubber (HTV) containing 0.07 percent of vinyl groups, produced by "Polish Silicones" Ltd. (Poland), reinforced by "Aerosil 200" fumed silica produced by "Evonik Industries" (Germany), an elastomer base. Boron oxide from "Alfa Aesar GmbH & Co KG" (Germany) we used as a fluxing agent, together with magnesium oxide "MagChem 50" obtained from "Martin Marietta Magnesia Specialties" (USA), applied for decreasing acidic character of the composition. Calcined kaolin "Polestar 200R" originated from "Imerys Minerals" Ltd (France), mica (phlogopite) "PW 150" from "Minelco Minerals" (Sweden), wollastonite "Termin 939-304" from "Quarzwerke Gruppe" (Germany), aluminum hydroxide "Martinal OL-107 LEO" from "Martinswerk" (Germany), and surface modified montmorillonite "Cloisite 20A" using dimethyl-dihydrogenatedtallow quaternary ammonium salt, produced by "Rockwood Additives Ltd." (UK) were applied as refractory filler. 50 percent paste of 2,4-dichlorobenzoyl peroxide originated from "Novichem" (Poland) was used for cross-linking of the silicone rubber composites.

7.2.2 SAMPLE PREPARATION

Due to quite large size of boron oxide particles, the first step was to grind them in a Pulverisette 5 planetary mill made by "Fritsch GmbH" (Germany) using bowls covered internally with agate (280 cm³ of capacity) and 20 mm of diameter agate balls. The mill operated with the speed of 350 rpm, during 30 min (Figure 7.5).

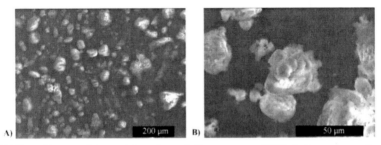

FIGURE 7.5 Boron oxide particles after grinding process, SEM photographs made under magnification of 150 x (**a**) and 1000 x (**b**).

Studied mixes containing boron oxide and other mineral fillers (Table 7.1) were prepared with a Brabender-Plasticorder laboratory mixer (Germany), whose rotors were operating with 20 rpm during components incorporation (5 min) and with 60 rpm during their homogenization (10 min). Composite samples were vulcanized in a steel mold, at 130°C during time t_{90} determined rheometrically according to ISO 3417.

TABLE 7.1 Composition of the ceramizable composites studied

Components \ Composites	C-KAO	WOL	ALOH	M-MMT	MIC
Silicone rubber	100	100	100	100	100
Fumed silica	50	50	50	50	50
B$_2$O$_3$	20	20	20	20	20
MgO	5	5	5	5	5
Calcined kaolin	10	-	-	-	-
Wollastonite	-	10	-	-	-
Aluminum hydroxide	-	-	10	-	-

Surface modified montmorillonite	-	-	-	10	-
Mica (phlogopite)	-	-	-	-	10
2,4-dichlorobenzoyl peroxide (50 % paste)	1.8	1.8	1.8	1.8	1.8

7.2.3 SAMPLE CERAMIZATION

Composite samples were ceramized in a laboratory furnace under four different regimes of heat treatment. Isothermal heating at 600, 800, or 1000°C during 20 min, and slow temperature speed ratio from room temperature up to 1,000°C during 2 hrs. Then they were cooled down in open air prior to testing.

7.2.4 TECHNIQUES

Vulcanization kinetics of the composite mixes were determined using a WG-02 vulcameter produced by Metalchem (Poland). Mechanical properties of the composites: elongation at break (EB), mechanical moduli at 100, 200, and 300 percent of elongation (SE_{100}, SE_{200}, SE_{300}), tensile strength (TS), and tear strength (TES) were measured with a Zwick Roell 1435 instrument (Germany). Hardness of vulcanized composites was determined using a Zwick Roell tester (Germany), applying force of 12,5 N.

Composites after heat treatment were tested for porosity, using a mercury porosimeter CarloErba 2000 (Italy) and endurance under compression using a Zwick Roell Z 2.5 tester (Germany). Compression tests were made on the samples of cylindrical geometry (Figure 7.6).

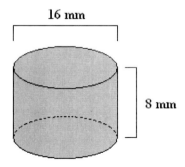

FIGURE 7.6 Cylindrical shape of samples ceramized in laboratory furnace.

For complementary microstructure analysis of the ceramized composite samples, a Hitachi S-4700 (Japan) scanning electron microscope equipped with a BSE detector was used

7.3 RESULTS AND DISCUSSION

7.3.1 VULCANIZATION KINETICS OF COMPOSITES

TABLE 7.2 Vulcameter's parameter of the composite mixes studied. Minimum (M_{min}), optimum (M_{opt}), and maximum (M_{max}) torque value, scorch time (t_{02}), and vulcanization time (t_{90})

Properties / Composites	C-KAO	WOL	ALOH	M-MMT	MIC
M_{min} [dNm]	59.7	61.3	62.3	45.5	52.8
M_{opt} [dNm]	136.2	140.0	145.5	87.1	132.8
M_{max} [dNm]	144.7	148.8	154.7	91.7	141.7
t_{02} [s]	29	34	22	50	34
t_{90} [s]	82	92	95	123	92

Vulcanization kinetics of the composites vulcanization are very similar regardless refractory filler type, excluding the mix containing surface modified montmorillonite (M-MMT) [Table 7.2]. In its case values of torque were significantly lower than for other samples: minimum (M_{min}), optimum (M_{opt}), and maximum (M_{max}) torque values respectively from 14 to 27 percent, from 34 to 40 percent and from 35 to 41 percent. This effect can be explained by plasticization of silicone matrix by quaternary ammonium salt present in M-MMT. Also values of scorch (t_{02}) and vulcanization (t_{90}) time were higher in the case of M-MMT sample: 47 to 127 percent and from 30 to 50 percent than other mixes studied for the scorch time and vulcanization time respectively. This fact could cause problems, because of high speed extrusion technology commonly used in cable industry.

7.3.2 MECHANICAL PROPERTIES OF COMPOSITES BEFORE CERAMIZATION

TABLE 7.3 Mechanical properties of the composites studied. Values of stress at 100 percent (SE_{100}), 200 percent (SE_{200}), and 300 percent (SE_{300}) of composites elongation and their tensile strength (TS), tear strength (TES), elongation at break (EB), and hardness (H).

Properties \ Composites	C-KAO	WOL	ALOH	M-MMT	MIC
SE_{100} [MPa]	1.9	1.7	1.7	1.3	2.0
SE_{200} [MPa]	2.8	2.6	2.6	1.5	2.7
SE_{300} [MPa]	3.7	3.5	3.4	1.8	3.4
TS [MPa]	3.9	3.9	3.7	3.1	3.9
EB [%]	322	350	338	641	357
TES [N/mm]	8.2	10.4	9.3	15.6	13.1
H [°ShA]	63.1	68.0	66.2	62.4	64.1

Values of stress at 100 percent (SE_{100}), 200 percent (SE_{200}), 300 percent (SE_{300}) of elongation, tensile strength (TS), and elongation at break (EB) of the composites are very similar, except M-MMT sample filled with surface modified montmorillonite (Table 7.3). In its case values of moduli at 100, 200, and 300 percent of elongation and tensile strength were slightly lower than for other composites, but its elongation at break was almost two times higher than determined for the other samples. Also M-MMT sample characterize itself by the best tear strength higher from 19 percent up to even 90 percent than composites filled with unmodified refractory powders. The highest value of hardness has the composite filled with wollastonite (WOL) whereas the lowest the composite containing surface modified montmorillonite (M-MMT). However the difference is not significant—it does not exceed ca 10 percent.

7.3.3 MECHANICAL PROPERTIES OF COMPOSITES AFTER HEAT TREATMENT

FIGURE 7.7 Photographs of the composite samples subjected to slow heating from room temperature to 1,000 °C during 2 hrs

TABLE 7.4 Force required to break samples ceramized under various conditions and changes to the sample diameter after slow ceramization (heating from room temperature to 1,000 °C during 2 hrs)

Properties / Composites	C-KAO	WOL	ALOH	M-MMT	MIC
600 °C 20min. [N]	20.0	34.4	33.8	17.8	53.9
800 °C 20min. [N]	16.2	27.3	17.4	17.2	28.0
1000 °C 20min. [N]	20.2	67.3	22.1	35.8	109.0
$T_R \to 1000$ °C [N]	478	548	505	613	601
Sample parameters					
Diameter of samples after $T_R \to 1000$ °C heating [mm]	13	15	11	11	18
Changes of samples diameter after $T_R \to 1000$ °C heating [mm]	-3	-1	-5	-5	+2

The best mechanical properties, after isothermal heating exhibit composite sample containing mica (phlogopite) (MIC) (Table 7.4). The weakest ceramic phases were obtained after isothermal heating of the composites at 800 °C, whereas the strongest structures were created after slow heating from room temperature to 1,000 °C. After this treatment the sample containing surface modified montmorillonite (M-MMT) and filled with mica flakes (MIC) exhibits the best mechanical strength (ca. 600 N).

All of the composites shrunk after slow heating treatment, except sample containing mica (MIC), which expanded in diameter by 2 mm. The largest shrinkage was detected for the samples filled with aluminum hydroxide (ALOH) and surface modified montmorillonite (M-MMT), which shrunk in diameter by 5 mm. Changes to shape of composites subjected to the action of high temperature is very important from the point of view of maintaining continuity of ceramic cover protecting copper wire of a cable on fire.

7.3.4 MORPHOLOGY OF COMPOSITES AFTER HEAT TREATMENT

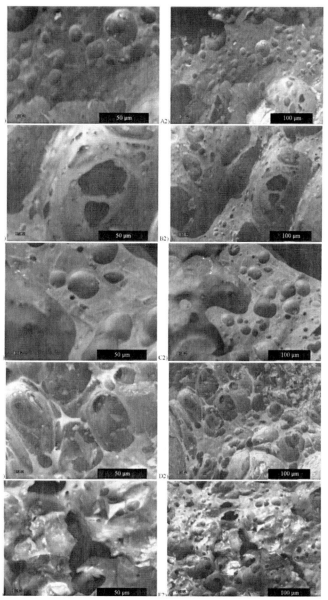

FIGURE 7.8 Micromorphology of composites after heating form room temperature to 1,000 °C during 2 hrs. SEM photographs of composites containing aluminum hydroxide (**a**), calcined kaolin (**b**), mica (**c**), surface modified montmorillonite (**d**), and wollastonite (**e**).

Scanning electron microscopy (SEM) pictures of composite samples ceramized slowly under heating from room temperature to 1,000 °C during 2 hrs demonstrate good adhesion between ceramic phase components in the composites filled with mica (MIC—C1,C2), wollastonite (WOL—E1,E2), and aluminum hydroxide (ALOH—A1,A2). Far worse adhesion is observed for the samples containing calcined kaolin (C-KAO—B1,B2) and surface modified montmorillonite (M-MMT—D1,D2) (Figure 7.8).

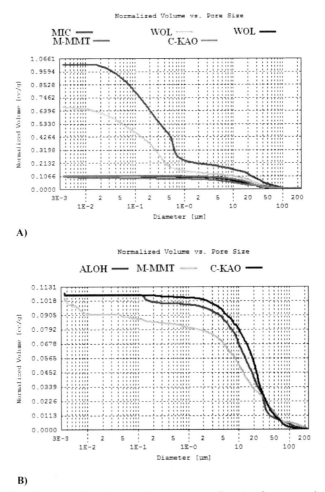

FIGURE 7.9 The normalized pore volume vs. pore diameter for composites subjected to slow heating from room temperature to 1,000 °C during 2 hrs. Graph B) presents precisely differences in the characteristics for samples containing aluminum hydroxide (ALOH), surface treated montmorillonite (M-MMT), and calcined kaolin (C-KAO).

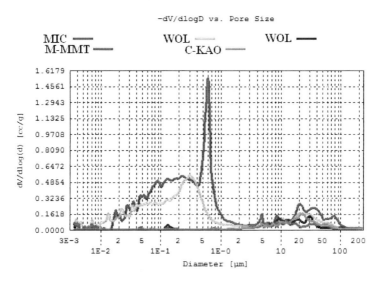

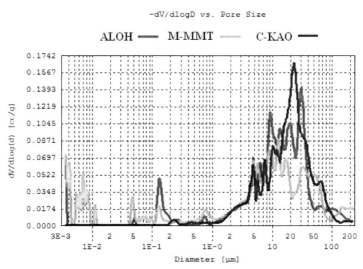

FIGURE 7.10 The normalized differential curve dV/log(d) vs. pore diameter for composites slowly ceramized by heating from room temperature to 1,000 °C during 2 hrs. Graph B) presents precisely differences in the characteristics for samples containing aluminum hydroxide (ALOH), surface treated montmorillonite (M-MMT), and calcined kaolin (C-KAO).

Porosimetry analysis demonstrates that samples containing mica and wollastonite have created nanoporous structure with high amount of pores (Figure 7.9a)). This kind of structure is expected to be the best from the point of view of mechanical properties as well as thermal insulation of the material. Amount of pores in samples filled with calcined kaolin (C-KAO), aluminum hydroxide (ALOH), and surface treated montmorillonie (M-MMT) is almost ten times lower in comparison to the composite containing mica (MIC) (Figure 7.9a)). Their pores are also significantly bigger (Figure 7.10a)).

7.4 SUMMARY AND CONCLUSIONS

Vulcanization kinetics of the composites studied are very similar, excluding sample filled with surface modified montmorillonite (M-MMT), for which torque values are lower, probably due to plasticization effect resulting from high concentration of organic modifier (quaternary ammonium salt). The M-MMT sample differs from other composites also in terms of the time of vulcanization. Both, scorch (t_{02}) and vulcanization (t_{90}) time of this sample are longer. It could be caused by acidic character of montmorillonite particles surface obtained as a result of Hoffmann beta-elimination of quaternary ammonium cation from the surface of particles [19, 25] during mixing of compounds. Acidic components affect the mechanism of peroxide cross-linking, leading to creation of nonactive ions instead of peroxide radicals [23–25].

Elongation at break (EB), tensile strength (TS), stress at 100 percent (SE_{100}), 200 percent (SE_{200}), and 300 percent (SE_{300}) of elongation of composites before ceramization were generally similar except the M-MMT sample. In its case moduli at elongation and tensile strength values are lower, probably due to the plasticization effect originated from the presence of organic modifier. Tear strength of samples differs from 8,2 N/mm for C-KAO to 15,6 N/mm for M-MMT composite. It is likely that the presence of organofilizer influences positively TES of composites due to decreasing stiffness of vulcanizates, allowing dissipation of energy during tear deformation. Hardness of composites depends not only on the presence of organofilizer but also on the type of mineral filler added. However the difference is not significant.

Mechanical strength of composites after ceramization strongly depends on the type of refractory filler added. The best mechanical properties after isothermal ceramization at 600, 800, and 1,000 °C present composite containing mica (MIC). After slow heating from room temperature to 1,000 °C the strongest showed to be the composites filled with surface modified montmorillonite (M-MMT), or mica (MIC). Samples subjected to slow heating were generally shrinking except the MIC one, which diameter increased by 2 mm. This fact can adversely affect cable covers, which are likely to lose continuity on fire.

Porosimetry and SEM analysis demonstrate that type of refractory filler effects strongly morphology of the composite after ceramization. Only two samples, either containing mica (phlogopite) or wollastonite (WOL and MIC) are able to create nanoporous structure which can strongly improve thermal insulation of ceramic phase, what is very important from the point of view of remaining functioning of electric installations. A ceramized layer should be able to protect copper wire from outside heating during fire.

This study has shown that the best composition of fillers promoting ceramization of silicone rubber-based composites consists of boron oxide (fluxing agent) and mica (refractory filler). Good mechanical properties and processability, in combination with very good mechanical properties and nanoporous structure after ceramization, give to this composite large industrial implementation capacity.

ACKNOWLEDGEMENTS

This work was supported by the EU Integrity Fund, project POIG 01.03.01-00-067/08-00.

KEYWORDS

- Ceramization
- Composites
- Fillers
- Morphology
- Silicone rubber

REFERENCES

1. Hamdani, S.; Longuet, C.; Lopez-Cuesta, J-M.; and Ganachaud, F.; Calcium and aluminium-based fillers as flame-retardant additives in silicone matrices. I. Blend preparation and thermal properties. *Polym. Degr. Stabil.* **2010**, *95*, 1911–1919.
2. Hamdani-Devarennes, S.; Pommier, A.; Longuet, C.; Lopez-Cuesta, J-M.; and Ganachaud, F.; Calcium and aluminium-based fillers as flame-retardant additives in silicone matrices. II. Analyses on composite residues from an industrial-based pyrolysis test, *Polym. Degr. Stabil.* **2011**, *96*, 1562–1572.
3. Hamdani-Devarennes, S.; Longuet, C.; Sonnier, R.; Ganachaud F.; and Lopez-Cuesta J-M. Calcium and aluminium-based fillers as flame-retardant additives in silicone matrices. III. Investigations on fire reaction. *Polym Degr Stabil,* **2013**, *98*, 2021-2032.
4. Mansouri, J.; Wood, C. A.; Roberts, K.; Cheng Y. B.; and Burford, R. P.; Investigation of the ceramifying process of modified silicone-silicate compositions. *J. Mater. Sci.* **2007**, *42*, 6046–6055.

5. Hanu, L. G.; Simon, G. P.; Mansouri, J.; Burford, R. P.; and Cheng, Y. B.; Development of polymer-ceramic composites for improved fire resistance. *J. Mater. Process. Tech.* **2004**, *153–154*, 401–407.
6. Bieliński, D. M.; Anyszka, R.; Pędzich, Z.; and Dul, J.; Ceramizable silicone rubber-based composites. *Int. J. Adv. Mater. Manuf. Charac.* **2012**, *1*, 17–22.
7. Hanu, L. G.; Simon, G. P.; and Cheng, Y. B.; Thermal stability and flammability of silicone polymer composites. *Polym. Degr. Stabil.* **2006**, *91*, 1373–1379.
8. Pędzich, Z.; Bukanska, A.; and Bieliński, D. M.; Anyszka R.; Dul, J.; and Parys, G.; Microstructure evolution of silicone rubber-based composites during ceramization at different conditions. *Int. J. Adv. Mater. Manuf. Charact.* **2012**, *1*, 29–35.
9. Pędzich Z.; and Bieliński, D. M.; Microstructure of silicone composites after ceramization. *Compos.* **2010**, *10*, 249–254.
10. Dul, J.; Parys, G.; and Pędzich, Z.; Bieliński, D. M.; Anyszka, R.; Mechanical properties of silicone-based composites destined for wire covers. *Int. J. Adv. Mater. Manuf. Charact.* **2012**, *1*, 23–28.
11. Hamdani, S.; Longuet, C.; Perrin, D.; Lopez-Cuesta J-M.; and Ganachaud, F.; Flame retardancy of silicone-based materials, *Polym. Degr. Stabil.* **2009**, *94*, 465–495.
12. Mansouri, J.; Burford, R. P.; Cheng, Y. B.; Hanu, L.; *Formation of strong ceramified ash from silicone-based composites*. *J. Mater. Sci.* **2005**, *40*, 5741–5749.
13. Mansouri, J.; Burford, R. P.; and Cheng, Y. B.; Pyrolysis behaviour of silicone-based ceramifying composites. *Mater. Sci. Eng.* **2006**, *425*, 7–14.
14. Hanu, L. G.; Simon, G. P.; Cheng, Y. B.; Preferential orientation of muscovite in ceramifiable silicone composites. *Mater. Sci. Eng. A.* **2005**, *398*, 180–187.
15. Xiong, Y.; Shen, Q.; Chen, F.; Luo, G.; and Yu, K.; Zhang, L.; High strength retention and dimensional stability of silicone/alumina composite panel under fire. *Fire. Mater.* **2012**, *36*, 254–263.
16. Pędzich, Z.; Anyszka, R.; Bieliński, D. M.; Ziąbka M.; Lach, R.; and Zarzecka-Napierała, M.; Silicon-basing ceramizable composites containing long fibers. *J. Mater. Sci. Chem. Eng.* **2013**, *1*, 43–48.
17. Bieliński, D. M.; Anyszka, R.; Pędzich, Z.; Parys, G.; and Dul, J.; Ceramizable silicone rubber composites. Influence of type of mineral on ceramization, *Compos,* **2012**, 12, 256–261.
18. Pędzich, Z.; Bukańska A.; Bieliński D. M.; Anyszka R.; Dul, J.; Parys G. Microstructure evolution of silicone rubber-based composites during ceramization in different conditions, *Compos,* **2012**, *12*, 251–255.
19. Morgan, A. B.; Chu, L. L.; and Harris, J. D.; A flammability performance comparsion between synthetic and natural clays in polystyrene nanocomposites. *Fire. Mater,* **2005**, *29*, 213–229.
20. Wang, T.; Shao H.; and Zhang Q.; Ceramifying fire-resistant polyethylene composites. *Adv. Compos. Lett.* **2010**, *19*, 175–179.
21. Shanks, R. A.; Al-Hassany Z.; and Genovese A.; Fire-retardant and fire-barrier poly(vinyl acetate) composites for sealant application. *Express. Polym. Lett.* **2010**, *4*, 79–93.
22. Thomson, K. W.; Rodrigo, P. D. D.; Preston, C. M.; and Griffin, G. J.; Ceramifying polymers for advanced fire protection coatings. In: *Proceedings of European Coatings Conference* 2006, 15th September, Berlin, Germany.
23. Heiner, J.; and Stenberg, B.; Persson, M.; Crosslinking of siloxane elastomers. *Polym. Test.* **2003**, *22*, 253–257.
24. Ogunniyi, S. D.; Peroxide vulcanization of rubber, *Prog. Rubber. Plast. Tech.* **1999**, *15*, 95–112.
25. Anyszka, R.; Bieliński, D. M.; Kowalczyk, M.; Influence of dispersed phase selection on ceramizable silicone composites cross-linking, *Elastom.* **2013**, *17*, 16–20.

CHAPTER 8

APPLICATION OF MICRO-DISPERSED SILICON CARBIDE ALONG WITH SLURRIES AS A FUNCTIONAL FILLER IN FIRE AND HEAT RESISTANT ELASTOMER COMPOSITIONS

V. S. LIPHANOV, V. F. KABLOV, S. V. LAPIN, V. G. KOCHETKOV, O. M. NOVOPOLTSEVA, and G. E. ZAIKOV

8.1 INTRODUCTION

The investigation of polymer materials for extreme operating conditions needs the new components that ensure the flow of physical and chemical transformations for increasing the operational resistance.

One of important components in elastomeric materials is the filler (carbon black, silica, etc.) The function of the filler is usually the improvement of mechanical properties (e.g., strength, hardness, etc.). In extreme conditions, when temperatures are near and above the temperature of the material performance functionally active fillers can play a stabilizing role during the destruction of the material because of high temperatures [1–2].

One effective solution is to use sloughing and fine fillers [5–8] and also highly dispersed silicon carbide [3–7].

Silicon carbide is one of the most promising materials. It has found application in many industries due to its high hardness and inertness to many corrosive environments. This material is now used to produce abrasive tools, as a filler for fire retardant materials, protecting covers of nuclear fuel, semiconductors, and refractory composites. It is an advanced material for highly integrated devices microwave electronics, operating at high temperatures, high electric fields, and high frequencies.

Silicon carbide (SiC) is a product of a chemical compound of carbon and silicon at high temperature[8]. It contains 70.04 percent of silicon and 29.96 percent carbon. The density is 3.1–3.2 g/cm3; microhardness 3,000–3,300 kgs/mm2; hardness on the Mohs scale of more than 9.

Chemically pure silicon carbide is colorless; technical one can be found in a variety of colors from black to green and has a metallic luster.

The material has a lot of structural polytypes. The silicon carbide atoms are in state of sp3-hybridization and form a bond of a tetrahedron. In the crystal lattice of silicon carbide, the short-range order is always the same but the long range can differ; that is why there are many polytypes of this material. The structural difference causes difference in physical and chemical properties (e.g., termal resistance, electrical and optical characteristics). It makes one or another polytype being more preferred for different applications.

Silicon carbide is a semiconductor material that is why it is a potential catalyst of thermal oxidation and pyrolysis processes. The silicon carbide particles have sharp corners and it allows to expect the appearance of physical and chemical activity in the processes of adsorption and chemical reactions. (due to the presence of unpaired electrons and excess surface energy). Silicon carbide also can be used as so called microbarrier because of its plastic forms on the surface layers of the material. But the usage of silicon carbide in elastomeric materials is poorly understood.

A cheap source of microfine silicon carbide is a sludge which is produced mainly while grinding with an abrasive tool based on silicon carbide. While grinding, the silicon carbide particles are crushed to smaller particles. Sludge also contains grinded metal microparticles and a small amount of surface-active substances.

8.2 THE PURPOSE OF THE RESEARCH

The purpose of the research is the investigation of the possibility to use the microfine silicon carbide—a basic material of a grinding sludge as a functionally active filler in the fire-resistant elastomeric materials.

8.3 MATERIALS AND METHODS

The object of the study is styrene-butadiene rubber vulcanizates SKMS-30ARKM 15 with sulfuric vulcanizing group [4–9]. Mixtures were prepared in laboratory rollers 160 × 320 mm. Vulcanization was carried out at a temperature of 145 °C.

The studied compositions are shown in Table 8.1.

TABLE 8.1 Test compounds

Ingredient Name	Weight Parts/100 parts of Rubber
Rubber SKMS-30ARKM-15	100,00
Carbon black P324	40

The research of the vulcanization kinetics of rubber compounds was carried out in accordance with GOST 12535-84 "Rubber compounds". The method for determining the cure characteristics with vulcametric obtained with Monsanto 100 rheometer. Physical and mechanical properties of the vulcanizates were measured using tensile testing machine IRI-60 according to GOST 270-75 75 "Rubber" The hardness was estimated according to GOST 263-75 "Rubber. Determination of Shore A hardness." The abrasion resistance was determined according to GOST 426-77 "Rubber. Method for the determination of abrasion resistance when sliding" on the Grasselli machine. Microscopic studies and determination of elemental composition were carried out on double-beam scanning electron microscope "Verse 3D".

8.4 THE RESULTS

Table 8.2 shows the rheological and vulcametrical parameters of the mixtures. As it follows from the table an optimal combination of rubber mixture is #3, it contains 20 weight parts of carbon black and 20 weight parts of grinding sludge. This combination of carbon black and silicon carbide increases the cure rate while increasing the induction period of vulcanization (time of vulcanization start). Increasing the amount of silicon carbide leads to faster curing that can be caused by the catalytic property of silicon carbide

TABLE 8.2 Vulcanization properties of rubber mixtures[a]

Properties	1	2	3	А	Б
The minimum torque (ML), N·m	1,37	1,23	1,03	0,75	0,96
Maximum torque (MH), N·m	8,15	7,60	6,64	5,41	6,23
The difference in torque (ΔM), N·m	6,78	6,37	5,61	4,66	5,27
Start time of vulcanization (t_s), min	5,7	5,7	6,37	10,1	5,0
Time to reach 50% of cure (τ_{50}), min	13,0	11,6	12,0	14,6	8,7
The optimum cure time (τ_{90}), min	25,0	21,0	20,1	21,5	15,0
Vulcanization speed indicator (R_v), min-1	5,18	6,54	7,28	8,77	10,0

[a]The vulcanization temperature is 145 °C

Physical and mechanical properties of vulcanizates are presented in Table 8.3. The results of flame resistance testing of the developed compositions are quiet interesting. The warm-up time of the back side of the sample to a temperature of 60 °C was determined by the shims with 50 mm diameter and 6 mm thickness.

The results of flame resistance testing of the developed compositions are quiet interesting. The warm-up time of the back side of the sample to a temperature of 60 °C was determined by the shims with 50 mm diameter and 6 mm thickness.

TABLE 8.3 Physical and mechanical properties of the vulcanized rubber

Parameter	1	2	3	А	Б
Vulcanization mode 145 °C × 25 min					
Apparent stress under 100% extension (f_{100}), MPa	2,06	2,20	1,44	0,97	0,97
Apparent stress under 300% extension, (f_{300}), MPa	9,4	8,8	4,3	1,56	1,40
Relative strength (fp), MPa	23,2	20,8	16,7	2,15	6,9
Elongation (ε), %	600	540	620	420	730
Permanent elongation (θ), %	20	12	13	4	20
Density, кг/м3	1110	1160	1170	1240	1600
Shore hardness, conventional unit.	51	49	42	33	37
Attritive, α, m3/TDzh	73,3	63,2	68,3	94,6	160,3

When the flame burns the sample with silicon carbone, the tight fire-resistant chark appears; it protects the sample from burning itself. The plastic form of silicon carbide particles generates the barrier layer that also protects sample from the flame.

The plastics of silicon carbide can be seen on the surface of the chark. As silicon carbide is a heat-resistant and a very hard to oxidize, so the barrier layer effectively protects the sample from burning through.

To estimate the heat resistance of the developed vulcanizates when heating with plasma torch the temperature of the nonheating surface was measured. After the silicon carbide was impregnated the time the sample reaches the temperature of 60 °C increased from 33 to 60 min.

8.5 CONCLUSIONS

Researches are showing that the grinding sludge with microfine silicon carbide can be used to increase the efficiency of fire-resistant elastomeric materials while decreasing their cost.

KEYWORDS

- Elastomers
- Fillers
- Fire resistance
- Silicon carbide

REFERENCES

1. Grishin, B.S.; *Materialy rezinovoy promyshlennosti (informatsionno-analiticheskaya baza dannykh). Ch. 2* [Rubber Materials Industry (information-analytical database)], Kazan, **2010**; 488 p.
2. Kornev, A. E.; *Tekhnologiya elastomernykh materialov* [Elastomeric materials technology] Moscow, Istek Publ, **2009**; 504 p
3. Kablov ,V. F.; and Novopoltseva, O. M.; *Vliyaniye napolnitelya perlit na teplostoykost rezin na osnove etilenpropilendiyenovogo kauchuka* [Effect of the heat-resistant filler perlite on ethylene-propylene rubbers] Modern problems of science and education, № 3, **2013**; p 444.
4. Kablov, V. F.; and Novopoltseva, O. M.; *Materialy i sozdaniye retseptur rezinovykh smesey dlya shinnoy i rezinotekhnicheskoy promyshlennosti.* [Materials and creating recipes rubber compounds for tire and rubber industry] *Tekhnologiya pererabotki plasticheskikh mass i elastomerov* (Processing technology of plastics and elastomers) Volgograd, **2009**; p 321.
5. Kablov, V. F.; and Novopoltseva, O. M.; *Razrabotka i issledovaniye ogneteplozashchitnykh materialov s vspuchivayushchimisya i mikrovoloknistymi napolnitelyami s elementorganicheskimi modifikatorami dlya ekstremalnykh usloviy ekspluatatsii* [Development and research of fire retardant intumescent materials and micro-fiber fillers with organoelement modifiers for extreme conditions] *Tezisy dokladov III-ey Vserossiyskoy konferentsii «Kauchuk i rezina – 2013: traditsii i novatsii»* (Abstracts of the III-rd Russian Conference "Caoutchouc and rubber"), Moscow, **2013**; pp 28–30.
6. Kablov, V. F.; and Novopoltseva, O. M.; *Teplozashchitnyye pokrytiya, soderzhashchiye perlit* [Thermal barrier coatings containing perlite] *Mezhdunarodnyy zhurnal prikladnykh i fundamentalnykh issledovaniy* (International Journal of Applied and Basic Research), № 1, **2012**; pp 174–175.
7. Novakov I. A.; and Kablov, V. F.; *Vliyaniye napolniteley, modifitsirovannykh metallami peremennoy valentnosti, na vysokotemperaturnoye stareniye rezin na osnove etilen-propilenovogo kauchuka* [Effect of fillers modified with transition metals at high temperature aging of rubbers based on ethylene-propylene rubber] *Izvestiya Volgogradskogo gosudarstvennogo tekhnicheskogo universiteta* (News.of the Volgograd State Technical University), Volgograd, 2., № 8, **2011**; pp. 102–105.
8. Knunyantsa, I. L.; *Khimicheskaya entsiklopediya. T. 2* [Chemical encyclopedia V.2], Moscow, Sovetskaya entsiklopediya Publ., **1990**; p 673.
9. Reznichenko, S. V.; and Morozova, Yu. L.; *Bolshoy spravochnik rezinshchika. Ch.1. Kauchuki i ingrediyenty* [Great reference book of rubber maker. P.1. Rubbers and ingredients], Moscow, Tekhinform Publ., **2012**; 744 p.

CHAPTER 9

THERMAL STABILITY OF ELASTIC POLYURETHANE

I. A. NOVAKOV, M. A. VANIEV, D.V. MEDVEDEV,
N. V. SIDORENKO, G. V. MEDVEDEV, and D. O. GUSEV

9.1 INTRODUCTION

Polyurethane elastomers (PUE) are of great practical importance in various fields [1]. In particular, in developing PUE of molding compositions for sports and roofing the liquid rubbers (oligomers) of diene nature with a molecular weight of 2,000–4,000 are widely used as a polyol component. Usually these are homopolymers of butadiene and isoprene, the products of copolymerization of butadiene with isoprene or butadiene with piperylen and isocyanate prepolymers based on these oligomers.

After curing, the materials exhibit good physical-mechanical, dynamic, and relaxation properties, and high hydrolytic stability [2, 3]. However, the disadvantage of these PUEs is their low resistance to thermal-oxidation aging, due to the presence of double bonds in the oligomer molecules. Under the effect of weather conditions, irreversible changes leading to partial or complete loss of the fundamental properties and materials reduced lifetime take place.

To minimize these negative effects the stabilizers and antioxidants are most commonly used. However, traditional methods of evaluating the effectiveness of a stabilizer within the PUE require lengthy field tests or the materials exposure to high air temperatures for a period of scores of hours and several days [4].

Modern methods of thermal analysis can significantly reduce the time of polymer tests, and informative results, their accuracy, and capability to forecast the coating lifetime are significantly improved and expanded [5–7]. In particular, the determination of oxidation induction time (OIT) and the oxidation onset temperature (OOT) by differential scanning calorimetry (DSC) is effective for the accelerated study of thermal-oxidation polymers stability. This rapid method has been recommended [8–10] and used for polyolefins [11–15], oils and hydrocarbons [16, 17], and PVC [18].

Information on the use of OIT method for polyurethanes is currently limited. There are only some patent data [19] and publications on the results of determination of OIT and OOT for automotive coating materials, derived from polyurethanes of simple and complex polyester structure [20]. There are actually no publications on the test techniques and results of evaluating the thermal-oxidation stability of PUE based on diene oligomers by using DSC. In addition, it should be noted that there are quite a number of manufacturers of commercial stabilizers in the market. Experience has proven that even with the same chemical structure their efficacy may vary. For this reason, when formulating the composition and selecting the stabilizer, or when making a decision on the feasibility and acceptability of direct replacement of one brand by another, one must use a modern rapid method that would quickly assess and predict the thermal stability of the coating material.

In view of the above, the purpose of the present research is to study, by using DSC, the oxidation stability of PUE samples derived from butadiene and isoprene copolymer, and comparative assessment of OIT performance in the presence of different brands of pentaerythritol tetrakis[3-(3′,5′-di-tert-butyl-4′-hydroxyphenyl)propionate] stabilizer.

9.2 EXPERIMENTAL

To obtain PUE we used an oligomer, which is a product of the anionic copolymerization of butadiene and isoprene in the ratio of 80:20. The molecular weight of 3200. Mass fraction of hydroxyl groups was 1 percent, and the oligomer functionality on them was 1.8 percent.

The compositions were being prepared in a ball mill from 12 to 15 hrs. Homogenization of the components was carried out to the degree of grinding equal to 65. All formulations contained the same amount of the following ingredients: the above oligomer, filler (calcium carbonate), plasticizer of a complex ester nature, desiccant (calcium oxide), organic red pigment (FGR CI 112, produced by Ter Hell & Co Gmbh.).

Stabilizing agents varied in the amounts of 0.2, 0.6, and 1.0 wt. parts to 100 oligomer weight parts. The compositions were numbered in accordance with Table 9.1. Comparison sample was the material under the code 0, which did not contain the stabilizer.

TABLE 9.1 Brands and content of the stabilizers used

PUE Sample Number	Stabilizer Brand and Content (Per Oligomer 100 Weight Parts)			
	Irganox 1010	Evernox 10	Songnox 1010	Chinox 1010
0	–	–	–	–
1	0,2	–	–	–
2	0,6	–	–	–

TABLE 9.1 (Continued)

PUE Number	Sample	Stabilizer Brand and Content (Per Oligomer 100 Weight Parts)			
		Irganox 1010	Evernox 10	Songnox 1010	Chinox 1010
3	1	–	–	–
4		–	0,2	–	–
5		–	0,6	–	–
6		–	1	–	–
7		–	–	0,2	–
8		–	–	0,6	–
9		–	–	1	–
10		–	–	–	0,2
11		–	–	–	0,6
12		–	–	–	1

The compositions were cured taking into account the general content of the hydroxyl groups in the system under the action of the estimated volume of Desmodur 44V20L polyisocyanate (Bayer MaterialScience AG). Mass fraction of isocyanate groups in the product was 32 percent. Chain branching agent was chemically pure glycerin, and the catalyst was dibutyl tin dilaurate (manufactured by "ACIMA Chemical Industries Limited Inc".).

Curing conditions: standard laboratory temperature and humidity, duration – 72 hrs.

Pentaerythritol tetrakis[3-(3′,5′-di-tert-butyl-4′-hydroxyphenyl)propionate] of Irganox 1010, Evernox 10, Songnox Chinox 1010, and 1010 trademarks was used as a stabilizer. Structural formula is shown in Figure 9.1.

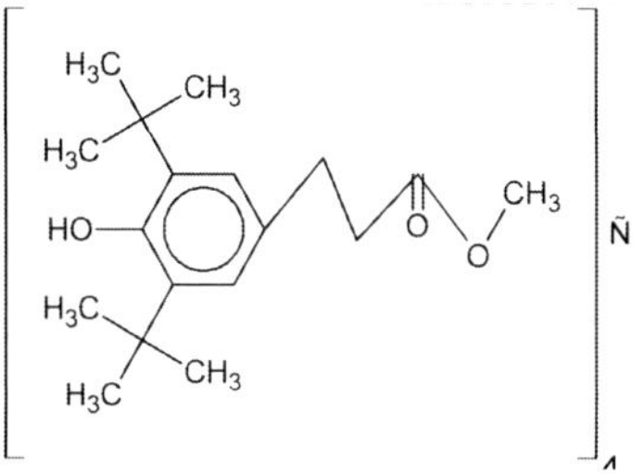

FIGURE 9.1 The structural formula of pentaerythritol tetrakis[3-(3′,5′-di-tert-butyl-4′-hydroxyphenyl)propionate] stabilizer.

Samples of cured PUE were tested by the Netzsch DSC 204 F1 Phoenix heat flow differential scanning calorimeter. Calibration was performed on an indium standard sample. Samples weighing 9–12 mg were placed in an open aluminum crucible. Test temperatures were attained at 10K/min rate under constant purging with an inert gas (argon). Upon reaching the target temperature the inert gas supply was stopped and the oxygen supply started at 50 ml/min rate. All data were recorded and processed using Netzsch Proteus special software in OIT registration mode.

9.3 RESULTS AND DISCUSSION

For polyolefins the tests to determine OIT are standardized in terms of both the recommended oxidizing gas (oxygen) flow and the isothermal segment temperature [8, 9]. There are no such standards for PUE. The authors [21] recommend that the test temperature be previously identified experimentally, and other settings be selected in accordance with the recommendations of the relevant ASTM or ISO. For this reason, we first determined the conditions of OIT fixation for the nonstabilized sample at two different temperatures (Figure 9.2).

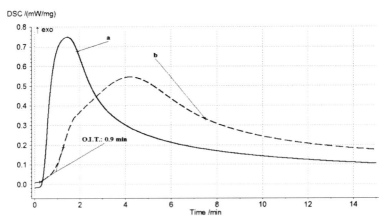

FIGURE 9.2 OIT determination for a nonstabilized PUE (number zero) at 200°C (**a**) and 180°C (**b**) by using atmosphere oxygen

As Figure 9.2 shows, isothermal mode at 200°C does not allow estimating the value of OIT for unstabilized sample. Under these conditions, snowballing oxidation degradation (curve a) starts almost immediately and oxidation induction time cannot be actually defined. As a result of reducing temperature to 180°C it is possible to fix the target parameter. For unstabilized sample the OIT value was 0.9 min (curve b). However, we found that tests at lower temperatures lead to the significant

Thermal Stability of Elastic Polyurethane

increase in test duration, especially for stabilized samples. This is undesirable, since the benefit of rapid OIT method in this case is partially lost. Thus, we found the necessary balance between time and temperature conditions that allow estimating OIT for the investigated objects at the recommended oxygen supply rate. In this regard, all subsequent tests were carried out at 180°C, and the results obtained are illustrated in each case, depending on the stabilizer type and content in comparison with the nonstabilized sample. The sample numbering and the stabilizer amount are consistent with Table 9.1.

Figure 9.3 shows the DSC curves for the samples containing Irganox 1010.

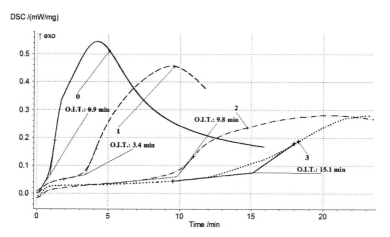

FIGURE 9.3 Isothermal DSC scans at 180°C for samples containing different amounts of Irganox 1010 stabilizer.

On DSC curves of the materials stabilized by Irganox 1010 antioxidant, the OIT value changes can be traced depending on the content of pentaerythritol tetrakis[3-(3′,5′-di-tert-butyl-4′-hydroxyphenyl)propionate] of this brand. Significant stabilizing effect can be observed even at proportion of 0.2 weight parts. When adding 0.6 and 1.0 weight parts of this product to PUE the OIT was 9.8 and 15.1 min respectively, which is 10 and 15 times as large as the corresponding parameter of reference sample.

Thereby it should be noted that the detected effects confirm and significantly specify the data obtained earlier [22] by using oligodiendiols and substituted phenol of this particular type, but using the classical evaluation method. The principal difference is that in the latter case, the implementation of the standard involves the need of thermostatic control of samples at higher air temperature (usually within 72 hrs), the subsequent physical and mechanical testing and correlation of properties before and after thermal aging. To assess the PUE sample oxidation stability by evaluating OIT by means of DSC the time expenditure is no more than 30 min and requires very little sample weight.

Materials containing Evernox 10, Songnox Chinox 1010, and 1010 were also investigated by this method. DSC data are shown in Figures 9.4, 9.5, and 9.6.

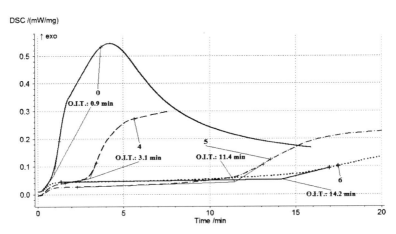

FIGURE 9.4 Isothermal DSC scans at 180 °C for samples containing different amounts of Evernox 10 stabilizer.

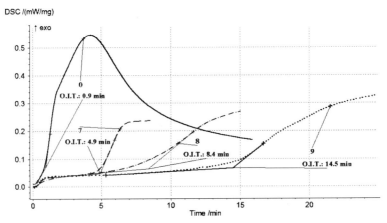

FIGURE 9.5 Isothermal DSC scans at 180°C for samples containing different amounts of Songnox 1010 stabilizer.

Thermal Stability of Elastic Polyurethane

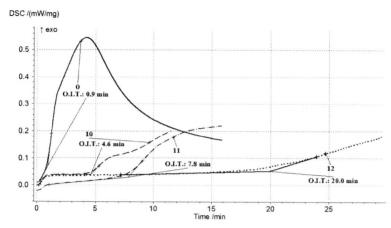

FIGURE 9.6 Isothermal DSC scans at 180°C for samples containing different amounts of Chinox 1010 stabilizer.

For general benchmarking the data obtained by processing the experimental array, are summarized in Table 9.2.

TABLE 9.2 OIT values for PUE, depending on the type and content of the stabilizer

Sample Number	Stabilizer Brand and Content (per 100 weight parts)				OIT, min
	Irganox 1010	Evernox 10	Songnox 1010	Chinox 1010	
0	–	–	–	–	0.9
1	0,2	–	–	–	3.4
2	0,6	–	–	–	9.8
3	1	–	–	–	15.1
4	–	0,2	–	–	3.1
5	–	0,6	–	–	11.4
6	–	1	–	–	14.2
7	–	–	0,2	–	4.9
8	–	–	0,6	–	8.4
9	–	–	1	–	14.5

TABLE 9.2 *(Continued)*

Sample Number	Stabilizer Brand and Content (per 100 weight parts)				OIT, min
	Irganox 1010	Evernox 10	Songnox 1010	Chinox 1010	
10	–	–	–	0,2	4.6
11	–	–	–	0,6	7.8
12	–	–	–	1	20

It follows from the OIT numerical values that regardless of the stabilizer brand, with an increase of its content in PUE within the investigated concentration range, the natural increase in the oxidation induction time is recorded. However, due to the high sensitivity of the method a significant difference in stabilizing effect of equal quantities of the single-type product manufactured by different vendors can be easily traced. Simple calculation shows that the deviation between the OIT maximum and minimum values for materials stabilized by 0.2 weight parts of Irganox 1010, Evernox 10, Songnox Chinox 1010, and 1010, is equal to 36.7 percent. With the content being 0.6 and 1.0 weight parts, this deviation is 20.4 and 31.5 percent, respectively. Apparently, this difference is due to the chemical purity and other factors that determine the protective capacity of the products used.

9.4 CONCLUSIONS

Thus, in the example of mesh polyurethane materials based on copolymer of butadiene and isoprene we show the high efficiency of using the option of OIT determination by DSC in order to carry out express tests on PUE thermal stability. The preferred temperature of the isothermal segment for accelerated test of this type of materials has been deduced from experiment.

Comparative evaluation of OIT indicators for PUE, stabilized by pentaerythritol tetrakis[3-(3',5'-di-tert-butyl-4'-hydroxyphenyl)propionate], depending on the manufacturer, ceteris paribus, revealed significant difference in terms of the protective effect of sterically hindered phenol. In practical terms, this means that prior to making the composition and selecting the stabilizer, as well as planning direct qualitative and quantitative replacement of one brand stabilizer by another in the PUE, one must take into consideration the potential significant differences in antioxidant efficacy of the products.

This work was supported by the Grant Council of the President of the Russian Federation, grant MK-4559.2013.3.

KEYWORDS

- DSC
- Material testing
- Oxidation induction time
- Polyurethane elastomers
- Stabilizers

REFERENCES

1. Prisacariu, C.; Polyurethane Elastomers. From Morphology to Mechanical Aspects. Springer-Verlag, Wien, **2011**.
2. Novakov, I. A.; Nistratov A. V.; Medvedev V. P.; Pyl'nov D. V.; Myachina, E. B.; and Lukasik, V. A.; et al.; Influence of hardener on physicochemical and dynamic properties of polyurethanes based on α,ω-di(2-hydroxypropyl)-polybutadiene Krasol LBH-3000. *Polym. Sci. Ser. D.* **2011**, *4*(2), 78–84.
3. Novakov, I. A.; Nistratov, A. V.; Pyl'nov, D. V.; Gugina, S. Y.; and Titova, E. N.; Investigation of the effect of catalysts on the foaming parameters of compositions and properties of elastic polydieneurethane foams. *Polym. Sci. Ser. D.* **2012**, *5*(2), 92–95.
4. ISO 188:2011 Rubber, vulcanized or thermoplastic. Accelerated ageing and heat resistance tests.
5. Menczel, Joseph D.; and Bruce Prime, R.; Thermal analysis of polymers. Fundamentals and Applications. John Wiley & Sons, Inc.: Hoboken, New Jersey; **2009**.
6. Gabbott, Paul; Principles and Applications of Thermal Analysis. Blackwell Pub: Oxford, Ames, Iowa; **2008**.
7. Pieichowski, J., and Pielichowski, K.: Application of thermal analysis for the investigation of polymer degradation processes. *J. Therm. Anal.* **1995**, *43*, 505–508.
8. ASTM D 3895-07: Standard Test Method for Oxidative-Induction Time of Polyolefins by Differential Scanning Calorimetry.
9. ISO 11357-6: Differential scanning calorimetry (DSC). Determination of oxidation induction time (isothermal OIT) and oxidation induction temperature (dynamic OIT).
10. ASTM E2009 – 08: Standard Test Method for Oxidation Onset Temperature of Hydrocarbons by Differential Scanning Calorimetry.
11. Gomory I.; and Cech, K.; A new method for measuring the induction period of the oxidation of polymers. *J. Therm. Anal.* **1971**, 3, 57–62.
12. Schmid, M.; Ritter, A.; Affolter, S.: Determination of oxidation induction time and temperature by DSC. *J. Therm. Anal. Cal.* **2006**, *83–2*, 367–371.
13. Woo, L.; Khare, A. R.; Sandford, C. L.; Ling, M. T. K.; and Ding, S. Y.; Relevance of high temperature oxidative stability testing to long term polymer durability. *J. Therm. Anal. Cal.* **2001**, *64*, 539–548.
14. Peltzer, M.; and Jimenez, A.; Determination of oxidation parameters by DSC for polypropylene stabilized with hydroxytyrosol (3,4-dihydroxy-phenylethanol). *J. Therm. Anal. Cal.* **2009**, *96*(I), 243–248.
15. Focke, Walter W.; and Westhuizen Isbe van der; Oxidation induction time and oxidation onset temperature of polyethylene in air. *J. Therm. Anal. Cal.* **2010**, *99*, 285–293.

16. Simon, P.; and Kolman, L.; DSC study of oxidation induction periods. *J. Therm. Anal. Cal.* **2001**, *64*, 813–820.
17. Conceicao Marta, M.; Dantas Manoel, B.; Rosenhaim, Raul, Fernandes Jr. Valter, J.; Santos Ieda, M. G.; and Souza Antonio, G.; Evaluation of the oxidative induction time of the ethylic castor biodiesel. *J. Therm. Anal. Cal.* **2009**, *97*, 643–646.
18. Woo, L.; Ding, S. Y.; Ling, M. T. K.; and Westphal, S. P.; Study on the oxidative induction test applied to medical polymers. *J. Therm. Anal.* **1997**, *49*, 131–138.
19. Dietmar MÄder (Oberursel, DE), inventors: Stabilization of polyol or polyurethane compostions against thermal oxidation, US20090137699, USA **2008**.
20. Simon, P.; Fratricova, M.; Schwarzer, P.; and Wilde H.-W.; Evaluation of the Residual Stability of Polyuretane Automotive Coatings by DSC. *J. Therm. Anal. Cal.* **2006**, *84*(3), 679–692.
21. Clauss, M.; Andrews, S. M.; Botkin, J. H.; Antioxidant Systems for Stabilization of Flexible Polyurethane Slabstock. *J. Cellul. Plas.* **1997**, *33*, 457.
22. Medvedev, V. P.; Medvedev, D. V.; Navrotskii, V. A.; Lukyanichev, V. V.; The study of oxidative aging polydieneurethane. *Polyurethane. Technol.* **2007**, *3*, 34–36.

CHAPTER 10

PAN/NANO–TIO$_2$–S COMPOSITES: PHYSICOCHEMICAL PROPERTIES

M. M. YATSYSHYN, A. S. KUN'KO, and O. V. RESHETNYAK

10.1 INTRODUCTION

Composites of conducting polymer (CP) and inorganic oxides are very attractive and perspective materials for different branches of science, namely for chemistry, physics, electronics, photonics, etc. due to synergetic effect which arises under the integration of the properties of oxide and CP [1, 2]. In the present time the titanium (IV) oxide (rutile and anatase crystalline modifications) is one among widely used inorganic oxides which is applied for the chemical synthesis of such nanocomposites [3-7].

Nanosized TiO$_2$ (anatase) is nontoxic, chemically inert, and inexpensive material. It is actively studied and widely used in the electronics, photonics [1,] and especially in the photocatalysis [8]. In recent years TiO$_2$ has been studied because it is also known as a photocatalyst [9] which accelerates the formation of the hydroxyl radicals under light. Hydroxyl radicals are powerful oxidizing agents that can disinfect and deodorize air, water, and surfaces in environmental–decontamination applications [10–13]. The features of nanosized TiO$_2$ induced with the presence of oxygen vacancies that leads to the occurrence of specific chemical and physical defects. However, the high energy gap width of the nano–TiO$_2$ (3.2 eV) renders impossible its effective operation under irradiation by sunlight. The decreasing of TiO$_2$ energy gap width requires the TiO$_2$ modification by different photosensitizers. For this purpose sulfur is frequently used [14-17], because their energy gap width is 2.9 eV [18].

Other widely used in last time photosensitizer for nano–TiO$_2$ is the polyaniline (PAn) with energy gap width 2.8 eV [19]. Polyaniline is organic metal with high conductivity in the doping form (~10^2 S×cm^{-1}) [20], good stability, nontoxic, inexpensive, and multifunctional polymer and also has the wide perspectives for usage [21-23]. These margins of polyaniline use develop considerably if the possibility of change of main polyaniline forms in the result of effect of different factors will be taken into consideration [21-23].

The combination of the properties of nano–TiO$_2$ and polyaniline enables to solve successfully the problems of the chemistry, physics, and electronics. Specific electronic structures of the nano–TiO$_2$ (as the n–type semiconductor) and polyaniline (as the electron's conductor in majority of the cases and as a p–type semiconductor under certain conditions) give the possibility to design the systems for different applications. For example, today such materials are equipped in the photocatalytic conversions of the different pollutants especially [14]. The modification of the surface of TiO$_2$ particles by polyanilines layers raises the catalytic activity of titanium (IV) oxide [5, 19]. Composite materials which have integrated properties of S–doped nano–TiO$_2$ and polyaniline layers can be effective in the photocatalytic processes especially.

PAn/TiO$_2$ composites are produced mainly during the aniline chemical oxidation in situ by different oxidants, for example (NH$_4$)$_2$S$_2$O$_8$ or Na$_2$S$_2$O$_8$, in the aqueous solutions of inorganic (HCl, H$_2$SO$_4$ etc.) or organic (for example salicylic, toluene sulfonic etc.) acids in the presence of produced previously micro– or nanosized TiO$_2$ particles [13, 22-26]. Polyaniline layers are applied on the surfaces of TiO$_2$ both without previous modification of oxide particles surface [13] and after such modification, for example by g–amino–propyl–triethoxysilane [27]. Synthesis of the hybrid PAn/nano–TiO$_2$ composites can be carried out also by one–step method [28-29], exactly in the micellar solutions of surfactants [30].

The composites particles on the basis of polyaniline and micro– or nanosized TiO$_2$ can be characterized by different structure, namely core–shell type [5], microrods [31], microspheres [1-3, 5, 13, 23, 32], nanowires [28], nanonets [33], which content the individual TiO$_2$ particles and their aggregates, and also the polymeric films with different thickness and morphology on the surface of the oxide nanoparticles [24]. The shape and size of the PAn/nano–TiO$_2$ composite particles are determined by synthesis conditions of the nanosized TiO$_2$ and composite on its base respectively.

The combination of polyaniline, nano–TiO$_2$ and other components [34-35] permits to produce the composites with improved physicochemical properties. Such nanocomposite materials are studied actively and are employed in the different branches of engineering and technics [31, 36], as cathodic materials in the chemical power sources [37], in the electronics [7, 38], chemo– and biosensors [39–40], and also as the components of corrosion protection coverages [41] or protective shades of different assignments [42].

It can be expected that the PAn/nano–TiO$_2$–S composite also has the attractive features and also arrives at a wide use in the modern life. Therefore the main aim of our work is the synthesis and study of physicochemical properties of such composite materials.

10.2 EXPERIMENTAL

10.2.1 MATERIALS

Aniline (Aldrich, 99.5 %) was distilled under the reduced pressure of 4 Torr and stored under argon. Other used in the work reagents were analytical grade and used without the additional purification. The solutions of chloride acid were prepared with use of the standard titers. For preparation of the all solutions the twice distilled water as used.

The nanorarticles of sulfur–doped TiO$_2$ (anatase) were synthesized by solid state method. In this case the initial metha–titanic acid H$_2$TiO$_3$ was dehydrated in the presence of sulfur (~1.5 mass percent) during its heating in the muffle furnace under the 500 °C over the time of 2 hrs. The synthesized sample was cooled then in the exiccator to the room temperature. An analysis of SEM–images of the synthesized TiO$_2$–S particles indicated that they are characterized by an aggregative structure with the size of aggregates, which form the spherical individual nanoparticles by ~20 nм size [33].

The samples of PAn and PAn/nano–TiO$_2$–S composites were produced by aniline oxidative polycondensation in the presence of sodium peroxydisulfate as an oxidant in the 0.5 M aqueous HCl solution under the temperature 2 ± 0.5 °C. Aniline: Na$_2$S$_2$O$_8$ molar ratio was 1 : 1.1. During the synthesis the 0.01 mole of aniline dissolved in the 50 cm^3 of 0.5 M aqueous HCl solution, introduced the batch of S–doped titanium (IV) oxide (0.25–2.0 g), exposure this mixture by ultrasonic machining for 10 min for the disaggregation of nano–TiO$_2$–S particles and stood over the time of 1 hr. In this solution under the mixing adds (by the ~0.5 cm^3 portions) 50 cm^3 solution of Na$_2$S$_2$O$_8$ (0.011 mole) in the cooled to the temperature 2 ± 0.5 °C 0.5 M HCl solution and mixes this reaction mixture further °C over the time of 1 hr at the same temperature. The reaction mixture further held at room temperature for 24 hrs, filtered, washed off by 250 cm^3 of distilled water to neutral value of pH and dried in the vacuum drier under the 60 ± 1 °C and underpressure 0.9 ± 0.05 kg×cm^{-2}. Preparation conditions of synthesized samples of PAn and composites shown on Table 10.1.

The reaction of the oxidative condensation of aniline in the HCL aqueous solutions can be presented by following scheme [43]:

TABLE 10.1 Preparation conditions of synthesized samples of PAn and PAn/nano–TiO_2–S composites

	SAMPLE	PAN	C2	C1	C0.75	C0.5	C0.25
REAGENTS	ANILINE / MOL	0.01	0.01	0.01	0.01	0.01	0.01
	HCL / M	0.5	0.5	0.5	0.5	0.5	0.5
	$NA_2S_2O_8$ / MOL	0.011	0.011	0.011	0.011	0.011	0.011
	NANO–TIO_2–S / G	-	2.0	1.0	0.75	0.50	0.25

The synthesized product was a powder by green color that indicates about production of the emeraldine form of polyaniline, exactly emeraldine salt PAn–HCl. The produced samples of PAn/nano–TiO_2–S composites had more light tinctures; their intensities were decreased with the increasing of the TiO_2 content in the initial reaction mixture.

10.2.2 INSTRUMENTAL

The physicochemical properties of the produced samples of PAn, nano–TiO_2–S and its composites were characterized with the use of the following experimental methods and instrumental supplies. X–ray phase analysis were carried out with use of DRON–3 diffractometer (Cu Kα radiation, l = 1.54 Å). FTIR spectra of synthesized samples were registered on the NICOLET IS 10 spectrometer (samples were pressed into a pellet together with KBr).

Paulic–Paulic–Erdei Q–1500 D derivatograph was used during studies of the thermal stability of synthesized samples. Measurements were carried out in the 20–900°C temperature interval (dynamic mode with a heating speed of 10 $^0K \times min^{-1}$) in an air atmosphere and with use of corundum pots. The sample's weight was 200 mg, reference substance—Al_2O_3.

Scanning electron microscopy (SEM) images and results of the energy dispersive X–ray (EDX) spectroscopy were obtained on the JEOL JSM–6400 microscope. The parallel beam of electrons with energy 3 kV was used for SEM–image acquisition of the samples surface, while under the qualitative and quantitative determination of elements in the samples the energy of electrons (EHT) was scanned in the 0-10 kV interval. In the first case the distance from electron source to samples surface was 3–4 mm, while during EDX studies—10 mm.

The cyclic voltammetry (CVA) measurements were carried out using a computer–controlled potentiostat / galvanostat ПИ–50–1 (Ukraine). Powders of synthesized polymer or composites were pressed into pellets with thickness ~2 mm and diameter 10 mm under the pressure 150 atm×cm^{-2} and temperature 20 °C over the time of 5 min and were used as the working electrodes. Platinum sheet was used as an counter electrode. The potential was scanned in the (-200 , (+600) mV interval with a 50 mV·s^{-1} scanning rate. All experimental potential values are referred to the saturated silver/silver chloride electrode.

Conductivity of produced samples was measured under the 20°C with use of sandwich–type cell. In this case the pellets, which were produced analogously as working electrodes for voltammetry, were used during the conductivity measurements.

10.4 RESULTS AND DISCUSSION

10.4.1 STRUCTURAL STUDIES OF PAN/NANO–TiO2–S COMPOSITES

The results of the X–ray phase analysis of the synthesized samples of nano–TiO$_2$–S, PAn and PAn/nano–TiO$_2$–S composites are shown on Figure 10.1. There are several sharp peaks, namely at 2q = 25.3, 38.1, 48.2, 54,9, 62.9 etc. on the diffractogram of nano–TiO$_2$–S sample (Figure 10.1, curve 1), which correspond to the crystalline structure of anatase [13, 28]. The diffraction peaks at 2q equal 9.6, 14.9, 21.4, 25.6, 27.6, 29.6 and 34.5 on the diffractogramm of polyaniline (Figure 10.1, curve 7) are typical for the high ordered structures of doped polyaniline PAn–HCl [30, 44] and corresponds to reflections of crystal planes (001), (011), (100), (110), (111) and (020) for PAn pseudoorthorhombic structure [45]. The presence of these seven peaks on the background of wide halo (Figure 10.1, curve 7) indicated that the synthesized sample of individual polyaniline has a mixed amorphous–crystalline structure [44]. The patterns exhibit sharp diffraction peaks at 2q = 25.6° and 21.4°, which can be ascribed to periodicity parallel and periodicity perpendicular to the polymer chain, respectively [46]. The peak at 2q=25.3° is most intensive on the TiO$_2$ diffractogramm (Figure 10.1, curve 1). It is agreed practically with PAn characteristic peak at 25,6°, which disappears during polymerization deposition of polyaniline on the surface of nano–TiO$_2$–S particles, but accrues peak at 2q=25.6° (Figure 10.1, curve 2-6). This fact indicates that majority of polyaniline molecules in the composites are deposited on the surface of TiO$_2$–S nanoparticles whose crystalline structure to bring influence on the crystallinity of PAn [28]. The observed high crystallinity and orientation of polyaniline macromolecules on the surface of nano–TiO$_2$ particles is promissory because such high–ordered systems shows higher electric conductivity [44].

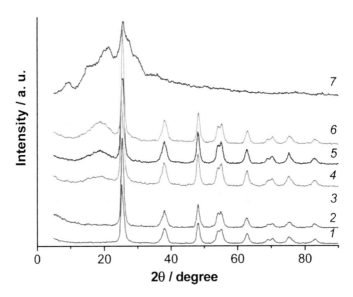

FIGURE 10.1 X–ray diffractogramms of nano–TiO$_2$–S (1), PAn (7) and PAn/nano–TiO$_2$–S composites: **C2** (2), **C1** (3), **C0.75** (4); **C0.5** (5), and **C0.25** (6) samples.

There are FTIR spectra of synthesized samples of individual PAn, nano–TiO$_2$–S and their composites on the Figure 10.2. The form of FTIR spectrum 7 in the 400–4,000 cm^{-1} interval and set observed characteristic bands, namely at 3400-2880, 1560, 1470, 1290, 1110, and 793 cm^{-1}, corresponds to polyaniline [28]. In particular, the peak at 3,400-2,880 cm^{-1} corresponds to so–called "H-band" which is a visualization of H–bonding imine– (–NH–) and protonated imine–group (–NH$^+$–) polyaniline macromolecules between itself [46]. Two characteristic bands at the 1560 and 1470 cm^{-1} are connected with vibrations of quinoid and benzenoid rings and they are characteristic features of the polyaniline. Intensive peaks at 1,290, 1,110 and 793 cm^{-1}, corresponds to the emeraldine salt and indicated on the high doping degree of PAn [46].

The slight shift of characteristic bands at 1,560 and 1,470 cm^{-1} observed for the samples of PAn/nano–TiO$_2$–S composites (Figure 10.2, spectra 2–5). It is a result of the interfacial interaction which takes place between PAn macromolecules and TiO$_2$–S nanoparticles. N-H…O and N…H-O H–bonding are realized in this case between imino–groups polymeric links and surface O= and HO- group of TiO$_2$, respectively [46]. The increasing of intensity of the wide band at 3 500-3 800 cm^{-1} with maximum at 3 700 cm^{-1} (Figure 10.2, spectra 2–5) is confirmation of this conclusion [3]. At once, the increasing of TiO$_2$–S content in the composite leads to intensive absorption of IR radiation in the 200–800 cm^{-1} interval

PAN/Nano–TiO$_2$–S Composites: Physicochemical Properties 161

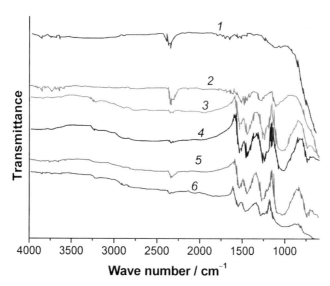

FIGURE 10.2 FTIR spectra of the of nano–TiO$_2$–S (1), PAn (6) and PAn/nano–TiO$_2$–S composites: **C1** (2), **C0.75** (3); **C0.5** (4), and **C0.25** (5) samples.

10.4.2 MORPHOLOGY OF NANO–TiO$_2$–S COMPOSITES

SEM images of the initial nano–TiO$_2$–S and synthesized composites (**C1** and **C2**) are shown on the Figure 10.3. The difference between images under the low TiO$_2$–S content is absent practically, while the morphology of samples under the high contents of mineral component is different. The analysis of SEM–image of synthesized TiO$_2$–S particles (Figure 10.3a-c) indicated that they are characterized by aggregative structure with size of aggregates from 200 to 800 нм. Analogous morphology is typical also for the synthesized PAn/nano–TiO$_2$–S composites, however the size of aggregates is slightly smaller—from 50 нм and they are looser (Figure 10.3d, f, g–i). The images which were obtained under the magnitude ´100.000 (Figure 10.3, b,c) confirms that aggregates of TiO$_2$–S are formed by individual nanoparticles with size ~20 нм [33]. Polyaniline in the **C1** composite (Figure 10.3e, f) forms 3D net structure which annexes individual TiO$_2$ particles and their fine aggregates. Under the higher TiO$_2$ content (Figure 10.3h) polymer deposits as a thin layer on the surface of individual mineral nanoparticles that favors disaggregation of the formed particles of PAn/nano–TiO$_2$–S composite. In this case the size of **C2** composite particles is ~25-30 nm (Figure 10.3i). It is obvious that TiO$_2$ nanoparticles has an influence on the formation of the polyaniline layer because they are the centers of the polymer-

10.4.3 EDX SPECTROSCOPY RESULTS

Spectra of energy dispersive X-ray spectroscopy in the chemistry of CP, in the case of nanostructured systems especially, are used for the confirmation of the composite formation and for the determination of the content of inorganic oxides or metallic particles components in them. EDX spectra of synthesized samples of nano–TiO_2–S and their composites with polyaniline are shown in the Figure 10.4. Intensive peaks at 4.55 and 4.95 keV and weak-intensive peak at 0.45 keV corresponds to titanium atoms while the peak at 0.52 keV to the atoms of oxygen (Figure 104a). The presence of sulfur as a doping agent in the TiO_2 confirms weak intensive peaks at 2.31 and 2.47 keV, respectively (Figure 10.4a). There are additional weak-intensive peaks at 0.27 and 2.62 keV in the EDX spectra of PAn/nano–TiO_2–S composites (Figure 10.4b, c), in comparison with spectrum of individual TiO_2, which corresponds to carbon and chlorine atoms (doping agent of PAn in the emeraldine form) respectively. Signal from nitrogen atoms (at 0,40 eV [47]) is absent in the measured spectra of composites. However, as it is known, it appears in the polyaniline EDX spectra not always [48–49], that is connected with special state of these atoms in the polyaniline macromolecules probably.

TABLE 10.2 Results of energy dispersive X-ray spectroscopy of synthesized samples (respective spectra are shown on Figure 10.4, a–c)

Element	nano–TiO_2–S Mass Fraction/Percent	nano–TiO_2–S Atomic Fraction/Percent	C1 Mass Fraction/Percent	C1 Atomic Fraction/Percent	C2 Mass fraction/percent	C2 Atomic fraction/percent
C K	-	-	34.31	49.36	18.93	32.74
O K	44.53	68.86	36.96	39.91	36.62	47.57
S K	1.31	0.71	1.33	0.72	1.19	0.77
Cl K	-	-	1.03	0.50	1.03	0.60
Ti K	53.52	30.43	26.37	9.51	42.23	18.32
S	100.00	100.00	100.00	100.00	100.00	100.00

The quantitative results of EDX spectra analysis (Table 10.2) permitted to conclude about the contents of samples. The juxtaposing of the results of X–ray phase analysis (Figure 10.1) and EDX spectra (Table 10.2) indicates that PAn/nano–TiO$_2$–S composites contains nanosized–TiO$_2$–S,

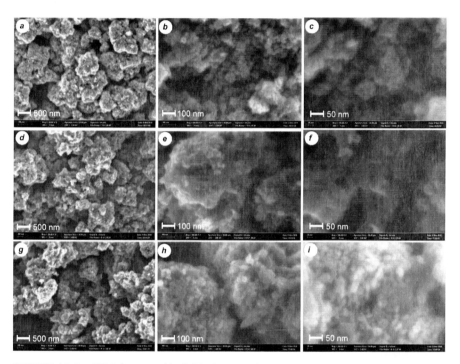

FIGURE 10.3 SEM–images of individual nano–TiO$_2$–S (a–c), **C1** (d–f) and **C2** (g–i) PAn/nano–TiO$_2$–S composites with ×20 000 (**a, d, g**), ×100.000 (**b, e, h**) and ×200.000 (**c, f, i**) magnification, respectively.

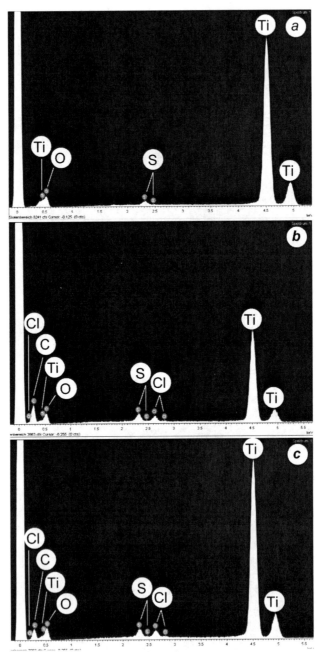

FIGURE 10.4 EDX spectra of individual nano–TiO_2–S (a), **C1** (b) and **C2** (c) PAn/nano–TiO_2–S composites.

PAN/Nano–TiO$_2$–S Composites: Physicochemical Properties   165

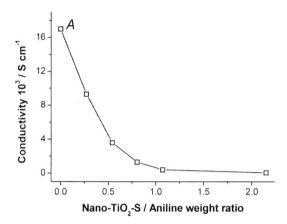

FIGURE 10.5 Dependence of the conductivity of synthesized samples of PAn/nano–TiO$_2$–S composites from the weight ratio nano–TiO$_2$–S: PAn in the reaction mixture during their synthesis polyaniline and Cl$^-$-anions as its doping agent. Decrease of the carbon content in the composite **C2** in comparison with **C1** even distribution is evidence of more even distribution of polymer in the composite in this case, exactly the formation thin polyaniline layer on the surface of mineral particles.

10.4.4 CONDUCTIVITY AND CYCLIC VOLTAMMETRY STUDIES

Results of conductivity studies of synthesized samples are shown on Figure 10.5. As it expected, the highest conductivity observes the polyaniline sample—17.0·10^{-3} S cm^{-1} (Figure 10.5, dot A). The introduction and further increasing of semiconductor TiO$_2$–S nanoparticles content in the composites leads to the exponential decreasing of the conductivity of samples.

The samples of PAn, **C0.25** and **C0.5** composites were tested as electrode materials. The studies were carried out in the 0.5 M sulfate acid aqueous solution under the potential scanning in the (-200)-(+600) mV interval. The cyclic volyammogramms for the first five cycles of potential scanning (Figure 10.6) demonstrates a good conductivity and reversibility of the produced samples. The higher conductivity, as it was noted above, characterizes the polyaniline sample. In this case, there is hysteresis loop on the cyclic voltammogamm (Figure 10.6a), which is decreased with the introduction of mineral component in mineral matrix (Figure 10.6b) and for the **C0.5** composite the voltammogramm corresponds to Ohm's law practically (Figure 10.6c). The values of cathodic and anodic currents are commensurate for the polyaniline (Figure 10.6a), while the higher values of anodic current in comparison with cathodic observes for the comopite's electrodes (Figure 10.6b-c), that can be explained by doping–dedoping processes of polyaniline component of composite during electrode cyclic polarization or less overvoltage of hydrogen evolution in the presence of TiO$_2$.

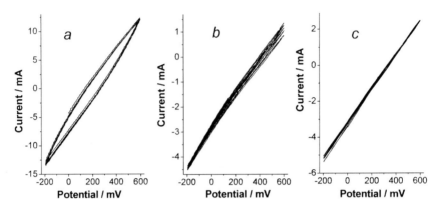

FIGURE 10.6 Cyclic voltammogramms (5 cycles of potential scanning) of the individual PAn (a), **C0.25** (b) and **C0.5** (c) PAn/nano–TiO$_2$–S composites in the 0.5 M H$_2$SO$_4$ aqueous solution.

10.4.4 THERMOGRAVIMETRIC STUDIES

The results of thermogravimetric studies of produced samples, namely individual nano–TiO$_2$–S and polyaniline, and its composite are shown on the Figures 10.7 and 10.8 and in Table 10.3. The character of obtained curves of mass loss Dm (TG), differential curves of mass loss (DTG) and differential thermal analysis (DTA) are different considerably (Figure 10.8).

The sample of nano–TiO$_2$–S is characterized by high thermal stability in the 20–900 °C temperature interval, where the total mass loss (Dm$_{900}$) is »5 percent (Figure 10.7, curve 7; Figure 10.8a, curve 1), while the total mass loss for the PAn sample is reached to » 95 percent (Figure 10.7, curve 1; Figure 10.8b, curve 1). Therefore, the main mass loss in the composites is connected with the thermal decomposition of polymeric component and values of total mass loss increases with increasing of PAn content in the samples (Figure 10.8, curves 2-6). In the same time, the increasing of TiO$_2$–S nanoparticles in the samples leads to the decreasing of thermodestruction rate of polyaniline component. It indicates the lower grade of TG curves in the 230–600°C temperature's interval. Besides, temperature of the termination of Pan component thermal destruction in the composites samples is shifted in the side of lower temperatures, whereas the start temperature of composites destruction (liberation of doping agent) is shifted in the side of the higher values in comparison with the individual polyaniline. Such shifts indicate on the strong interaction between the particles of titanium (IV) oxide and the polyaniline molecules in the composites [50].

Three stages can be marks out on the thermogravimetric curves (Figure 10.8a-c, curves 1) during the samples heating (results for **C1** sample are presented as an example of thermal characteristics of synthesized composites). Results of the quantitative analysis of these data are shown in Table 10.3. The first stage of mass loss (section A-B on the curves 1, Figure 10.8) is connected with desorption of physically fixed water, while the second stage (interval B-C)—with liberation of the water, which H–bonded with PAn macromolecules, doping agent (HCl) and hydration water of doping Cl$^-$-anions [51-55]. The limited sulfur burning from TiO$_2$–S particles takes place for sample of individual mineral component (Figure 10.8a) in the in the C-D interval on the TG curves, while oxidative destruction of polymer observes mainly in the same interval for the Pan and composites samples. The mass losses of the individual polyaniline sample for the noted intervals averages ≈ 11,5 percent (T ≈ 50–132 °C), ≈ 2,5 percent (132–250 °C) and ≈ 73,5 percent (250–900 °C), respectively.

Analysis of the results presented in Table 10.3 indicates that the increasing of TiO$_2$–S nanoparticles content in the composites leads to the certain decreasing of mass losses on the all marked out sections of curves 1 (Figure 10.8b, c), that confirms above noted explanations about nature of mass losses during oxidative destruction of Pan and composites samples. Temperature T_1 (see Figure 10.8) of first maximum on DTG curves decreases from 98 °C to 66 °C from PAn to nano–TiO$_2$–S individual samples. In the same time the value of temperatures T_2 of second maximum increases from 285 to 350 °C under the introduction of mineral component in polymer matrix and with its further increasing of its content. The several discrepancies of T_2 values can be stipulated by inorganic oxide influence on the processes on the second stage of thermal destruction of samples. In particular, the hydrated state of the surface of TiO$_2$–S nanoparticles and, as a result, the participation of terminated HO–groups in formation H–bonds with polymer macromolecules must be taken into consideration. Therefore the value T_2 = 129°C (Table 10.3) for the individual TiO$_2$–S samples can be connected with the process of dehydration of the surface of nanoparticles.

The temperature T_3 is decreased with the increasing of TiO$_2$ content in composites, but for the **C0.5–C2** samples it is constant practically. Such character of T_3 change, in our opinion,

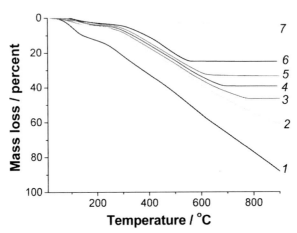

FIGURE 10.7 Thermogravimetric (TG) curves of the polyaniline (1), nano–TiO$_2$–S (7) and their composites: **C0.25** (2), **C0.5** (3), **C0.75** (4), **C1** (5) and **C2** (6)

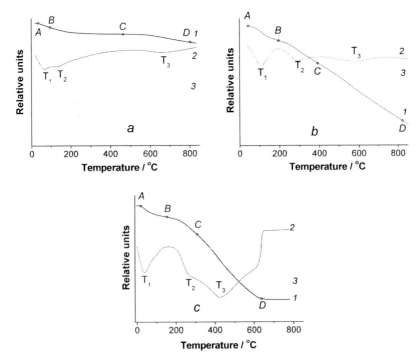

FIGURE 10.8 TG (1), DTA (1) and DTG (1) curves for the synthesized samples of nano–TiO$_2$–S (a); PAn (b) and **C1** composite (c)

PAN/Nano–TiO$_2$–S Composites: Physicochemical Properties 169

TABLE 10.3 Results of the analysis of TG and DTG curves of studied samples

Samples	Mass Loss of Samples on the Section of TG curve/Percent			Total Mass Loss Dm$_{900}$ / percent	Temperatures that Corresponds to Maxima of mass Losses / (±1.0) °C		
	A - B^a	B - C^a	C – D^a		T_1^a	T_2^a	T_3^a
PAn	11.5	2.5	73.5	95.2	98	285	554
C0.25	4.4	2.3	54.7	61.4	95	300	500
C0.5	4.2	2.7	39.2	46.7	85	325	485
C0.75	4.0	2.9	32.3	42.4	82	342	479
C1	3.4	3.2	28.0	34.6	81	290	470
C2	2.8	3.7	19.5	26.0	78	350	483
TiO$_2$-S	2.1	0.6	2.25	5.0	66	129	670

^{a}See Figure 10.8a, b, c and comments in the text

is connected with the changes in morphology of produced samples—from aggregative (3D net structure in the **C0.25** and **C0.5** samples) to thin polyaniline layers on the surface of mineral nanoparticles (for the **C0.5–C2** samples), that assured fuller oxidative destruction of polymeric component of composite under lower temperatures. However, the more detailed analysis of DTG curves of produced samples with high content of nano–TiO$_2$–S indicated that the third temperature interval (C–D, see Figure 10.8) contains several successive steps of polymer destruction. Besides main step of polyaniline oxidative destruction, it are the steps of oligomeric fraction of polyaniline liberation, supplementary finish liberation of doping agent, final carbonization of intermediate chemicals [56] and limited sulfur burning from TiO$_2$–S particles (with T$_3$ = 670°C, see Table 10.3), that confirms the complicacy of the destruction of composites samples.

10.5 CONCLUSIONS

The PAn/nano–TiO$_2$–S composites were synthesized in the 0.5 M HCl aqueous solutions by oxidative oxidation of aniline in the presence of nanoparticles of mineral component with size ~20 nm. The results of X–ray diffraction phase analysis, FTIR and EDX spectroscopy and thermogravimetric studies, indicate that polyaniline is deposited on the surface of mineral component particles and is characterized by sufficiently ordering of macromolecules. The change of samples morphology from aggregative (3D net structure) to thin polymeric layers on the surface of TiO$_2$–S nanoparticles with increasing of mineral component content was detected under the analysis of SEM images.

The formation of composite, but not mechanical mixture of components, confirms the results of EDX and FTIR spectroscopy. Moreover, the shift of characteristic bands in the FTIR spectra of composites samples confirms the presence of strong interfacial interaction between polymeric macromolecules and surface of TiO$_2$–S nanoparticles in the result of H–bonding.

It was determined, that the increasing of content of the mineral component leads to the increasing of the thermal stability of the PAn/nano–TiO$_2$–S composite in comparison with individual polyaniline, but to the decreasing of electric conductivity of samples.

It is expected that the synthesized composites with the high contents of TiO$_2$–S, when the thin polyaniline layer deposited on the mineral nanoparticles, will be effective sensitizers in the processes of the photooxidation in consequence of the synergetic effect occurrence under the integration of the properties of S–doped TiO$_2$ and polyaniline.

KEYWORDS

- **Conductivity**
- **Crystallinity**
- **Interfacial interaction.**
- **Nanocomposite**
- **Polyaniline**
- **Surface morphology**
- **Thermal stability**
- **Titanium (IV) oxide**

REFERENCES

1. Korotcenkov, G.; *Mater. Sci. Eng. B.* **2007**, *139*, 1–23.
2. Wan, M.; Conducting Polymers with Micro or Nanometer Structure. *Tsinghua University Press/Springer–Verlag GmbH,* Berlin–Heidelberg/Beijing, 312 p. (2010).
3. Xu, J.–C. Liu, W.–M.; and Li, H.–L.; *Mater. Sci. Eng. C.* **2005**, *25*, 444–447.
4. Parvatikar, N.; Jain, S.; and Khashim, S.; et al. *Sens. Actuat. B.* **2006**. *114*, 599–603.
5. Wang, X.; Tang, S.; Zhou, C.; Liu, J.; and Feng, W.; *Synth. Met.* **2009**, 159, 1865–1869.
6. Shi, L.; Wang, X.; and Lu, L.; et al. *Synth. Met.* **2009**, *159*, 2525–2529.
7. Bian, C.; Yu, A.; and Wu, H.; *Electrochem. Commun.* **2009**. *11*, 266–269.
8. Sánchez, J.; and Peral, X.; Doménech. *Electrochim. Acta.* **1997**, *42*, 1872–1882.

9. Su, S.-J.; and Kuramoto, N.; *Synth. Met.* **2000**, *114*, 147–153.
10. Zhang, H.; Zong, R. L.; Zhao, J. C.; and Zhu, Y. F.; *Environ. Sci. Technol.* **2008**, *42*, 3803-3807.
11. Mohammad, R. K.; Jeong, H. Y.; Mu, S. L.; and Kwon, T. L.; *React. Funct. Polym.*, **2008**, *68*, 1371-1376.
12. Zhang, L.; Liu, P.; and Su, Z.; **2006**, *Polym. Degrad. Stabil.* **91**, 2213-2219.
13. Zhang, L.; Wan, M.; and Wei, Y.; *Synth. Met.* **2005**, *151*, 1–5.
14. The, C. M.; and Mohamed, A. R.; *J. Alloys Compd.* **2011**, *509*, 1648-1660.
15. Koca, A.; and Sahin, M.; *Int. J. Hydrogen. Energy.* **2002**, *27*, 363–367.
16. Umebayashi, T.; Yamaki, T.; Itoh, H.; and Asai, K.; *Appl. Phys. Lett.* **2002**, *81*, 454–456.
17. Tian, H.; Ma, J.; Li, K.; and Li, J.; *Ceram. Int.* **2009**. *35*, 1289–1292.
18. Umebayashi, T.; Yamaki, S.; Yamamoto, A.; Miyashita, S.; Tanaka, T.; Sumita, K.; and Asai J.; *Appl. Phys.* **2003**, *93*, 5156–5160.
19. Li, X.; Wang, D.; Cheng, G.; Luo, Q.; An, J.; and Wang, Y; *Appl. Catal. B: Environ.* **2008**, *81*, 267–273.
20. Kim, S.–C.; Kim, D.; Lee, J.; Wang, Y.; Yang, K.; Kumar, J.; Bruno, F. F.; and Samuelson, L. A.; *Macromol. Rapid Commun.* **2007**, *28*, 1356-1360.
21. Malinauskas, A.; *Polymer.* **2001,** *42*, 3957-3972.
22. Koval'chuk, Ye.; Yatsyshyn, M.; and Dumanchuk, N.; *Proc. Shevchenko Sci. Soc. Chem. Biochem.* **21**, 108-122.
23. Eftekhari, A. (Ed.) Nanostructured Conducting Polymers. **2010**, *John Wiley & Sons, Ltd.*: Chichester, UK, 800.
24. Tai, H.; Jiang, Y.; Xie, G.; Yu, J.; and Chen, X.; *Sens. Actuat. B.* **2007** *125*, 644–650.
25. Kim, B.–S., Lee, K–T.; Huh, P.–H.; Lee, D.–H.; and Lee, J.–O.; *Synth. Met. 159.* **2009**, 1369–1372.
26. Oh, M.; and Kim, S.; *Electrochim. Acta.* **2012**. *78*, 279-285.
27. Chuang, F.–Y., and Yang, S.-M. *Synth. Met.* **2005**, *152*, 361–364.
28. Bian, C.; Yu, Y.; and Xue, G.; *Appl. Polym. Sci.* **2007**, *104*, 21–26.
29. Savitha, K. U.; and Prabu, H. G.; *Mater. Chem. Phys.*, **2011**, *130*, 275-279.
30. Yang, J, Ding, Y.; and Zhang, J.; *Mater. Chem. Phys.* **2008**, *112*, 322-324.
31. Karim, M. R.; Yeum, J. H.; Lee, M. S.; and Lim, K. T.; *React. Funct. Polym.* **2008**, *68*, 1371–1376, ().
32. Mo, T.–C.; Wang, H.–W.; Chen, S.–Y.; and Yeh, Y.–C.; *Ceram. Int.* **2008**, *34*, 1767–1771.
33. Yatsyshyn, M. M.; Koval'chuk, A. S.; Kun'ko, V. S.; and Besaga, Kun'ko, Kh. *Mat. Sci. Nanostruct.* **2011**, *1*, 81-86.
34. Zhou, H.; Zhang, C.; Wang, X.; Li, H.; and Du, Z.; *Synth. Met.*, **2011**, *161*, 2199-2205.
35. Çetin, H.; Boyarbay, B.; Akkaya, A.; Uygun, A.; and Ayyildiz, E.; *Synth. Met.* **2011**, *161*, 2384-2389.
36. Ameen, S.; Akhtar, M. S.; Kim, G.–S.; Kim, Y. S.; Yang, O.–B.; and Shin, H.–S.; *J. Alloys Compd.* **2009**, *487*, 382–386.
37. Gurunathan, K.; Amalnerkar, D. P., and Trivedi, D. C.; *Mater. Lett.*, **57**, 1642–1648.
38. Li, Y.; Hagen, J.; and Haarer, D.; *Synth. Met.*, *94*, 273–277.
39. Zheng, J.; Li, G.; Ma, X.; Wang, Y.; Wu, G.; and Cheng, Y.; *Sens. Actuators B*, **2008**, *133*, 374–380.
40. Tai, H.; Jiang, Y.; Xie, G.; Yu, J.; Chen, X.; and Ying, Z.; *Sens. Actuat. B*, 2008. *129*, 319–326.
41. Radhakrishnan, S.; Siju, C. R.; Mahanta, D.; Patil, S.; and Madras, G.; *Electrochim. Acta.* **2009**, *54*, 1249–1254.
42. Phang, S. W.; Tadokoro, M.; Watanabe, J.; and Kuramoto, N.; *Curr. Appl. Phys.* 8, 391–394.
43. Sapurina, I.; and Stejskal, J.; *Polym. Int.* **2008**, *57*, 1295–1325.
44. Rahy, A.; and Yang, D. J.; *Mater. Lett.* **2008**, *62*, 4311–4314.
45. Pouget, J. P.; Jòzefowicz, M. E.; and Epstein, A. J.; et al. *Macromolecules.* **1991**, *24*, 779–789.
46. Li, X.; *Electrochim. Acta.* **2009**, *54*, 5634–5639

47. Karim, M. R.; Lee, H. W.; Cheong, I. W.; Park, S. M.; Oh, W., and Yeum, J. H. *Polym. Composite.* **31**, 83-88 (2010).
48. Liu, X.-X.; Zhang, L.; Li, Y.-B.; Bian, L.-J.; Huo, Y.-Q.; and Su, Z.; *Polymer Bull.* **2006**, *57*, 825–832.
49. Karthikeyan, M.; Satheeshkumar, K. K.; and Elango, K. P.; *J. Hazardous Mater.* **2009**, *163*, 1026-1032.
50. Deng, J.; He, C. L.; Peng, Y.; Wang, J.; Long, X.; Li, P.; and Chan, A. S. C.; *Synth. Met.* **2003**, *139*, 295–301.
51. Pielichowski, K.; *Solid. State. Ionics.* **1997**, *104*, 123-132.
52. Tsocheva, D.; Zlatkov, T.; and Terlemezyan, L.; *J. Therm. Anal.*, 1998, *53*, 895–904.
53. Rodrigues, P. C.; de Souza, G. P.; Neto, J. D. D. M.; and Akcelrud, L.; *Polymer*, *43*, 5493–5499.
54. Lee, I. S.; Lee, J. Y.; Sung, J. H.; and Choi, H. J. *Synth. Met.* **2005**, *152*, 173–176.
55. Salahuddin, N., Ayad, M. M.; and Ali, M.; *J. Appl. Polymer Sci.* **2008**, *107*, 1981–1989.
56. Bhadra, S.; Chattopadhyay, S.; Singha, N. K.; and Khastgir, D.; *J. Appl. Polymer Sci.* **2008**, *108*, 57–64.

CHAPTER 11

VISCOELASTIC PROPERTIES OF THE POLYSTYRENE

YU. G. MEDVEDEVSKIKH, O. YU. KHAVUNKO, L. I. BAZYLYAK, and G. E. ZAIKOV

11.1 INTRODUCTION

The viscosity η of polymeric solutions is an object of the numerous experimental and theoretical investigations generalized in Refs. [1-4]. This is explained both by the practical importance of the presented property of polymeric solutions in a number of the technological processes and by the variety of the factors having an influence on the η value, also by a wide diapason (from 10^{-3} to 10^2 $Pa{\times}s$) of the viscosity change under transition from the diluted solutions and melts to the concentrated ones. The all above said gives a great informational groundwork for the testing of different theoretical imaginations about the equilibrium and dynamic properties of the polymeric chains.

It can be marked three main peculiarities for the characteristic of the concentrated polymeric solutions viscosity, *namely*:

1. Measurable effective viscosity η for the concentrated solutions is considerable stronger than the η for the diluted solutions and depends on the velocity gradient *g* of the hydrodynamic flow or on the shear rate.

It can be distinguished [4] the initial η_0 and the final η_∞ viscosities ($\eta_0 > \eta_\infty$), to which the extreme conditions $g \to 0$ and $g \to \infty$ correspond respectively.

Due to dependence of *η* on *g* and also due to the absence of its theoretical description, the main attention of the researches [4] is paid into the so–called most newton (initial) viscosity η_0, which is formally determined as the limited value at $g{\to}0$. Exactly this value η_0 is estimated as a function of molar mass, temperature, and concentration (in solutions).

The necessity of the experimentally found values of effective viscosity extrapolation to «zero» shear stress doesn't permit to obtain the reliable value of η_0. This leads to the essential and far as always easy explained contradictions of the experimental results under the critical comparison of data by different authors.

2. Strong power dependence of η on the length N of a polymeric chain and on the concentration ρ (g/m^3) of a polymer in solution exists: $\eta \sim \rho^\alpha N^\beta$ with the indexes $\alpha = 5 \div 7$, $\beta = 3{,}3 \div 3{,}5$, as it was shown by authors [4].
3. It was experimentally determined by authors [1, 5] that the viscosity η and the characteristic relaxation time t^* of the polymeric chains into concentrated solutions and melts are characterized by the same scaling dependence on the length of a chain:

$$\eta \sim t^* \sim N^\beta \qquad (11.1)$$

with the index $\beta = 3{,}4$.

Among the numerous theoretical approaches to the analysis of the polymeric solutions viscosity anomaly, i. e. the dependence of h on g, it can be marked the three main approaches. The first one connects the anomaly of the viscosity with the influence of the shear strain on the potential energy of the molecular kinetic units transition from the one equilibrium state into another one and gives the analysis of this transition from the point of view of the absolute reactions rates theory [6]. However, such approach hasn't take into account the specificity of the polymeric chains; that is why, it wasn't win recognized in the viscosity theory of the polymeric solutions. In accordance with the second approach the polymeric solutions viscosity anomaly is explained by the effect of the hydrodynamic interaction between the links of the polymeric chain; such links represent by themselves the "beads" into the "necklace" model. Accordingly to this effect the hydrodynamic flow around the presented "bead" essentially depends on the position of the other "beads" into the polymeric ball. An anomaly of the viscosity was conditioned by the anisotropy of the hydrodynamic interaction which creates the orientational effect [7, 8]. High values of the viscosity for the concentrated solutions and its strong gradient dependence cannot be explained only by the effect of the hydrodynamic interaction.

That is why the approaches integrated into the conception of the structural theory of the viscosity were generally recognized. In accordance with this theory the viscosity of the concentrated polymeric solutions is determined by the quasi–net of the linkages of twisted between themselves polymeric chains and, therefore, depends on the modulus of elasticity E of the quasi–net and on the characteristic relaxation time t^* [1-2]:

$$\eta = E \cdot t^* \qquad (11.2)$$

It is supposed, that the E is directly proportional to the density of the linkages assemblies and is inversely proportional to the interval between them along the same chain. An anomaly of the viscosity is explained by the linkages assemblies' density decreasing at their destruction under the action of shear strain [9], or by the change of the relaxation spectrum [10], or by the distortion of the polymer chain links distribution function relatively to its center of gravity [11]. A gradient dependence of the viscosity is described by the expression [11]:

$$(\eta - \eta_\infty)/(\eta_0 - \eta_\infty) = f(gt^*) \qquad (11.3)$$

It was greatly recognized the universal scaling ratio [1, 5]:

$$\eta = \eta_0 \cdot f(gt^*) \qquad (11.4)$$

in which the dimensionless function $f(gt^*) = f(x)$ has the asymptotes $f(0) = 1$, $f(x)\underset{x\gg1}{=} x^{-g}$, $g = 0{,}8$.

Hence, both expressions (11.3) and (11.4) declare the gradient dependence of h by the function of the one nondimensional parameter gt^*. However, under the theoretical estimation of h and t^* as a function of N there are contradictions between the experimentally determined ratio (11.1) and b = 3,4. Thus, the analysis of the entrainment of the surrounding chains under the movement of some separated chain by [12] leads to the dependencies $\eta \sim N^{3.5}$ but $t^* \sim N^{4.5}$. At the analysis [13] of the self–coordinated movement of a chain enclosing into the tube formed by the neighboring chains it was obtained the $\eta \sim N^3$, $t^* \sim N^4$. The approach in [14] which is based on the conception of the reptational mechanism of the polymeric chain movement gives the following dependence $\eta \sim t^* N^3$. So, the index b = 3,4 in the ratio (11.1) from the point of view of authors [2] remains by one among the main unsolved tasks of the polymers' physics.

Summarizing the above presented short review, let us note, that the conception about the viscosity–elastic properties of the polymeric solutions accordingly to the *Maxwell's* equation should be signified the presence of two components of the effective viscosity, *namely*: the *frictional* one, caused by the friction forces only, and the *elastic* one, caused by the shear strain of the conformational volume of macromolecules. But in any among listed above theoretical approaches the shear strain of the conformational volumes of macromolecules was not taken into account. The sustained opinion by authors [3-4] that the shear strain is visualized only in the strong hydrodynamic flows whereas it can be neglected at little g, facilitates to this fact. But in this case the inverse effect should be observed, *namely* an increase of h at the g enlargement.

These contradictions can be overpassed, if to take into account [15, 16], that, although at the velocity gradient of hydrodynamic flow increasing the external action leading to the shear strain of the conformational volume of polymeric chain is increased, but at the same time, the characteristic time of the external action on the rotating polymeric ball is decreased; in accordance with the kinetic reasons this leads to the decreasing but not to the increasing of the shear strain degree. Such analysis done by authors [15-17] permitted to mark the *frictional* and the *elastic* components of the viscosity and to show that *exactly the elastic component* of the viscosity *is the gradiently dependent value.* The elastic properties of the conformational volume of polymeric chains, in particular shear modulus, were described early by authors [18-19] based on the self–avoiding walks statistics (*SAWS*).

Here presented the experimental data concerning to the viscosity of the concentrated solutions of styrene in toluene and also of the melt and it is given their interpretation on the basis of works [15-19].

11.2 EXPERIMENTAL DATA AND STARTING POSITIONS

In order to obtain statistically significant experimental data we have studied the gradient dependence of the viscosity for the concentrated solution of polystyrene in toluene at concentrations $0,4\times10^5$; $0,5\times10^5$ and $0,7\times10^5$ g/m^3 for the four fractions of polystyrene characterizing by the apparent molar weights $M = 5,1\times10^4$; $M = 4,1\times10^4$; $M = 3,3\times10^4$ and $M = 2,2\times10^4$ g/mole. For each pair of values r and M the gradient dependence of the viscosity has been studied at fourth temperatures 25°C, 30°C, 35°C, and 40°C.

The viscosity for the polystyrene melt were investigated using the same fractions at 210°C. Temperature dependence of the polystyrene melt was investigated for the fraction with average molecular weight $M = 2,2\times10^4$ g/mole under three temperatures, namely 190, 200, and 210°C.

The experiments have been carried out with the use of the rotary viscometer *RHEOTEST* 2.1 equipped by the working cylinder having two rotary surfaces by diameters $d_1 = 3,4\times10^{-2}$ and $d_2 = 3,9\times10^{-2}$ m in a case of the concentrated solutions of polystyrene investigation and using the device by "cone-plate" type equipped with the working cone by 0,3° angle and radius $r = 1,8\times10^{-2}$ m in a case of the polystyrene melt investigation.

11.3 CONCENTRATED SOLUTIONS

11.3.1 INITIAL STATEMENTS

Typical dependences of viscosity η of solution on the angular velocity ω *(turns/s)* of the working cylinder rotation are represented on Figures 11.1, 11.2, and 11.3. Generally it was obtained the 48 curves of $\eta(\omega)$.

For the analysis of the experimental curves of $\eta(\omega)$ it was used the expression [15, 20]:

$$\eta = \eta_f + \eta_e\left(1-\exp\{-b/\omega\}\right)/\left(1+\exp\{-b/\omega\}\right) \tag{11.5}$$

Viscoelastic Properties of the Polystyrene

in which η is the measured viscosity of the solution at given value ω of the working cylinder velocity rate; η_f and η_e are frictional and elastic components of η;

$$b/\omega = t_v^* / t_m^* \qquad (11.6)$$

where t_m^* is the characteristic time of the shear strain of the conformational volume for m–ball of intertwined polymeric chains; t_v^* is the characteristic time of the external action of gradient rate of the hydrodynamic flow on the m–ball.

The notion about the m–ball of the intertwined polymeric chains will be considered later.

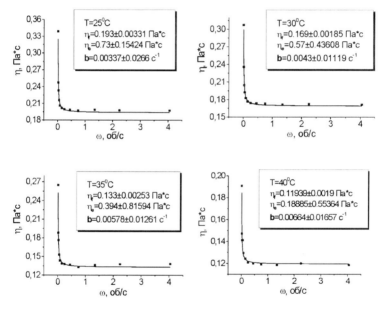

FIGURE 11.1 Experimental (points) and calculated in accordance with the Eq. (11.5) (curves) dependencies of the effective viscosity on the rotation velocity of the working cylinder: $\rho = 4.0 \cdot 10^5$ g/m^3, $M = 4.1 \cdot 10^4$ g/mole, $T = 25 \div 40°C$.

The shear strain of the conformational volume of m–ball and its rotation is realized in accordance with the reptational mechanism presented in Ref. [2], i. e. via the segmental movement of the polymeric chain, that is why t_m^* is also the characteristic time of the own, i. e. without the action g, rotation of m–ball [17].

The expression (11.5) leads to the two asymptotes:

$$\eta = \eta_f + \eta_e \text{ at } b/\omega \gg 1$$

$$\eta = \eta_f \text{ at } b/\omega \ll 1$$

So, it is observed a general regularity of the effective viscosity dependence on the rotation velocity ω of the working cylinder for diluted, concentrated solutions and melts. Under condition, that $b/\omega \gg 1$, that is at $\omega \to 0$, the effective viscosity is equal to a sum of the frictional and elastic components of the viscosity, and under condition $\omega \to \infty$ the measurable viscosity is determined only by a frictional component of the viscosity.

In accordance with Eq. (11.5) the effective viscosity $\eta(\omega)$ is a function on three parameters, *namely* η_f, η_e and b. They can be found on a basis of the experimental values of $\eta(\omega)$ via the optimization method in program *ORIGIN* 5.0. As an analysis showed, the numerical values of η_f are easy determined upon a plateau on the curves $\eta(\omega)$ accordingly to the condition $b/\omega \ll 1$ (see Figures 11.1, 11.2 and 11.3).

However, the optimization method gave not always the correct values of η_e and b. There are two reasons for this. Firstly, in a field of the $\omega \to 0$ the uncertainty of $\eta(\omega)$ measurement is sharply increased since the moment of force registered by a device is a small. Secondly, in very important field of the curve transition $\eta(\omega)$ from the strong dependence of η on ω to the weak one the parameters η_e and b are interflowed into a composition $\eta_e b$, i. e. they are by one parameter. Really, at the condition $b/\omega \ll 1$ decomposing the exponents into (11.5) and limiting by two terms of the row $\exp\{-\frac{b}{\omega}\} \approx 1 - \frac{b}{\omega}$, we will obtained $\eta = \eta_f + \eta_e b/2$. Due to the above-mentioned reasons the optimization method gives the values of η_e and b depending between themselves but doesn't giving the global minimum of the errors functional. That is why at the estimation of η_e and b parameters it was necessary sometimes to supplement the optimization method with the "manual" method of the global minimum search varying mainly by the numerical estimation of η_e.

Viscoelastic Properties of the Polystyrene

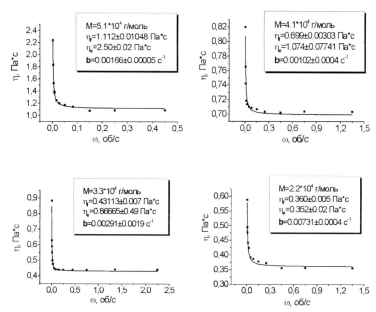

FIGURE 11.2 Experimental (points) and calculated in accordance with the Eq. (11.5) (curves) dependencies of the effective viscosity on the rotation velocity of the working cylinder: $\rho = 5.0 \cdot 10^5$ g/m^3, $M = 5.1 \div 2.2 \cdot 10^4$ g/mole, $T = 25°C$.

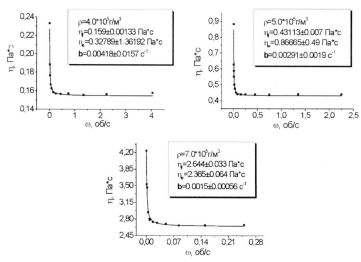

FIGURE 11.3 Experimental (points) and calculated in accordance with the Eq. (11.5) (curves) dependencies of the effective viscosity on the rotation velocity of the working cylinder: $\rho = 4.0 \cdot 10^5 \div 7.0 \cdot 10^5$ g/m^3, $M = 3.3 \cdot 10^4$ g/mole, $T = 25°C$.

As we can see from the Figures 11.1, 11.2, and 11.3, calculated curves $\eta(\omega)$ accordingly to the Eq. (11.5) and found in such a way parameters η_f, η_e and b, are described the experimental values very well.

The results of η_f, η_e and b numerical estimations for the all 48 experimental curves $\eta(\omega)$ are represented in Table 11.1. The mean–square standard deviations of the η_f, η_e and b calculations indicated on the Figures.

A review of these data shows that the all three parameters are the functions on the concentration of polymer into solution, on the length of a chain and on the temperature. But at this, the η_e and η_f are increased at the ρ and M increasing and are decreased at the T increasing whereas the b parameter is changed into the opposite way. The analysis of these dependencies will be represented in Sections 3.2, 3.3 and 3.4. Here let us present the all needed for this analysis determinations, notifications, and information concerning to the concentrated polymeric solutions.

Investigated solutions of the polystyrene in toluene were concentrated; since the following condition was performing for them:

$$\rho \geq \rho^*, \tag{11.7}$$

where ρ^* is a critical density of the solution per polymer corresponding to the starting of the polymeric chains conformational volumes overlapping having into diluted solution $(\rho \leq \rho^*)$ the conformation of *Flory* ball by the radius

$$R_f = aN^{3/5}, \tag{11.8}$$

here a is a length of the chain's link. It's followed from the determination of ρ^*

$$\rho^* = M / N_A R_f^3 = M_0 N / N_A R_f^3 \tag{11.9}$$

where M_0 is the molar weigh of the link of a chain. Taking into account the Eq. (11.8) and (11.9) we have:

$$\rho^* = \rho_0 N^{-4/5}, \tag{11.10}$$

where

$$\rho_0 = M_0 / a^3 N_A \tag{11.11}$$

can be called as the density into volume of the monomeric link.

Viscoelastic Properties of the Polystyrene

TABLE 11.1 Optimization parameters η_f, η_e and b in Eq. (11.5)

$\rho \cdot 10^{-5}$, g/m³		4,0				5,0				7,0			
T, °C	$M \cdot 10^4$ g/mole	5,1	4,1	3,3	2,2	5,1	4,1	3,3	2,2	5,1	4,1	3,3	2,2
25	η_f, Pa·s	0,35	0,19	0,16	0,06	1,11	0,69	0,43	0,36	6,50	2,66	2,64	0,86
	η_e, Pa·s	1,40	0,73	0,33	0,09	2,50	1,10	0,87	0,35	7,60	3,75	2,37	1,50
	$b \cdot 10^3$, s⁻¹	1,15	3,37	4,20	32,3	1,66	1,02	2,91	7,31	0,36	0,76	1,50	2,44
30	η_f, Pa·s	0,31	0,17	0,14	0,05	1,00	0,62	0,36	0,24	4,95	2,11	2,03	0,68
	η_e, Pa·s	0,95	0,57	0,25	0,06	1,30	0,76	0,52	0,32	4,05	2,21	1,86	1,00
	$b \cdot 10^3$, s⁻¹	1,38	4,30	5,90	35,0	2,23	1,80	3,14	8,69	0,72	0,83	1,70	2,65
35	η_f, Pa·s	0,19	0,13	0,11	0,04	0,68	0,50	0,26	0,19	4,07	1,85	1,45	0,43
	η_e, Pa·s	0,60	0,39	0,21	0,05	0,90	0,35	0,23	0,22	3,50	1,80	1,59	0,79
	$b \cdot 10^3$, s⁻¹	3,67	5,80	6,37	49,0	2,41	3,56	4,60	9,10	0,88	0,96	1,93	3,20
40	η_f, Pa·s	0,17	0,12	0,10	0,04	0,56	0,42	0,22	0,17	2,91	1,46	0,98	0,27
	η_e, Pa·s	0,40	0,19	0,13	0,03	0,65	0,29	0,15	0,12	2,01	1,39	1,19	0,57
	$b \cdot 10^3$, s⁻¹	5,35	6,60	6,90	73,9	2,67	5,60	5,60	16,8	1,33	1,41	2,27	4,24

In accordance with the *SARWS* [19] the conformational radius R_m of the polymeric chain into concentrated solutions is greater than into diluted ones and is increased at the polymer concentration ρ increasing. Moreover, not one, but m macromolecules with the same conformational radius are present into the conformational volume R_m^3. This leads to the notion of twisted polymeric chains m–ball for which the conformational volume R_m^3 is general and equally accessible. Since the m–ball is not localized with the concrete polymeric chain, it is the virtual, i. e. by the mathematical notion.

It is followed from the *SARWS* [19]:

$$R_m = R_f \cdot m^{1/5} \qquad (11.12)$$

$$m^{1/5} = (\rho/\rho^*)^{1/2} \text{ at } \rho \geq \rho^*, \qquad (11.13)$$

thus, it can be written

$$R_m = aN(\rho/\rho_0)^{1/2} \tag{11.14}$$

The shear modulus μ for the m–ball was determined by the expression [19]:

$$\mu = 1.36\frac{RT}{N_A a^3}\left(\frac{\rho}{\rho_0}\right)^2 \tag{11.15}$$

and, as it can be seen, doesn't depend on the length of a chain into the concentrated solutions.

Characteristic time t_m^* of the rotary movement of the m–ball and, respectively its shear, in accordance with the prior work [17] is equal to

$$t_m^* = \frac{4}{7}N^{3.4}\left(\frac{\rho}{\rho_0}\right)^{2.5} L_m \tau_m \tag{11.16}$$

Let us compare the t_m^* with the characteristic time t_f^* of the rotary movement of *Flory* ball into diluted solution [17]:

$$t_f^* = \frac{4}{7}N^{1.4}L_f \tau_f. \tag{11.17}$$

In these expressions τ_m and τ_f are characteristic times of the segmental movement of the polymeric chains and L_m and L_f are their form factors into concentrated and diluted solutions respectively. Let us note also, that the expressions (11.16) and (11.17) are self–coordinated since at $\rho = \rho^*$ the expression (11.16) transforms into the Eq. (11.17). The form factors L_m and L_f are determined by a fact how much strong the conformational volume of the polymeric chain is strained into the ellipsoid of rotation, flattened, or elongated one as it was shown by author [21].

11.3.2 FRICTIONAL COMPONENT OF THE EFFECTIVE VISCOSITY

In accordance with the data of Table 11.1 the frictional component of the viscosity η_f strongly depends on a length of the polymeric chains, on their concentration and on the temperature. The all spectrum of η_f dependence on N, ρ and T we will be considered as the superposition of the fourth movement forms giving the endowment into the frictional component of the solution viscosity. For the solvent such movement form is the *Brownian* movement of the molecules, i. e. their translation freedom degree: the solvent viscosity coefficient η_s will be corresponding to this translation freedom degree. The analog of the *Brownian* movement of the solvent molecules is the segmental movement of the polymeric chain which is responsible for its translation and rotation movements and also for the shear strain. The viscosity

Viscoelastic Properties of the Polystyrene

coefficient η_{sm} will be corresponding to this segmental movement of the polymeric chain.

Under the action of a velocity gradient g of the hydrodynamic flow the polymeric m–ball performs the rotary movement also giving the endowment into the frictional component of the viscosity. In accordance with the superposition principle the segmental movement and the external rotary movement of the polymeric chains will be considered as the independent ones. In this case the external rotary movement of the polymeric chains without taking into account the segmental one is similar to the rotation of m–ball with the frozen equilibrium conformation of the all m polymeric chains represented into m–ball. This corresponds to the inflexible *Kuhn's* wire model [22]. The viscosity coefficient η_{pm} will be corresponding to the external rotating movement of the m–ball under the action of g. The all listed movement forms are enough in order to describe the diluted solutions. However, in a case of the concentrated solutions it is necessary to embed one more movement form, *namely*, the transference of the twisted between themselves polymeric chain one respectively another in m–ball. Exactly such relative movement of the polymeric chains contents into itself the all possible linkages effects. Accordingly to the superposition principle the polymeric chains movement does not depend on the above–listed movement forms if it doesn't change the equilibrium conformation of the polymeric chains in m–ball. The endowment of such movement form into η_f let us note via η_{pm}.

Not all the listed movement forms give the essential endowment into the η_f, however for the generality let us start from the taking into account of the all forms. In such a case the frictional component of a viscosity should be described by the expression:

$$\eta_f = \eta_s(1-\phi) + \left(\eta_{sm} + \eta_{pm} + \eta_{pz}\right)\phi, \tag{11.18}$$

or

$$\eta_f - \eta_s = \left(\eta_{sm} + \eta_{pm} + \eta_{pz} - \eta_s\right)\phi, \tag{11.19}$$

here φ is the volumetric part of the polymer into solution. It is equal to the volumetric part of the monomeric links into m–ball; that is why it can be determined by the ratio:

$$\phi = \bar{V}N/N_A R_m^3, \tag{11.20}$$

in which $\bar{V}$ is the partial–molar volume of the monomeric link into solution.

Combining the eq. Eqs. (11.9), (11.10), (11.11), (11.12), (11.13), and (11.14) and Eq. (11.20) we will obtain:

$$\phi = \bar{V}\rho/M_0 . \qquad (11.21)$$

The ratio of $M_0/\bar{V}$ should be near to the density ρ_m of the liquid monomer. Assuming of this approximation, $M_0/\bar{V} \approx \rho_m$ we have:

$$\phi = \rho/\rho_m . \qquad (11.22)$$

At the rotation of m–ball under the action of g the angular rotation rate for any polymeric chain is the same but their links depending on the remoteness from the rotation center will have different linear movement rates. Consequently, in m–ball there are local velocity gradients of the hydrodynamic flow. Let g_m represents the averaged upon m–ball local velocity gradient of the hydrodynamic flow additional to g. Then, the tangential or strain shear σ formed by these gradients g_m and g at the rotation movement of m–ball in the medium of a solvent will be equal to:

$$\sigma = \eta_s(g + g_m) . \qquad (11.23)$$

However, the measurable strain shear correlates with the well–known external gradient g that gives another effective viscosity coefficient:

$$\sigma = \eta_{pm}g \qquad (11.24)$$

Comparing the Eqs. (11.23) and (11.24) we will obtain

$$\eta_{pm} - \eta_s = \eta_s g_m / g . \qquad (11.25)$$

Noting

$$\eta_{pm}^0 = \eta_s \cdot g_m / g \qquad (11.26)$$

instead of the Eq. (11.19) we will write

$$\eta_f - \eta_s = \left(\eta_{sm} + \eta_{pm}^0 + \eta_{pz}\right)\phi \qquad (11.27)$$

The endowment of the relative movement of twisted polymeric chains in m–ball into the frictional component of the viscosity should be in general case depending on a number of the contacts between monomeric links independently to which polymeric chain these links belong. That is why we assume:

$$\eta_{pz} \sim \phi^2 . \qquad (11.28)$$

Viscoelastic Properties of the Polystyrene

The efficiency of these contacts or linkages let us estimate comparing the characteristic times of the rotation (shear) of m–ball into concentrated solution t_m^* and polymeric ball into diluted solution t_f^* determined by the expressions (11.16) and (11.17).

Let's note that in accordance with the determination done by author [17] t_m^* is the characteristic time not only for m–ball rotation, but also for each polymeric chain in it. Consequently, t_m^* is the characteristic time of the rotation of polymeric chain twisted with others chains whereas t_f^* is the characteristic time of free polymeric chain rotation. The above–said permits to assume the ratio t_m^*/t_f^* as a measure of the polymeric chains contacts or linkages efficiency and to write the following in accordance with the (11.16) and (11.17):

$$\eta_{pz} \sim t_m^*/t_f^* = N^2 (\rho/\rho_0)^{2.5} (L_m/L_f) \tag{11.29}$$

Taking into account the (11.22) and combining the (11.28) and (11.29) into one expression we will obtain:

$$\eta_{pz} = \eta_{pz}^0 N^2 \left(\frac{\rho}{\rho_0}\right)^{2.5} \left(\frac{\rho}{\rho_m}\right)^2 \tag{11.30}$$

Here the coefficient of proportionality η_{pz}^0 includes the ratio $L_m \tau_m / L_f \tau_f$, which should considerably weaker depends on ρ and N that the value η_{pz}.

Substituting the (11.30) into (11.27) with taking into account the (11.22) we have:

$$\eta_f - \eta_s - \left[\eta_{sm} + \eta_{pm}^0 + \eta_{pz}^0 N^2 \left(\frac{\rho}{\rho_0}\right)^{2.5} \left(\frac{\rho}{\rho_m}\right)^2\right] \frac{\rho}{\rho_m} \tag{11.31}$$

Let us estimate the endowment of the separate terms in Eq. (11.31) into η_f. In accordance with Table 11.1 under conditions of our experiments the frictional component of the viscosity is changed from the minimal value $\approx 4 \times 10^{-2}$ $Pa \times s$ to the maximal one ≈ 6.5 $Pa \times s$. Accordingly to the reference data the viscosity coefficient η_s of the toluene has the order 5×10^{-4} $Pa \times s$. The value of the viscosity coefficient η_{sm} representing the segmental movement of the polymeric chains estimated by us upon η_f of the diluted solution of polystyrene in toluene consists of the value by 5×10^{-3} $Pa \times s$ order. Thus, it can be assumed $\eta_{sm}, \eta_s \ll \eta_f$ and it can be neglected the respective terms in Eq. (11.31). With taking into account of this fact, the Eq. (11.31) can be rewritten in a form convenient for the graphical test:

$$\eta_f \frac{\rho_m}{\rho} = \eta_{pm}^0 + \eta_{pz}^0 N^2 \left(\frac{\rho}{\rho_0}\right)^{2.5} \left(\frac{\rho}{\rho_m}\right)^2. \tag{11.32}$$

On Figure 11.4 it is presented the interpretation of the experimental values of η_f into coordinates of the Eq. (11.32).

At that, it were assumed the following values: $M = 104,15$ g/mole, $a = 1,86 \times 10^{-10}$ m under determination of P_0 accordingly to Eq. (11.11) and $\rho_m = 0,906 \cdot 10^6$ g/m³ for liquid styrene. As we can, the linear dependence is observed corresponding to Eq. (11.32) at each temperature; based on the tangent of these straight lines inclination (see the regression equations on Figure 11.4) it were found the numerical values of η_{pz}^0, the temperature dependence of which is shown on Figure 11.5 into the *Arrhenius*' coordinates.

It is follows from these data, that the activation energy E_{pz} regarding to the movement of twisted polymeric chains in toluene is equal to 39,9 kJ/mole.

It can be seen from the Figure 11.4 and from the represented regression equations on them, that the values η_{pm}^0 are so little (probably, $\eta_{pm}^0 \ll 0,1$ $Pa \times s$) that they are located within the limits of their estimation error. This, in particular, didn't permit us to found the numerical values of the ratio g_m/g.

So, the analysis of experimental data, which has been done by us, showed that the main endowment into the frictional component of the effective viscosity of the concentrated solutions "polystyrene in toluene" has the separate movement of the twisted between themselves into m–ball polymeric chains. Exactly this determines a strong dependence of the η_f on concentration of polymer into solution $\left(\eta_f \sim \rho^{5,5}\right)$ and on the length of a chain $\left(\eta_f \sim N^2\right)$.

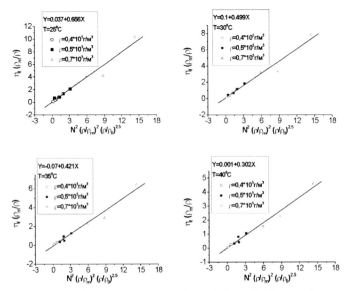

FIGURE 11.4 An interpretation of the experimental data of η_f in coordinates of the Eq. (11.32).

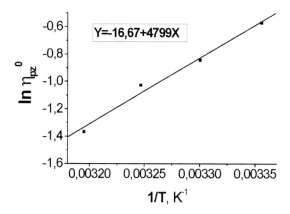

FIGURE 11.5 Temperature dependence of the viscosity coefficient η_{pz}^0 in coordinates of the *Arrhenius* equation.

11.3.3 ELASTIC COMPONENT OF THE EFFECTIVE VISCOSITY

It is follows from the data of Table 11.1, that the elastic component of viscosity η_e is a strong increasing function on polymer concentration ρ, on a length of a chain N and a diminishing function on a temperature T.

The elastic properties of the conformational state of the m–ball of polymeric chains are appeared in a form of the resistance to the conformational volume deformation under the action of the external forces. In particular, the resistance to the shear is determined by the shear modulus μ, which for the m–ball was determined by the expression (11.15). As it was shown by author [17], the elastic component of the viscosity is equal to:

$$\eta_e = \mu t_m^* L_m . \tag{11.33}$$

The factor of form L_m depends on the deformation degree of the conformational volume of a ball [17, 21].

Combining the (11.15) and (11.16) into (11.33) and assuming $\frac{4}{7} \cdot 1,36 \approx 1$ we will obtain

$$\eta_e = \frac{RT}{M_0} N^{3.4} \rho \left(\frac{\rho}{\rho_0}\right)^{3.5} L_m \tau_m . \tag{11.34}$$

Comparing the (11.16) and (11.34) we can see, that the known from the reference data ratio $\eta_e \sim t_m^* \sim N^{3.4}$ is performed but only for the elastic component of a viscosity.

It is follows from the expression (11.34), that the parameters L_m and τ_m are inseparable; so, based on the experimental values of η_e (see Table 11.1) it can be found the numerical values only for the composition $L_m \cdot \tau_m$. The results of $(L_m\tau_m)\eta$ calculations are represented in Table 11.2. In spite of these numerical estimations scattering it is overlooked their clear dependence on T, but not on ρ and N.

TABLE 11.2 Calculated values $L\tau$, τ/L, τ and L based on the experimental magnitudes η_e and b

$\rho \cdot 10^{-5}$, g/m³		4,0				5,0				7,0				$\tilde{\tau} \cdot 10^{10}$, s $\tilde{L}$
T,°C	$M \cdot 10^4$, g/mole	5,1	4,1	3,3	2,2	5,1	4,1	3,3	2,2	5,1	4,1	3,3	2,2	
25	$(L\tau)_{\eta_e} \cdot 10^{10}$, s	2,63	3,14	2,72	2,99	1,71	1,72	2,61	4,25	1,15	1,29	1,57	4,00	
	$(\tau/L)_b \cdot 10^{10}$, s	3,25	1,81	2,54	0,89	1,17	3,43	1,91	2,06	1,98	1,86	1,38	2,29	
	$\tau \cdot 10^{10}$, s	2,92	2,38	2,63	1,63	1,41	2,43	2,23	2,96	1,51	1,61	1,47	3,03	2,19
	L	0,90	1,32	1,03	1,83	1,21	0,71	1,17	1,44	0,76	0,86	1,07	1,32	1,13
30	$(L\tau)_{\eta_e} \cdot 10^{10}$, s	1,75	2,41	2,03	1,96	0,88	1,17	1,54	3,83	0,60	0,75	1,21	2,63	
	$(\tau/L)_b \cdot 10^{10}$, s	2,10	1,56	1,81	0,82	0,87	1,94	1,39	1,73	1,00	1,62	1,22	2,11	
	$\tau \cdot 10^{10}$, s	2,17	1,94	1,92	1,27	0,88	1,51	1,46	2,57	0,78	0,98	1,21	2,56	1,59
	L	0,81	1,24	1,00	1,55	1,00	0,78	1,05	1,49	0,78	0,60	1,00	1,12	1,04
35	$(L\tau)_{\eta_e} \cdot 10^{10}$, s	1,09	1,62	1,67	1,61	0,60	0,53	0,67	2,58	0,51	0,60	1,02	2,04	
	$(\tau/L)_b \cdot 10^{10}$, s	1,01	1,16	1,67	0,59	0,79	0,98	1,21	1,65	0,81	1,35	1,09	1,75	
	$\tau \cdot 10^{10}$, s	1,05	1,37	1,67	0,97	0,70	0,72	0,90	2,06	0,64	0,90	1,05	1,89	1,16
	L	1,04	1,18	1,00	1,65	0,87	0,73	0,74	1,25	0,79	0,67	0,97	1,08	1,00

… Viscoelastic Properties of the Polystyrene

40	$(L\tau)_{\eta e} \cdot 10^{10}, s$	0,72	0,78	1,03	0,96	0,43	0,44	0,43	1,40	0,29	0,46	0,75	1,46	
	$(\tau/L)_b \cdot 10^{10}, s$	0,70	1,01	1,54	0,39	0,73	0,62	1,00	0,90	0,54	0,92	0,91	1,31	
	$\tau \cdot 10^{10}, s$	0,71	0,89	1,26	0,61	0,56	0,52	0,66	1,12	0,40	0,65	0,83	1,38	0,80
	L	1,01	0,88	0,82	1,57	0,77	0,84	0,66	1,25	0,73	0,71	0,91	1,06	0,93

11.3.4 PARAMETER B

In accordance with the determination (11.6), the b parameter is a measure of the velocity gradient of hydrodynamic flow created by the working cylinder rotation, influence on characteristic time t_v^* of g action on the shear strain of the m–ball and its rotation movement. Own characteristic time t_m^* of m–ball shear and rotation accordingly to (11.16) depends only on ρ, N and T via τ_m.

It is follows from the experimental data (see Table 11.1) that the b parameter is a function on the all three variables ρ, N and T, but, at that, is increased at T increasing and is decreased at ρ and N increasing. In order to describe these dependences let us previously determine the angular rate ω_m^0 (s^{-1}) of the strained m–ball rotation with the effective radius $R_m L_m$ of the working cylinder by diameter d contracting with the surface:

$$\omega_m^0 = \pi d\omega / R_m L_m \quad (11.35)$$

Here π is appeared due to the difference in the dimensionalities of ω_m^0 and ω.

Let us determine the t_v^0 as the reverse one ω_m^0:

$$t_v^0 = R_m L_m / \pi d\omega \quad (11.36)$$

Accordingly to (11.36) t_v^0 is a time during which the m–ball with the effective radius $R_m L_m$ under the action of working cylinder by diameter d rotation will be rotated on the angle equal to the one radian. Let us note, that the t_m^* was determined by authors [17] also in calculation of the m–ball turning on the same single angle.

Since in our experiments the working cylinder had two rotating surfaces with the diameters d_1 and d_2, the value ω_m^0 was averaged out in accordance with the condition $d = (d_1 + d_2)/2$; so, respectively, the value t_v^0 was averaged out too:

$$t_v^0 = 2R_m L_m / \pi (d_1 + d_2)\omega \quad (11.37)$$

So, t_v^0 is in inverse proportion to ω; therefore through the constant device it is in inverse proportion to g: $t_v^0 \sim g^{-1}$. However, as it was noted in Section 2.2, in m–ball due to the difference in linear rates of the polymeric chains links it is appeared the hydrodynamic interaction which leads to the appearance of the additional to g local averaged upon m–ball velocity gradient of the hydrodynamic flow g_m. This local gradient g_m acts not on the conformational volume of the m–ball but on the monomeric framework of the polymeric chains (the inflexible Kuhn's wire model [22]). That is why the endowment of g_m into characteristic time t_v^* depends on the volumetric part φ of the links into the conformational volume of m–ball, i. e. $t_v^* \sim (g + g_m \varphi)^{-1}$.

Therefore, it can be written the following:

$$\frac{t_v^*}{t_v^0} = \frac{g}{g + g_m \varphi}, \qquad (11.38)$$

that with taking into account of Eq. (11.37) leads to the expression

$$t_v^* = \frac{2 R_m L_m}{\pi (d_1 + d_2) \omega} \bigg/ \left(1 + \frac{g_m}{g} \frac{\rho}{\rho_m}\right). \qquad (11.39)$$

Combining the (11.16) and (11.39) into (11.6) we will obtain

$$b = \frac{7a}{2\pi(d_1 + d_2)} \cdot \frac{L_m}{\tau_m} \bigg/ N^{2.4} \left(\frac{\rho}{\rho_0}\right)^2 \left(1 + \frac{g_m}{g} \frac{\rho}{\rho_m}\right). \qquad (11.40)$$

As we can see, here the parameters L_m and τ_m are also inseparable and can not be found independently one from another. That is why based on the experimental data presented in Table 11.1 it can be found only the numerical values of the ratio $(\tau_m / L_m)_b$. After the substitution of values $a = 1{,}86 \times 10^{-10}$ m, $d_1 = 3{,}4 \times 10^{-2}$ m, $d_2 = 3{,}3 \times 10^{-2}$ m we have

$$\left(\frac{\tau_m}{L_m}\right)_b = 2.84 \cdot 10^{-9} \bigg/ N^{2.4} \left(\frac{\rho}{\rho_0}\right)^2 \left(1 + \frac{g_m}{g} \frac{\rho}{\rho_m}\right) b. \qquad (11.41)$$

As it was marked in Section 2.2, we could not estimate the numerical value of g_m / g due to the smallness of the value η_{pm}^0 lying in the error limits of its measuring. That is why, we will be consider the ratio g_m / g as the fitting parameter starting from the consideration that the concentrated solution for polymeric chains is more ideal than the diluted one and, moreover, the m–ball is less strained than the single polymeric ball. That is why, g_m / g was selected in such a manner that the factor of form L_m was near to the 1. This lead to the value $g_m / g = 25$.

The calculations results of $(\tau_m / L_m)_b$ accordingly to Eq. (11.41) with the use of experimental values from Table 11.1 and also the values $g_m / g = 25$ are represented in Table 11.2. They mean that the $(\tau_m / L_m)_b$ is a visible function on a temperature but not on a ρ and N.

Viscoelastic Properties of the Polystyrene

On a basis of the independent estimations of $(\tau_m/L_m)_\eta$ and $(\tau_m/L_m)_b$ it was found the values of τ_m and L_m, which also presented in Table 11.2. An analysis of these data shows that with taking into of their estimation error it is discovered the clear dependence of τ_m and L on T, but not on P and N. Especially clear temperature dependence is visualized for the values $\tilde{\tau}_m$, obtained via the averaging of τ_m at giving temperature for the all values of P and N (Table 11.2). The temperature dependence of $\tilde{\tau}_m$ into the coordinates of the *Arrhenius'* equation is presented on Figure 11.6.

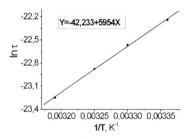

FIGURE 11.6 Temperature dependence of the average values of the characteristic time $\tilde{\tau}$ of the segmental movement of polymeric chain in coordinates of the *Arrhenius* equation.

11.4 POLYSTYRENE'S MELT

11.4.1 EXPERIMENTAL DATA

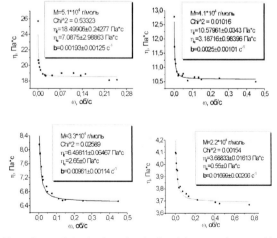

FIGURE 11.7 Experimental (points) and calculated in accordance with the equation (5) (curves) dependencies of the effective viscosity on the velocity of the working cone rotation. $T = 210°C$.

Typical dependencies of the melt viscosity η on the angular rate ω (rotations per second) of the working cone rotation are represented on Figures 11.7 and 11.8.

In order to analyze the experimental curves of $\eta(\omega)$, the Eq. (11.5) with the same remarks as to the numerical estimations of parameters η_e, η_f and b was used.

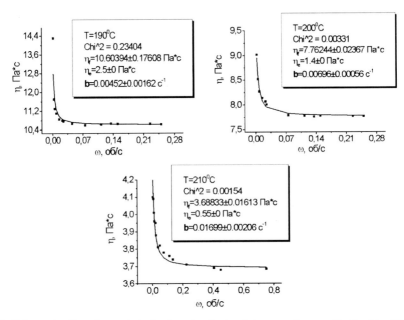

FIGURE 11.8 Experimental (points) and calculated in accordance with the Eq. (11.5) (curves) dependencies of the effective viscosity on the velocity of the working cone rotation. $M = 2{,}2 \cdot 10^4$ g/mole.

As it can be seen from the Figures 11.7 and 11.8, calculated curves of $\eta(\omega)$ accordingly to the Eq. (11.5) with the founded parameters η_f, η_e and b describe the experimental data very well.

Results of the numerical estimations of η_f, η_e та b on a length of the polymeric chain at 210°C are represented in Table 11.3 and the temperature dependencies are represented in Table 11.4. Review of these data shows, that the all three parameters are the functions on the length of a chain and on temperature. But at this, η_e and η_f are increased at N increasing and are decreased at T increasing, whereas b parameter is changed into the opposite way.

Viscoelastic Properties of the Polystyrene

TABLE 11.3 Optimization parameters η_f, η_e and b obtained from the experimental data at $T = 210°C$

$M \cdot 10^{-4}$, g/mole	η_f, Pa·s	η_e, Pa·s	b, s^{-1}
5.1	18.49	7.09	0.0019
4.1	10.58	3.19	0.0025
3.3	6.50	2.65	0.0096
2.2	3.69	0.55	0.0169

TABLE 11.4 Optimization parameters η_f, η_e and b obtained from the experimental data for polystyrene with $M = 2{,}2 \cdot 10^4$ g/mole

T, °C	η_f, Pa·s	η_e, Pa·s	b, s^{-1}
190	10.60	2.50	0.0045
200	7.76	1.40	0.007
210	3.69	0.55	0.0169

11.4.2 FRICTIONAL COMPONENT OF THE EFFECTIVE VISCOSITY OF MELT

Results represented in Tables 11.3 and 11.4 show that the frictional component of the viscosity η_f very strongly depends on the length of the polymeric chains and on the temperature. The whole spectrum of the dependence of η_f on N and T will be considered as the superposition of the above earlier listed three forms of the motion which make the endowment into the frictional component of the viscosity of melt, *namely* the frictional coefficients of viscosity η_{sm}, η_{pm} and η_{pz} (see Section 2.2).

Not the all listed forms of the motion make the essential endowment into η_f, however for the generalization let us start from the taking into account of the all forms. So, the frictional component of the viscosity should be described by the expression (11.19), but for the melt it is necessary to accept that $\eta_s = 0$, and $\varphi = 1$.

$$\eta_f = \eta_{sm} + \eta_{pm} + \eta_{pz}. \tag{11.42}$$

In a case of the melts η_{pz} is determined by the Eq. (11.30), but at $\varphi \cong \rho/\rho_m = 1$

$$\eta_{pz} = \eta_{pz}^0 N^2 \left(\frac{\rho}{\rho_0}\right)^{2.5}. \tag{11.43}$$

Here the coefficient of the proportionality η_{pz}^0 contains the ratio L_m/L_f, however, since the melt is the ideal solution for polymer, that is why it can be assumed that $L_m = 1$.

By substituting of the (11.43) into (11.40), we will obtain

$$\eta_f = \eta_{sm} + \eta_{pm} + \eta_{pz}^0 N^2 \left(\frac{\rho}{\rho_0}\right)^{2.5} \tag{11.44}$$

Let's estimate an endowment of the separate components into η_f. The results represented in Table 11.3 show, that under conditions of our experiments the frictional component of viscosity is changed from the minimal value ≈ 3,7 $Pa \cdot s$ till the maximal one equal to ≈ 18,5 $Pa \cdot s$. The value of the viscosity coefficient η_{sm}, which represents the segmental motion of the polymeric chains and estimated [20] on the basis of η_f for diluted solution of polystyrene in toluene is equal approximately $5 \cdot 10^{-3}$ $Pa \cdot s$. Thus, it can be assumed that $\eta_{sm} \ll \eta_f$ and to neglect by respective component in (11.44). Taking into account of this fact, the expression (11.44) let's rewrite as follow:

$$\eta_f = \eta_{pm} + \eta_{pz}^0 N^2 \left(\frac{\rho}{\rho_0}\right)^{2.5}. \tag{11.45}$$

For the melts $\rho/\rho_0 = const$, that is why the interpretation of the experimental values of η_f as the function of N^2 is represented on Figure 11.9. It can be seen from the Figure 11.9, that the linear dependence corresponding to the Eq. (11.45) is observed, and the numerical value of $\eta_{pz}^0 \left(\frac{\rho}{\rho_0}\right)^{2.5}$ was found upon the inclination tangent of the straight line; under the other temperatures this coefficient was found using the experimental data from Table 11.4. For the estimation of η_{pz}^0 it was assumed that $\rho = 1.05 \cdot 10^6$ g/m³, ρ_0 was calculated in accordance with (11.11) at $M_0 = 104{,}15$ g/mole, $a = 1{,}86 \cdot 10^{-10}$ m.

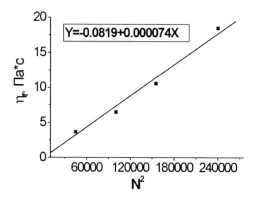

FIGURE 11.9 Interpretation of the experimental values of η_f in the coordinates of the Eq. (45) at $T = 210$ °C.

Temperature dependence of η_{pz}^0 into coordinates of the *Arrhenius* equation is represented on Figure 11.10.

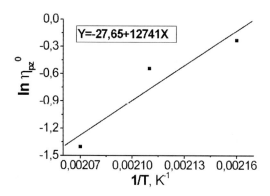

FIGURE 11.10 Temperature dependence of the numerical estimations of η_{pz}^0 in the coordinates of the *Arrhenius* equation.

So, an activation energy E_{pz} of the relative motion of intertwined polymeric chains into polystyrene's melt consists of 106 ± 35 kJ/mole.

It can be seen from the presented Figure 11.9 and from the regression equation, that the values of η_{pz}^0 are very little and are within the ranges of their estimation error; this cannot give the possibility to estimate the value of g_m/g.

So, as the analysis of the experimental data showed, the main endowment into the frictional component of the effective viscosity of the polystyrene's melt has the relative motion of the intertwining between themselves into m–ball polymeric chains. Exactly this determines the dependence of η_f on the length of a chain ($\eta_f \sim N^2$).

11.4.3 ELASTIC COMPONENT OF THE EFFECTIVE VISCOSITY OF MELT

It can be seen from the Tables 11.3, and 11.4, that the elastic component of the viscosity η_e is strongly growing function on a length of a chain N and declining function on temperature T.

The elastic component of the viscosity is described by the Eq. (11.43), but at $L_m = 1$.

$$\eta_e = \mu t_m^* \qquad (11.46)$$

Correspondingly, instead of the (11.34) we obtained

$$\eta_e = \frac{RT}{M_0} N^{3.4} \rho \left(\frac{\rho}{\rho_0}\right)^{3.5} \tau_m \qquad (11.47)$$

Using the expression (11.47) and the experimental values of η_e (see Tables 11.3 and 11.4) it was found the numerical values of the characteristic time of the segmental motion τ_m. The results of calculation $(\tau_m)_{\eta_e}$ are represented in Tables 11.5 and 11.6. Despite the disagreement in numerical estimations, it is observed their dependence on T, but not on the N; this fact is confirmed by the expression (11.47).

TABLE 11.5 Characteristic times of the segmental motion calculated based on the experimental values of η_e and b ($M = 2.2 \cdot 10^4$ g/mole)

T, °C	$(\tau_m)_{\eta_e} \cdot 10^{11}$, s	$(\tau_m)_b \cdot 10^{11}$, s	$\tilde{\tau}_m \cdot 10^{11}$, s
190	6.86	5.50	6.18
200	3.76	3.58	3.67
210	1.45	1.48	1.47

Viscoelastic Properties of the Polystyrene 197

TABLE 11.6 Characteristic times of the segmental motion calculated based on the experimental values of η_e та b ($T = 210°C$)

M·10⁻⁴, g/mole	$(\tau_m)_{\eta e} \cdot 10^{11}$, s	$(\tau_m)_b \cdot 10^{11}$, s	$\tilde{\tau}_m \cdot 10^{11}$, s
5.1	1.11	1.70	
4.1	1.05	2.25	
			1.48
3.3	1.84	0.99	
2.2	1.45	1.48	

11.4.4 PARAMETER B

In accordance with the determination (11.6), the b parameter is a measure of the velocity gradient of hydrodynamic flow created by the working cone rotation, influence on characteristic time t_v^* of g action on the shear strain of the m–ball and its rotation movement. Own characteristic time t_m^* of m–ball shear and rotation accordingly to (11.16) depends only on N and T via τ_m.

It is follows from the experimental data (see Tables 11.3 and 11.4) that the b parameter is a function of N and T, but at this it is increased at T increasing and is decreased at N growth. In order to describe of these dependencies let's previously determine the angular rate ω_m^0 (s^{-1}) of the rotation of m–ball with the effective radius R_m, which contacts with the surface of the working cone with radius r

$$\omega_m^0 = \pi r \omega / R_m \qquad (11.48)$$

Here π is appeared as a result of the different units of the dimention $\omega_m^0(s^{-1})$ and ω, (rot/s).

Let's determine the t_v^0 as the reverse one to the ω_m^0

$$t_v^0 = R_m / \pi r \omega \qquad (11.49)$$

In Eq. (11.49) t_v^0 is a time during which the m–ball with the conformational radius R_m under the action of working cone rotation with radius r will be rotated on the angle equal to the one radian. Let us note, that the t_m^* was determined by authors [17] also in calculation of the m–ball turning on the same single angle.

Thereby, t_v^0 is inversely proportional to ω, so via the constant of the device is inversely proportional to g: $t_v^0 \sim g^{-1}$. However, into m–ball as a result of the difference in the linear rates of the links of polymeric chains under their rotation the hydrodynamic interaction is appeared, which leads to the appearance of the additional to the g local averaged upon m–ball gradient velocity of the hydrodynamic flow g_m. This local gradient g_m acts not on the conformational volume of m–ball, but on the monomeric frame of the polymeric chains (the inflexible *Kuhn's* wire model [22]). That is why the endowment of g_m into characteristic time t_v^* depends on the volumetric part φ of the links into the conformational volume of m–ball, i. e. $t_v^* \sim (g + g_m\varphi)^{-1}$. Into the melt $\varphi = 1$, therefore, it can be written the following:

$$t_v^*/t_v^0 = g/(g + g_m), \qquad (11.50)$$

that leads with taking into account of the (11.49), to the expression

$$t_v^* = \frac{R_m}{\pi r \omega} \bigg/ \left(1 + \frac{g_m}{g}\right). \qquad (11.51)$$

By combining of the (11.16) and (11.51) in (11.6), we obtained

$$b = \frac{7a}{4\pi r \tau_m} \bigg/ N^{2.4} \left(\frac{\rho}{\rho_0}\right)^2 \left(1 + \frac{g_m}{g}\right). \qquad (11.52)$$

As we can see, using the experimental values of b parameter (*see* Tables 11.3 and 11.4) it can be calculated $(\tau_m)_b$. After the substitution of the values $a = 1.86 \cdot 10^{-10}$ m, r = $1.8 \cdot 10^{-2}$ m, we obtained

$$(\tau_m)_b = 3.78 \cdot 10^{-6} \bigg/ N^{2.4} \left(\frac{\rho}{\rho_0}\right)^2 \left(1 + \frac{g_m}{g}\right) b. \qquad (11.53)$$

Numerical value of the ratio g_m/g was considered as a parameter, which selected in such a way, that calculated accordingly to (11.53) values of $(\tau_m)_b$ corresponded to the calculated values $(\tau_m)_{\eta_e}$ accordingly to (11.47). So, the obtained value of g_m/g = 39.

The results of calculations of $(\tau_m)_b$ and $(\tau_m)_{\eta_e}$, are compared in Tables 11.5 and 11.6. As the results show, $(\tau_m)_b$ and $(\tau_m)_{\eta_e}$ is visible function on the temperature but not on the N. That is why based on the data of Tables 11.5 and 11.6 it were calculated the averaged values of the $\tilde{\tau}_m$ of the characteristic time of segmental motion of the macromolecule.

Temperature dependence of $\tilde{\tau}_m$ into coordinates of the *Arrhenius* equation is represented on Figure 11.11.

Viscoelastic Properties of the Polystyrene

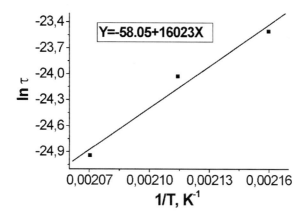

FIGURE 11.11 Temperature dependence of the averaged values of characteristic time $\tilde{\tau}$ of the segmental motion of polymeric chain into coordinates of the *Arrhenius* equation

11.5 CHARACTERISTIC TIME OF THE SEGMENTAL MOTION OF POLYSTYRENE IN SOLUTIONS AND MELT

The presentation of values $\tilde{\tau}_m$ into *Arrhenius'* coordinates equation (Figure 11.6) and (Figure 11.11) permitted to obtain the expressions for concentrated solutions and melt, respectively:

$$\ln \tilde{\tau}_m = -42.23 + 5950/T, \qquad (11.54)$$

$$\ln \tilde{\tau}_m = -58.05 + 16020/T, \qquad (11.55)$$

For diluted solution of polystyrene in toluene it was early obtained [20]:

$$\ln \tilde{\tau}_f = -44.07 + 6660/T, \qquad (11.56)$$

On a basis of the Eqs. (11.54), (11.55) and (11.56) it was calculated the activation energies of the segmental motion of polystyrene in diluted, concentrated solutions and melt, which consists of 55,4; 49,4 and 133.0 *kJ/mole*, respectively.

Characteristic time τ can be obtained by equation of the theory of absolute reactions rates [16]:

$$\tau = \frac{2h}{kT}\exp\left\{-\frac{\Delta S^*}{R}\right\}\exp\left\{\frac{\Delta H^*}{RT}\right\} = \tau_0 \exp\{E/RT\}, \qquad (11.57)$$

where $\Delta H^* = E$ is an activation energy of the segmental motion; ΔS^* is an activation entropy of the segmental motion.

By comparing the expression (11.57) and experimental data (11.54), (11.55) and (11.56) it was found for solutions at $t = 30$ °C and for melt at $t = 200$ °C the values of the activation entropy $\Delta S^*/R = 15,3$; 13,5 and 28,8 respectively.

As we can see, the difference between energies and entropies of activation in diluted and concentrated solutions is little and is in a range of the error of their estimation. At the same time, indicated parameters into melt of the polystyrene is approximately in two times higher. Besides, the growth of the activation entropy does not compensate the activation energy growth; as a result, the characteristic time of the segmental motion into melt is on 2–3 orders higher, than into the solutions (at the extrapolation of τ on general temperature).

Let's compare the values of the activation energies E with the evaporation heats ΔH_{evap} of styrene (–43,94 kJ/mole) and toluene (–37,99 kJ/mole). So, independently on fact, which values of ΔH_{evap} were taken for styrene or toluene, it is observed a general picture: E_{tm}, $E_{rf} > \Delta H_{eun}$. It is known [23], that for the low–molecular liquids, viscosity of which is determined by the Brownian or translational form of the molecules motion, the activation energy of the viscous flow is in 3–4 times less than the evaporation heats. This points on fact, that the segmental motion which is base of the reptation mechanism of the polymeric chains motion, is determined by their deformation–vibrational freedom degrees.

However, let's mark another circumstance. During the study of the bimolecular chains termination kinetics [24] which is limited by their diffusion, in polymeric matrixes of the dimethacrylate *TGM–3* (triethylenglycole dimethacrylate), monomethacrylate *GMA* (2,3–-epoxypropylmethacrylate) and their equimolar mixture *TGM–3* : *GMA* = 1 : 1 in the temperature range 20 ÷ 70°C it were obtained the following values of the activation energies: 122,2, 142,3 and 131,0 kJ/mole. Since the diffusion coefficient of the macroradical is also determined by the characteristic time of the segmental motion, it can be stated that the presented above activation energies of the segmental motion in melt and polymeric matrix are good agreed between themselves. A sharp their difference from the activation energy into solutions points on: *firstly*, a great influence of the solvent as a factor activating the segmental motion of polymeric chain, and *secondly*, on fact, that the dynamic properties of the polymeric chains in melt are very near to their dynamic properties in polymeric matrix.

11.6 DYNAMIC PROPERTIES OF POLYSTYRENE IN SOLUTIONS AND MELT

Dynamical properties of the polymeric chains are determined by characteristic times of their translational motion (t_t^*) and rotation (t_r^*) motions. As it was noted earlier, the characteristic time of the shear strain is also equal to t_r^*. Since the monomeric links connected into a chain, the all of these types of motion are realized exceptionally in accordance with the reptation mechanism, that is via the segmental motion with the characteristic time τ_s. That is why, let's analyze and generalize once more the obtained experimental data of the characteristic times of the segmental motion of the chains of polystyrene in solutions and melts, which were estimated based on elastic component of the viscosity η_e and parameter b. Besides, let's add to this analysis the characteristic times of the segmental motion, estimated based on coefficient of the frictional component of viscosity of diluted solution (η_{sm}), concentrated solution and melt ($\eta_{pz}°$).

The values τ_s will be used in the sequel for the estimation of the characteristic time of the translation motion t_t^* and of the coefficient of the diffusion D of the polystyrene chains into solutions and melt. Accordingly to the experimental data the temperature dependence τ_s, estimated based on the elastic component of the viscosity η_e and parameter b, is described by the equations:

in diluted solution (temperature range 20 – 35 °C)

$$\ln \tau_s = -44.07 + 6660/T, \quad (11.58)$$

in concentrated solution (temperature range 25–40°C)

$$\ln \tau_s = -42.23 + 5950/T, \quad (11.59)$$

in melt (temperature range 190–210°C)

$$\ln \tau_s = -58.05 + 16020/T \quad (11.60)$$

Let's write also the temperature dependencies of the coefficients of a frictional component of the viscosity:
    in diluted solution

$$\ln \eta_{sm} = -29.04 + 7300/T, \quad (11.61)$$

in concentrated solution

$$\ln \eta_{pz}° = -16.67 + 4800/T, \quad (11.62)$$

in melt

$$\ln \eta^\circ_{p\bar{z}} = -27.65 + 12740/T. \qquad (11.63)$$

Next, let's use the proposed earlier expression for characteristic time of the segmental motion in the following form

$$\ln \tau_s = \ln 2 \frac{h}{kT} - \frac{\Delta S_s}{R} + \frac{E_s}{RT}, \qquad (11.64)$$

where $\ln 2 \frac{h}{kT} = -28{,}78$ and $-29{,}22$ at $T = 303\ K$ and $T = 473\ K$ correspondingly.

Using of these values and comparing (11.64) and (11.58)–(11.60), we will obtain the numerical estimations for the activation entropy of the segmental motion $\Delta S_s/R$, which represented in Table 11.7.

TABLE 11.7 Characteristic parameters of segmental motion of polystyrene in solutions and melt

System	E_s, kJ/mole	$\Delta S_s/R$	E_{pz}, kJ/mole	$\Delta S_p/R$	T = 303 K τ_s, s	τ_{pz}, s	T = 473 K τ_s, s	τ_{pz}, s
Diluted solutions	55,3	15,3	—	—	2.5×10^{-10}	—	—	—
Concentrated solutions	49,4	13,5	39,9	6,0	1.5×10^{-10}	6.0×10^{-9}	—	1.9×10^{-11}
Melt	133,1	28,8	105,9	17,0	$5.6\times10^{-3*}$	$1.5\times10^{-2*}$	3.1×10^{-11}	4.0×10^{-9}

Note: *Data found by the extrapolation accordingly to the equations (60) and (71) in the field of the glass–like state of melt

In Table 11.7 also the activation energies E_s of the segmental motion and the value τ_s at $T = 303\ K$ and $T = 473\ K$ are represented too. Values τ_s at $T = 303\ K$ in melt were obtained by the extrapolation of expression (11.60) on given temperature, at which melt is in the solid glass–like state.

It can be seen from the Table 11.7, that the numerical values both of τ_s, and the thermodynamic characteristics ($\Delta S_s/R$ and E_s) of the segmental motion into diluted and concentrated solutions are differed only within the limits of the experimental error of their estimations. In melt these values are essentially differed. At this, the growth of the activation energy (approximately from 55 kJ/mole till 133 kJ/mole) of the segmental motion is not compensated by the growth of the activation entropy (till $\Delta S_s/R \approx 29$); as a result, the values of τ_s in melt are on two orders greater than in solutions (at $T = 473\ K$) and on six

Viscoelastic Properties of the Polystyrene

orders greater at $T = 303$ K.

Let's assume, that the coefficients of the frictional component of the viscosity of polymeric chains are described by as same general expression [16], as the coefficients of the viscosity of low–molecular solution. At that time it can be written:

$$\eta_{sm} = 3\frac{RT}{V}\tau_{sm}, \tag{11.65}$$

$$\eta_{pz}^° = 3\frac{RT}{V}\tau_{pz}, \tag{11.66}$$

where V is the partial–molar volume of the monomeric link of a chain.

τ_{sm} and τ_{pz} are per sense, the characteristic times of the segmental motion of free polymeric chain into diluted solution and overlapping one with others polymeric chains into the concentrated solution and melt taking into account the all possible gearing effects, correspondingly.

Since the partial–molar volume V of the monomeric link of the polystyrene is unknown, then for the next calculations it can be assumed without a great error to be equal to the molar volume of the monomeric link into the melt:

$$V = \rho/M_0, \tag{11.67}$$

where $\rho = 1.05 \cdot 10^6$ g/m^3 is a density of the polystyrene melt; $M_0 = 104.15$ $g/mole$ is the molar mass of styrene. Let us write (11.65) and (11.66) in general form:

$$\ln \eta = \ln 3RT \frac{\rho}{M_0} + \ln \tau. \tag{11.68}$$

At this $\ln 3RT \frac{\rho}{M_0} = 18,15$ and $18,59$ at $T = 303$ K and $T = 473$ K, correspondingly.

Taking into account of this value and comparing the (11.68) and (11.61) and (11.62) and (11.63), we will obtain the temperature dependences τ_{sm} and τ_{pz}:

for diluted solution

$$\ln \tau_{sm} = -47.15 + 7300/T, \tag{11.69}$$

for concentrated solution

$$\ln \tau_{pz} = -34.82 + 4800/T, \tag{11.70}$$

for melt

$$\ln \tau_{pz} = -46.24 + 12740/T. \tag{11.71}$$

On the basis of two last ones expressions the τ_{pz} have been calculated at T = 303 K and T = 473 K. Taking into account a general Eq. (11.64) it has been found also the value of the activation entropy $\Delta S_{pz}/R$ (see Table 11.7).

Comparing the parameters of the Eq. (11.58) for τ_s and (11.69) for τ_{sm}, it can be seen, that the difference between them is adequately kept within the error limits of their estimation. The values of τ_s and τ_{sm} at $T = 303$ K, equal to $2.5 \cdot 10^{-10}$ s and $1.0 \cdot 10^{-10}$ s correspondingly prove of this fact. Thus, it can be assumed that $\tau_s \equiv \tau_{sn}$, and that is why the coefficient of the frictional component of the viscosity η_{sm} of the polymeric chains can be described by as same general expression (11.65) as for the coefficient of the low–molecular solution. The values of τ_{pz} calculated accordingly to the expressions (11.71) and (11.72) for concentrated solution at $T = 303$ K and melt at T = 473 K correspondingly (see Table 11.7), are essentially differed from τ_s : $\tau_{pz} > \tau_s$, approximately on two orders. An analysis of the parameters of the Eqs. (11.59), (11.60) and (11.70), (11.71) showed that the difference between τ_s and τ_{pz} is caused by two factors, which abhorrent the one of the other: by insignificant decreasing of the activation energy ($E_{pz} < E_s$), that should be decreased the τ_{pz}, and by a sharp decreasing of the activation entropy $(\Delta S_{pz} < \Delta S_s)$, that increases of τ_{pz}.

As it was said, the coefficient of the frictional component of viscosity η_{pz} in concentrated solutions and melt caused by the motion of the overlapping between themselves polymeric chains relatively the one of the other and characterizes the efficiency of the all possible gearings. However, the mechanism of this motion is also reptational that is realized via the segmental motion. Correspondingly, between the times τ_s and τ_{pz} the some relationship should be existing. Let's assume the thermodynamical approach for the determination of this relationship as a one among the all possible.

Let's determine the notion «gearing» as the thermodynamical state of a monomeric link of the chain, at which its segmental motion is frozen. This means, that under the relative motion of the intertwining between themselves polymeric chains the reptational mechanism of the transfer at the expense of the segmental motion takes place, but under condition that the part of the monomeric links of a chain is frozen.

Let the ΔG_z° is a standard free energy of the monomeric link transfer from a free state into the frozen one. Then the probability of the frozen states formation or their part should be proportional to the value $\exp\{-\Delta G_z^\circ/RT\}$. That is why, if the k_s is a constant rate of the free segmental transfer, and k_{pz} is the rate constant of the frozen segmental transfer, then between themselves the relationship should be existing:

$$k_{pz} = k_s \exp\{-\Delta G_z^\circ/RT\}. \qquad (11.72)$$

Then k_{pz}, additionally to k_s, has a free activation energy equal to the standard free defrosting energy of the frozen state.

Since $k_s = \tau_s^{-1}, k_{pz} = \tau_{pz}^{-1}$, we obtained

Viscoelastic Properties of the Polystyrene

$$\tau_{pz} = \tau_s \exp\{\Delta G_z^\circ / RT\}. \qquad (11.73)$$

By assigning

$$\Delta G_z^\circ = \Delta H_z^\circ - T\Delta S_z^\circ \qquad (11.74)$$

and taking into account the experimentally determined ratios $\tau_{pz} > \tau_s$, $E_{pz} < E_s$ and $\Delta S_{pz} < \Delta S_s$, we conclude, that in (11.74) $\Delta G_z^\circ > 0$, $\Delta H_z^\circ < 0$ and $\Delta S_z^\circ < 0$, and besides the entropy factor $T\Delta S_z^\circ$ should be more upon the absolute value than the enthalpy factor ΔH_z°. These ratios per the physical sense are sufficiently probable. A contact of the links under the gearing can activates a weak exothermal effect $(\Delta H_z^\circ < 0)$ at the expense of the intermolecular forces of interaction, and the frosting of the segmental movement activates a sharp decrease of the entropy of monomeric link $\Delta S_z^\circ < 0$, but at this $|T\Delta S_z^\circ| > |\Delta H_z^\circ|$. Let's rewrite the (11.73) with taking into account of (11.74) in a form

$$\ln \tau_{pz} = \ln \tau_s + \frac{\Delta H_z^\circ}{RT} - \frac{\Delta S_z^\circ}{R}. \qquad (11.75)$$

Comparing the expressions (11.59) and (11.70), (11.60) and (11.72) and taking into account (11.75) we obtained:

for concentrated solution

$\Delta G_z^\circ = 9.0$ kJ/mole, $\Delta H_z^\circ = -9.6$ kJ/mole, $\Delta S_z^\circ / R = -7.4$,

for melt

$\Delta G_z^\circ = 15.0$ kJ/mole, $\Delta H_z^\circ = -31.4$ kJ/mole, $\Delta S_z^\circ / R = -11.8$.

In connection with carried out analysis the next question is appeared: why in the concentrated solutions and melt the gearing effect hasn't an influence on the elastic component of viscosity η_e°, and determined based on this value characteristic time of the segmental motion is τ_s; at the same time, the gearings effect strongly influences on the frictional component of viscosity, on the basis of which the τ_{pz} is estimated. Probably, the answer on this question consists in fact that the elastic component of the viscosity is determined by the characteristic time of the shear which is equal to the characteristic time of rotation. Accordingly to the superposition principle the rotation motion of the m-ball of the intertwining between themselves polymeric chains can be considered independently on their mutual relocation, that is as the rotation with the frozen conformation. As a result, the gearings effects have not an influence on the characteristic time of the rotation motion. Free segmental motion gives a contribution in a frictional component of viscosity, but it is very little and is visible only in the diluted solutions. That is why even a little gearing effect is determining for the frictional component of viscosity in concentrated solutions and melts.

Let's use the obtained numerical values of the characteristic times of the segmental movement τ_s for the estimation of dynamical properties of the polystyrene

chains that is their characteristic time of the translational movement t_t^* and coefficient of diffusion D into solutions and melt. Accordingly to [25], the values t_t^* and D are determined by the expressions:
in diluted solutions

$$t_t^* = N^{8/5} \tau_s,$$ (11.76)

$$D = \frac{a^2}{2\tau_s} N^{-3/5}.$$ (11.77)

in concentrated solutions and melt

$$t_t^* = N^{3.4} \left(\frac{\rho}{\rho_0}\right)^{2.5} \tau_s,$$ (11.78)

$$D = \frac{a^2}{2\tau_s} \Big/ \left(\frac{\rho}{\rho_0}\right)^{2.5} N^{2.4}.$$ (11.79)

In order to illustrate the dynamic properties of the polystyrene in solutions and melt in Table 11.8 are given the numerical estimations of the characteristic times of segmental τ_s and translational t_t^* motions of the polystyrene and diffusion coefficients D. It was assumed for the calculations $a = 1.86 \cdot 10^{-10}$ m, $N = 10^3$ and $\rho = 0.5 \cdot 10^6$ g/m³ for concentrated solution and melt correspondingly. As we can see, the characteristic time of the translational motion t_t^* of the polystyrene chains is on 4 and 6 orders higher than the characteristic time of their segmental motion; this is explained by a strong dependence of t_t^* on the length of a chain. The coefficients of diffusion weakly depend on the length of a chain, that is why their values into solutions is on 2–3 order less, than the coefficients of diffusion of low–molecular substances, which are characterized by the order 10^{-9} m²/s.

TABLE 11.8 Dynamic characteristics of polystyrene in solutions and melt

System	T = 303 K			T = 473 K		
	τ_s, s	t_t^*, s	D, m²/sc	τ_s, s	t_t^*, s	D, m²/s
Diluted solutions	$2.0 \cdot 10^{-10}$	$1.3 \cdot 10^{-6}$	$1.4 \cdot 10^{-12}$	—	—	—
Concentrated solutions	$2.0 \cdot 10^{-10}$	$2.9 \cdot 10^{-4}$	$1.0 \cdot 10^{-13}$	—	—	—
Melt	$5.0 \cdot 10^{-3}$*	$7.2 \cdot 10^{3}$*	$7.3 \cdot 10^{-22}$*	$3.0 \cdot 10^{-11}$	$4.3 \cdot 10^{-5}$	$1.2 \cdot 10^{-13}$

Note: *Data found by the extrapolation in a field of the glass–like state of melt.

A special attention should be paid into a value of the diffusion coefficient at $T = 303\ K$ in a field of the glass–like state of melt $D = 7 \cdot 10^{-22}\ m^2/s$. Let's compare of this value D with the diffusion coefficients of the macroradicals in polymeric matrixes $TGM-3$, $TGM-3-GMA$ and GMA which estimated experimentally [24] based on the kinetics of macroradicals decay, which under the given temperature consist of $10^{-21} \div 10^{-22}\ m^2/s$.

Thus, carried out analysis shows, that the studies of the viscosity of polymeric solutions permits sufficiently accurately to estimate the characteristic times of the segmental and translational movements, on the basis of which the coefficients of diffusion of polymeric chains into solutions can be calculated.

***

Investigations of a gradient dependence of the effective viscosity of concentrated solutions of polystyrene and its melt permitted to mark its frictional η_f and elastic η_e components and to study of their dependence on a length of a polymeric chain N, on concentration of polymer ρ in solution and on temperature T. It was determined that the main endowment into the frictional component of the viscosity has the relative motion of the intertwined between themselves in m–ball polymeric chains. An efficiency of the all possible gearings is determined by the ratio of the characteristic times of the rotation motion of intertwined between themselves polymeric chains in m–ball t_m^* and Flory ball t_f^*. This lead to the dependence of the frictional component of viscosity in a form $\eta_f \sim N^2 \rho^{5.5}$ for concentrated solutions and in a form $\eta_f \sim N^2$ for melt, which is agreed with the experimental data.

It was experimentally confirmed the determined earlier theoretical dependence of the elastic component of viscosity for concentrated solutions $\eta_e \sim N^{3.4} \rho^{4.5}$, and for the melt $\eta_e \sim N^{3.4}$, that is lead to the well known ratio $\eta_e \sim t_m^* \sim N^{3.4}$, which is true, however, only for the elastic component of the viscosity. On a basis of the experimental data of η_e and b it were obtained the numerical values of the characteristic time τ_m of the segmental motion of polymeric chains in concentrated solutions and melt. As the results showed, τ_m doesn't depend on N, but only on temperature. The activation energies and entropies of the segmental motion were found based on the average values of $\tilde{\tau}_m$. In a case of a melt the value of E and $\Delta S^*/R$ is approximately in twice higher than the same values for diluted and concentrated solutions of polystyrene in toluene; that points on a great activation action of the solvent on the segmental motion of the polymeric chain, and also notes the fact that the dynamical properties of the polymeric chains in melt is considerably near to their values in polymeric matrixes, than in the solutions.

An analysis which has been done and also the generalization of obtained experimental data show, that as same as in a case of the low-molecular liquids, an investigation of the viscosity of polymeric solutions permits sufficiently accurately to estimate the characteristic time of the segmental motion on the basis of which the diffusion coefficients of the polymeric chains in solutions and melt can be calculated; in other words, to determine their dynamical characteristics.

KEYWORDS

- Elastic component
- Frictional component
- Polystyrene
- Reptational mechanism

REFERENCES

1. Ferry, J. D.; Viscoelastic Properties of Polymers. John Wiley and Sons: NewYork; **1980**, 641 p.
2. De Gennes, P. G.; Scaling Concepts in Polymer Physics. Cornell University Press: Ithaca, **1979**, 300 p.
3. Tsvietkov, V. N.; The Structure of Macromolecules in Solutions. In V. N. Tsvietkov; V. E. Eskin; S. Ya. Frenkel – M.: *"Nauka"*, **1964**, 700 p. (in Russian)
4. Malkin, A. Ya.; Rheology: Conceptions, Methods, Applications. In A. Ya. Malkin; and A. I. Isayev – M.: "Proffesiya" **2010**, 560 p. (in Russian)
5. Grassley, W. W.; The Entanglement Concept in Polymer Rheology. *Adv. Polym. Sci.* **1974**, *16*, 1–8.
6. Eyring, H.; Viscosity, Plasticity, and Diffusion as Examples of Absolute Reaction Rates. *J. Chem. Phys.* **1936**, *4*, 283–291.
7. Peterlin, A.; Gradient Dependence of the Intrinsic Viscosity of Linear Macromolecules. A. Peterlin, M. Čopic. *J. Appl. Phys.* **1956**, *27*, 434–438.
8. Ikeda, Ya.; On the effective diffusion tensor of a segment in a chain molecule and its application to the non-newtonian viscosity of polymer solutions. *J. Phys. Soc. Japan*, **1957**, *12*, 378–384.
9. Hoffman, M.; Strukturviskositat und Molekulare Struktur von Fadenmolekulen. M. Hoffman; and R. Rother. *Macromol. Chem.* **1964**, *80*, 95–111.
10. Leonov, A. I.; Theory of Tiksotropy. A. I. Leonov, and G. V. Vynogradov. Reports of the Academy of Sciences of USSR. **1964**, *155*(2), 406-409.
11. Williams, M. C.; Concentrated Polymer Solutions: Part II. Dependence of Viscosity and Relaxation Time on Concentration and Molecular Weight. *A. I. Ch. E. Journal*. **1967**, *13*(3), 534-539.
12. Bueche, F.; Viscosity of Polymers in Concentrated Solution. *J. Chem. Phys.* **1956**, *25*, 599–605.
13. Edvards, S. F.; The Effect of Entanglements of Diffusion in a Polymer Melt. S. F. Edvards; and J. W. Grant. *J. Phys.* **1973**, *46*, 1169-1186.
14. De Gennes, P. G.; Reptation of a Polymer Chain in the Presence of Fixed Obstacles. *J. Chem. Phys.* **1971**, *55*, 572-579.
15. Medvedevskikh, Yu. G.; Gradient Dependence of the Viscosity for Polymeric Solutions and Melts. In Yu. G.; Medvedevskikh, A. R.; Kytsya, L. I.; Bazylyak, G. E.; Zaikov. Conformation

of Macromolecules. Thermodynamic and Kinetic Demonstrations. Nova Sci. Publishing: New York, **2007**, 145-157.
16. Medvedevskikh, Yu. G.; Phenomenological Coefficients of the Viscosity of Low-Molecular Simple Liquids and Solutions. Medvedevskikh Yu. G.; Khavunko O. Yu. Collection Book: Shevchenko' Scientific Society Reports; **2011**, vol. 28, pp 70 – 83 (in Ukrainian).
17. Medvedevskikh, Yu. G.; Viscosity of Polymer Solutions and Melts. Yu. G. Medvedevskikh. Conformation of Macromolecules Thermodynamic and Kinetic Demonstrations. Nova Sci. Publishing: New York, **2007**, pp 125-143.
18. Medvedevskikh, Yu. G.; Statistics of Linear Polymer Chains in the Self-Avoiding Walks Model. Yu. G. Medvedevskikh. *Condens. Matter. Phys.* **2001**, *2*, (26), 209-218.
19. Medvedevskikh, Yu. G.; Conformation and Deformation of Linear Macromolecules in Dilute Ideal Solution in the Self–Avoiding Random Walks Statistics. *J. Appl. Polym. Sci.* **2008**, *109*. (4).
20. Medvedevskikh, Yu.; Frictional and Elastic Components of the Viscosity of Polysterene-Toluene Diluted Solutions. Yu. Medvedevskikh, O. Khavunko. *Chem. Chem. Technol.* **2011**, *5*(3), 291–302.
21. Medvedevskikh, Yu. G.; Conformation of Linear Macromolecules in the Real Diluted Solution. Yu. G. Medvedevskikh., L. I. Bazylyak, A. R. Kytsya. *Conformation of Macromolecules Thermodynamic and Kinetic Demonstrations – N. Y.: Nova Sci. Publishing*, 2007. - p. p. 35–53.
22. Kuhn, H.; Effects of Hampered Draining of Solvent on the Translatory and Rotatory Motion of Statistically Coiled Long-Chain Molecules in Solution. Part II. Rotatory Motion, Viscosity, and Flow Birefringence. H. Kuhn; W. Kuhn, *J. Polymer. Sci.* **1952**, *9*, 1-33.
23. Tobolsky, A. V.; Viscoelastic Properties of Monodisperse Polystyrene. A. V. Tobolsky, J. J. Aklonis, G. Akovali. *J. Chem. Phys.* **1965**, *42*(2), 723-728.
24. Medvedevskikh Yu. G.; Kinetics of Bimolecular Radicals Decay in Different Polymeric Matrixes. Yu. G. Medvedevskikh, A. R. Kytsya, O. S. Holdak, G. I. Khovanets, L. I. Bazylyak, G. E. Zaikov. Conformation of Macromolecules Thermodynamic and Kinetic Demonstrations, Nova Sci. Publishing: New York, **2007**, 139-209.
25. Medvedevskikh Yu. G.; Diffusion Coefficient of Macromolecules into Solutions and Melts . Conformation of Macromolecules. Thermodynamic and Kinetic Demonstrations. Nova Sci. Publishing: NewYork, **2007**, 107-123.

CHAPTER 12

NANOSTRUCTURED POLYMERIC COMPOSITES FILLED WITH NANOPARTICLES

A. K. MIKITAEV, A. YU. BEDANOKOV, and M. A MIKITAEV

12.1 INTRODUCTION

The polymeric nanocomposites are the polymers filled with nanoparticles which interact with the polymeric matrix on the molecular level in contrary to the macro-interaction in composite materials. Mentioned nanointeraction results in high adhesion hardness of the polymeric matrix to the nanoparticles [1, 52].

Usual nanoparticle is less than 100 nanometers in any dimension, 1 nanometer being the billionth part of a meter [1, 2].

The analysis of the reported studies tells that the investigations in the field of the polymeric nanocomposite materials are very promising.

The first notion of the polymeric nanocomposites was given in patent in 1950 [3]. Blumstain pointed in 1961 [4] that polymeric clay—based nanocomposites had increased thermal stability. It was demonstrated using the data of the thermogravimetric analysis that the polymethylmetacrylate intercalated into the Na^+ - methylmetacrylate possessed the temperature of destruction 40–50°C higher than the initial sample.

This branch of the polymeric chemistry did not attract much attention until 1990 when the group of scientists from the Toyota Concern working on the polyamide – based nanocomposites [5–9] found two—times increase in the elasticity modulus using only 4.7weight percent of the inorganic compound and 100°C increase in the temperature of destruction, both discoveries widely extending the area of application of the polyamide. The polymeric nanocomposites based on the layered silicates began being intensively studied in state, academic, and industrial laboratories all over the world only after that.

12.2 STRUCTURE OF THE LAYERED SILICATES

The study of the polymeric nanocomposites on the basis of the modified layered silicates (broadly distributed and well—known as various clays) is of much interest. The natural layered inorganic structures used in producing the polymeric nanocomposites are the montmorillonite [10, 11, 12], hectorite [13], vermiculite [14], kaolin, saponine [15], and others. The sizes of inorganic layers are about 220 and 1 nanometers in length and width, respectively [16, 17].

The perspective ones are the bentonite breeds of clays which include at least 70 percent of the minerals from the montmorillonite group.

Montmorillonite $(Na,K,Ca)(Al,Fe,Mg)[(Si,Al)_4O_{10}](OH)_2 \cdot nH_2O$, named after the province Montmorillion in France, is the high—dispersed layered aluminous silicate of white or gray color in which appears the excess negative charge due to the nonstoichiometric replacements of the cations of the crystal lattice, charge being balanced by the exchange cations from the interlayer space. The main feature of the montmorillonite is its ability to adsorb ions, generally cations, and to exchange them. It produces plastic masses with water and may enlarge itself 10 times. Montmorillonite enters the bentonite clays (the term "bentonite" is given after the place Benton in USA).

The inorganic layers of clays arrange the complexes with the gaps called layers or galleries. The isomorphic replacement within the layers (such as Mg^{2+} replacing Al^{3+} in octahedral structure or Al^{3+} replacing Si^{4+} in tetrahedral one) generates the negative charges which electrostatically are compensated by the cations of the alkali or alkali-earth metals located in the galleries (Figure 12.1) [18]. The bentonite is very hydrophilic because of this. The water penetrates the interlayer space of the montmorillonite, hydrates its surface and exchanges cations what results in the swelling of the bentonite. The further dilution of the bentonite in water results in the viscous suspension with bold tixotropic properties.

The more pronounced cation—exchange and adsorption properties are observed in the bentonites montmorillonite of which contains predominantly exchange cations of sodium.

Nanostructured Polymeric Composites

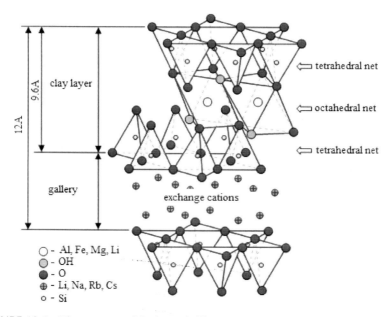

FIGURE 12.1 The structure of the layered silicate.

12.3 MODIFICATION OF THE LAYERED SILICATES

Layered silicates possess quite interesting properties—sharp drop of hardness at wetting, swelling at watering, dilution at dynamical influences and shrinking at drying.

The hydrophility of aluminous silicates is the reason of their incompatibility with the organic polymeric matrix and is the first hurdle need to be overridden at producing the polymeric nanocomposites.

One way to solve this problem is to modify the clay by the organic substance. The modified clay (organoclay) has at least two advantages: (1) it can be well dispersed in polymeric matrix [19] and (2) it interacts with the polymeric chain [13].

The modification of the aluminous silicates can be done with the replacement of the inorganic cations inside the galleries by the organic ones. The replacing by the cationic surface—active agents like bulk ammonium and phosphonium ions increases the room between the layers, decreases the surface energy of clay and makes the surface of the clay hydrophobic. The clays modified such a way are more compatible with the polymers and form the layered polymeric nanocomposites [52]. One can use the nonionic modifiers besides the organic ones which link themselves to the clay surface through the hydrogen bond. Organoclays produced with help of non— ionic modifiers in some cases become more chemically stable than the organoclays produced with help of cationic modifiers (Figure 12.2a) [20].

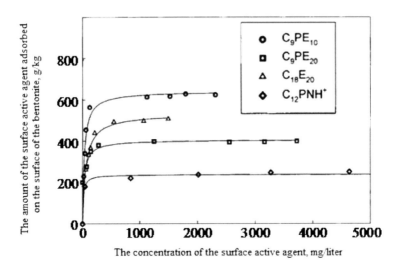

FIGURE 12.2A The adsorption of different modifiers on the clay surface

The least degree of desorption is observed for non—ionic interaction between the clay surface and organic modifier (Figure 2b). The hydrogen bonds between the ethylenoxide grouping and the surface of the clay apparently make these organoclays more chemically stable than organoclays produced with nonionic mechanism.

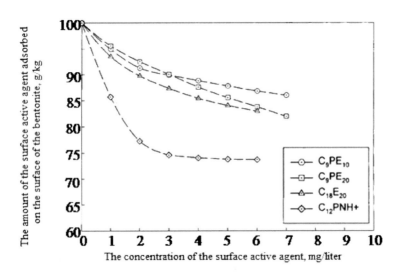

FIGURE 12.2B The desorption of different modifiers from the clay surface:

$C_9PE_{10} - C_9H_{19}C_6H_4(CH_2 CH_2O)_{10}OH$;
$C_9PE_{20} - C_9H_{19}C_6H_4(CH_2 CH_2O)_{20}OH$;
$C_{18}E_{20} - C_{18}H_{37}(CH_2 CH_2O)_{20}OH$;
$C_{12}PNH^+ - C_{12}H_{25}C_6H_4NH^+Cl^-$.

12.4 STRUCTURE OF THE POLYMERIC NANOCOMPOSITES ON THE BASIS OF THE MONTMORILLONITE

The study of the distribution of the organoclay in the polymeric matrix is of great importance because the properties of composites obtained are in the direct relation from the degree of the distribution.

According Giannelis [21], the process of the formation of the nanocomposite goes in several intermediate stages (Figure 12.3). The formation of the tactoid happens on the first stage—the polymer surrounds the agglomerations of the organoclay. The polymer penetrates the interlayer space of the organoclay on the second stage. Here the gap between the layers may reach 2–3 nanometers [22]. The further separation of the layers, third stage, results in partial dissolution and disorientation of the layers. Exfoliation is observed when polymer shifts the clay layers on more than 8–10 nanometers.

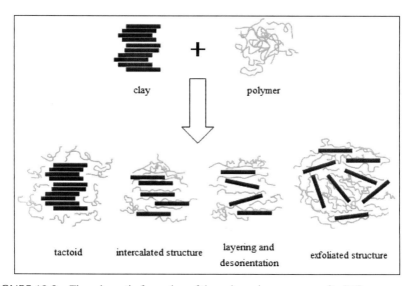

FIGURE 12.3 The schematic formation of the polymeric nanocomposite [24].

All mentioned structures may be present in real polymeric nanocomposites in dependence from the degree of distribution of the organoclay in the polymeric matrix. Exfoliated structure is the result of the extreme distribution of the organoclay. The excess of the organoclay or bad dispersing may born the agglomerates of the organoclays in the polymeric matrix what finds experimental confirmation in the X-ray analysis [11, 12, 21, 23].

In the following subsections we describe a number of specific methods used at studying the structure of the polymeric nanocomposites.

12.4.1 DETERMINATION OF THE INTERLAYER SPACE

The X-ray determination of the interlayer distance in the initial and modified layered silicates as well as in final polymeric nanocomposite is one of the main methods of studying the structure of the nanocomposite on the basis of the layered silicate. The peak in the small—angle diapason ($2\theta = 6-8°C$) is characteristic for pure clays and responds to the order of the structure of the silicate. This peak drifts to the smaller values of the angle 2θ in organomodified clays. If clay particles are uniformly distributed in the bulk of the polymeric matrix then this peak disappears, what witnesses on the disordering in the structure of the layered silicate. If the amount of the clay exceeds the certain limit of its distribution in the polymeric matrix, then the peak reappears again. This regularity was demonstrated on the instance of the polybutylenterephtalate (Figure 12.4) [11].

The knowledge of the angle 2θ helps to define the size of the pack of the aluminous silicate consisting of the clay layer and interlayer space. The size of such pack increases in a row from initial silicate to polymeric nanocomposite according to the increase in the interlayer space. The average size of that pack for montmorillonite is 1.2–1.5 nanometers but for organomodified one varies in the range of 1.8–3.5 nanometers (Figures 12.5, 12.6, and 12.7)

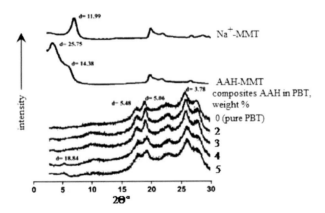

FIGURE 12.4 The data of the X-ray analysis for clay, organoclay, and nanocomposite PBT/organoclay.

Nanostructured Polymeric Composites

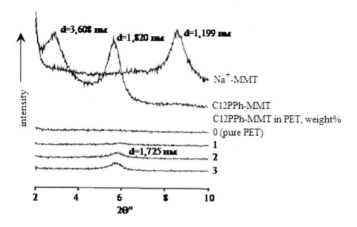

FIGURE 12.5 The data of the X-ray analysis for clay, organoclay, and nanocomposite PET/organoclay.

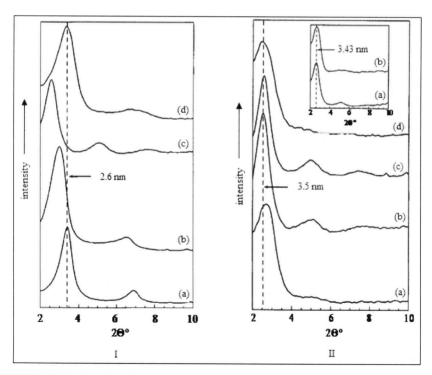

FIGURE 12.6 The data of the X-ray analysis for:

I. (a)—dimetyldioctadecylammonium (DMDODA)—hectorite;
   (b)—50 percent polystyrene (PS) / 50 percent DMDODA—hectorite;
   (c)—75 percent polyethylmetacrylate (PEM)/25 percent DMDODA;
   (d)—50 percent PS/ 50 % DMDODA—hectorite after 24 hrs of etching in cyclohexane.
II. Mix of PS, PEM, and organoclay:
   (a)—23.8 percent PS / 71.2 percent PEM / 5 percent DMDODA – hectorite;
   (b)—21.2 percent PS / 63.8 percent PEM / 15 percent DMDODA – hectorite;
   (c)—18.2 percent PS / 54.8 percent PEM / 27 percent DMDODA – hectorite;
   (d)—21.2 percent PS / 63.8 percent PEM / 15 percent DMDODA – hectorite after 24 hrs of etching in cyclohexane.

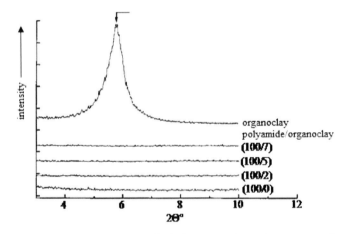

FIGURE 12.7 The data of the X-ray analysis for organoclay and nanocomposite polyamide acid/organoclay.

Summing up we conclude that comparing the data of the X-ray analysis for the organoclay and nanocomposite allows for the determination of the optimal clay amount need be added to the composite. The data from the scanning tunneling (STM) and transmission electron (TEM) microscopes [27, 28] can be used as well.

12.4.2 THE DEGREE OF THE DISTRIBUTION OF THE CLAY PARTICLES IN THE POLYMERIC MATRIX

The two structures, namely the intercalated and exfoliated ones, could be distinguished with the respect to the degree of the distribution of the clay particles, Figure

Nanostructured Polymeric Composites 219

12.8. One should note that clay layers are quite flexible though they are shown straight in the figure. The formation of the intercalated or exfoliated structures depends on many factors, for example the method of the production of the nanocomposite or the nature of the clay etc [29].

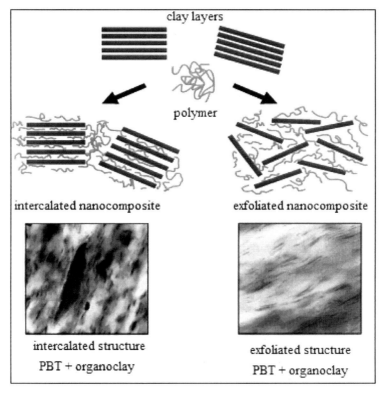

FIGURE 12.8 The formation of the intercalated and exfoliated structures of the nanocomposite

The TEM images of the surface of the nanocomposites can help to find out the degree of the distribution of the nanosized clay particles, see plots (a) to (d) in the Figure 12.9.

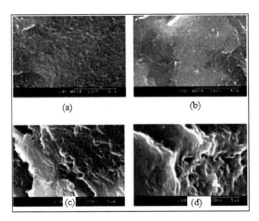

FIGURE 12.9 The images from scanning electron microscope for the nanocomposite surfaces:
(a)—pure PBT;
(b)—3weight% of organoclay in PBT;
(c)—4weight% of organoclay in PBT;
(d)—5weight% of organoclay in PBT.

The smooth surface tells about the uniform distribution of the organoclay particles. The surface of the nanocomposite becomes deformed with the increasing amount of the organoclay, see plots (a) to (d) in the Figure 12.10. Probably, this is due to the influence of the clay agglomerates [30, 31].

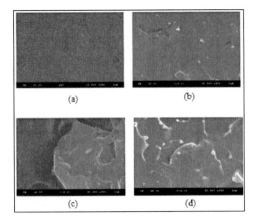

FIGURE 12.10 The images from scanning electron microscope for the nanocomposite surfaces:
(a)—pure PET;
(b)—3weight% of organoclay in PET;
(c)—4weight% of organoclay in PET;
(d)—5weight% of organoclay in PET.

Nanostructured Polymeric Composites

Also one can use the STM images to judge on the degree of the distribution of the organoclay in the nanocomposite, Figures 12.11 and 12.12. If the content of the organoclay is 2–3weight% then the clay layers are separated by the polymeric layer of 4–10 nanometers width, Figure12.11. If the content of the organoclay reaches 4–5weight% then the majority of the clay becomes well distributed however the agglomerates of 4–8 nanometers may appear.

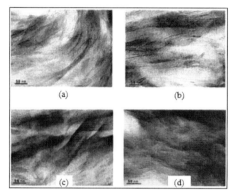

FIGURE 12.11 The images from tunneling electron microscope for the nanocomposite surfaces:
(a)—2weight% of organoclay in PBT;
(b)—3weight% of organoclay in PBT;
(c)—4weight% of organoclay in PBT;
(d)—5weight% of organoclay in PBT.

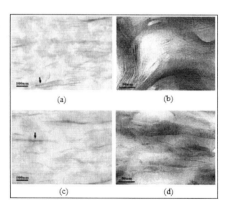

FIGURE 12.12 The images from tunneling electron microscope for the nanocomposite surfaces:
(a)—1weight% of organoclay in PET;
(b)—2weight% of organoclay in PET;
(c)—3weight% of organoclay in PET;
(d)—4weight% of organoclay in PET.

So the involvement of the X-ray analysis and the use of the microscopy data tell that the nanocomposite consists of the exfoliated clay at the low content (below 3weight%) of the organoclay.

12.5 PRODUCTION OF THE POLYMERIC NANOCOMPOSITES ON THE BASIS OF THE ALUMINOUS SILICATES

Different groups of authors [32–35] offer following methods for obtaining nanocomposites on the basis of the organoclays: (1) in the process of the synthesis of the polymer [33, 36, 37], (2) in the melt [38, 39], (3) in the solution [40–46] and (4) in the sol-gel process [47–50].

The most popular ones are the methods of producing in melt and during the process of the synthesis of the polymer.

The producing of the polymeric nanocomposite in situ is the intercalation of the monomer into the clay layers. The monomer migrates through the organoclay galleries and the polymerization happens inside the layers [19, 51], Figure 12.13.

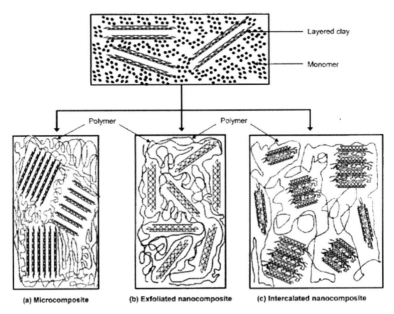

FIGURE 12.13 The production of the nanocomposite in situ:
(a)—microcomposite;
(b)—exfoliated nanocomposite;
(c)—intercalated nanocomposite [51].

The polymerization may be initiated by the heat, irradiation or other source. Obviously, the best results on the degree of the distribution of the clay particles in the polymeric matrix must emerge if using given method. This is associated with the fact that the separation of the clay layers happens in the very process of the inclusion of the monomer in the interlayer space. In other words, the force responsible for the separation of the clay layers is the growth of the polymeric chain whereas the main factor for reaching the necessary degree of the clay distribution in solution or melt is just satisfactory mixing. The most favorable condition for synthesizing the nanocomposites is the vacuuming or the flow of the inert gas. Besides, one has to use the fast speeds of mixing for satisfactory dispersing of the organoclay in the polymeric matrix.

The method of obtaining the polymeric nanocomposites in melt (or the method of extrusion) is the mixing of the polymer melted with the organoclay. The polymeric chains lose the considerable amount of the conformational entropy during the intercalation. The probable motive force for this process is the important contribution of the enthalpy of the interaction between the polymer and organoclay at mixing. One should add that polymeric nanocomposite on the basis of the organoclays could be successively produced by the extrusion [22]. The advantage of the extrusion method is the absence of any solvents what excludes the unnecessary leaks. Moreover the speed of the process is several times more and the technical side is simpler. The extrusion method is the best one in the industrial scales of production of the polymeric nanocomposites what acquires the lesser source expenses and easier technological scheme.

If one produces the polymer—silicate nanocomposite in solution then the organosilicate swells in the polar solvent such as toluene or N,N-dimethylformamide. Then the added is the solution of the polymer which enters the interlayer space of the silicate. The removing of the solvent by means of evaporation in vacuum happens after that. The main advantage of the given method is that 'polymer—layered silicate' might be produced from the polymer of low polarity or even nonpolar one. However this method is not widely used in industry because of much solvent consumption [52].

The sol-gel technologies find application at producing nanocomposites on the basis of the various ceramics and polymers. The initial compounds in these technologies are the alcoholates of specific elements and organic oligomers.

The alcoholates are firstly hydrolyzed and obtained hydroxides being polycondensated then. The ceramics from the inorganic 3D net is formed as a result. Also the method of synthesis exists in which the polymerization and the formation of the inorganic glass happen simultaneously. The application of the nanocomposites on the basis of ceramics and polymers as special hard defensive coverages and like optic fibers [53] is possible.

12.6 PROPERTIES OF THE POLYMERIC NANOCOMPOSITES

Many investigations in physics, chemistry, and biology have shown that the jump from macroobjects to the particles of 1–10 nanometers results in the qualitative transformations in both separate phases and systems from them [54].

One can improve the thermal stability and mechanical properties of the polymers by inserting the organoclay particles into the polymeric matrix. It can be done by means of joining the complexes of properties of both the organic and inorganic substances, that is combining the light weight, flexibility, and plasticity of former and durability, heat stability and chemical resistance of latter.

Nanocomposites demonstrate essential change in properties if compared to the nonfilled polymers. So, if one introduces modified layered silicates in the range of 2–10weight% into the polymeric matrix then he observes the change in mechanical (tensile, compression, bending, and overall strength), barrier (penetrability and stability to the solvent impact), optical and other properties. The increased heat and flame resistance even at low filler content is among the interesting properties too. The formation of the thermal isolation and negligible penetrability of the charred polymer to the flame provide for the advantages of using these materials.

The organoclay as a nanoaddition to the polymers may change the temperature of the destruction, refractoriness, rigidity, and rupture strength. The nanocomposites also possess the increased rigidity modulus, decreased coefficient of the heat expansion, low gas-penetrability, increased stability to the solvent impact and offer broad range of the barrier properties [54]. In Table 12.1 we gather the characteristics of the nylon-6 and its derivative containing 4.7weight% of the organomodified montmorillonite.

TABLE 12.1 The properties of the nylon-6 and composite based on it [54]

	Rigidity modulus, GPa	Tensile strength, MPa	Temperature of the deformation, °C	Impact viscosity, kJ/m^2	Water consumption, weight %	Coefficient of the thermal expansion (x,y)
Nylon-6	1.11	68.6	65	6.21	0.87	13×10^{-5}
Nanocomposite	1.87	97.2	152	6.06	0.51	6.3×10^{-5}

It is important that the temperature of the deformation of the nanocomposite increases on 87 °C.

The thermal properties of the polymeric nanocomposites with the varying organoclay content are collected in Table 12.2.

TABLE 12.2 The main properties of the polymeric nanocomposites

Property	Polybutyleneterephtalate + AAX-montmorillonite					Polyethyleneterephtalate + C_{12}PPh-montmorillonite			
	Organoclay content, %								
	0	2	3	4	5	0	1	2	3
Viscosity, dliter/g	0.84	1.16	0.77	0.88	0.86	1.02	1.26	0.98	1.23
T_g, °C	27	33	34	33	33	---	---	---	---
T_m, °C	222	230	230	229	231	245	247	245	246
T_d, °C	371	390	388	390	389	370	375	384	386
W_{tR}^{600c}, %	1	6	7	7	9	1	8	15	21
Strength limit, MPa	41	50	60	53	49	46	58	68	71
Rigidity modulus, GPa	1.37	1.66	1.76	1.80	1.86	2.21	2.88	3.31	4.10
Relative enlargement, %	5	7	6	7	7	3	3	3	3

The inclusion of the organoclay into the polybutyleneterephtalate leads to the increase in the glass transition temperature (T_g) from 27 to 33 degrees centigrade if the amount of the clay raises from zero to 2weight%. That temperature does not change with the further increase of the organoclay content. The increase in T_g may be the result of two reasons [56–59]. The first is the dispersion of the small amount of the organoclay in the free volume of the polymer, and the second is the limiting of the mobility of the segments of the polymeric chain due to its interlocking between the layers of the organoclay.

The same as the T_g, the melting temperature T_m increases from the 222 to 230°C if the organoclay content raises from 0 to 2weight% and stays constant up to 5weight%, see Table 12.2. This increase might be the consequence of both complex multilayer structure of the nanocomposite and interaction between the organoclay and polymeric chain [60, 61]. Similar regularities have been observed in other polymeric nanocomposites also.

The thermal stability of the nanocomposites polybutyleneterephtalate, briefly PBT, (or polyethyleleterephtalate)/organoclay determined by the thermogravimetric analysis is presented in Table 12.3 and in Figures 12.14 and 12.15 [11, 12].

TABLE 12.3 The thermal properties of the fibers from PET with varying organoclay content

Organoclay content, weight %	h_{inh} [a]	T_m (°C)	DH_m [b] (J/g)	T^i_D [c] (°C)	Wt_R^{600} [d] (%)
0 (pure PET)	1.02	245	32	370	1
1	1.26	247	32	375	8
2	0.98	245	33	384	15
3	1.23	246	32	386	21

a—viscosities were measured at 30°C using 0.1 gram of polymer on 100 milliliters of solution in mix phenol / tetrachlorineethane (50/50);
b—change in enthalpy of melting;
c—initial temperature of decomposition;
d—weight percentage of the coke remnant at 600°C.

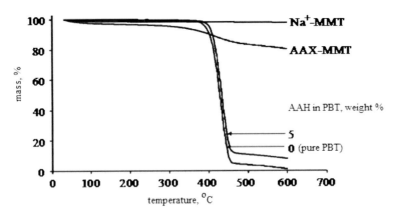

FIGURE 12.14 The thermogravimetrical curves for the montmorillonites, PBT, and nanocomposites PBT/organoclay

Nanostructured Polymeric Composites

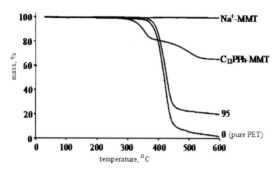

FIGURE 12.15 The thermogravimetrical curves for the montmorillonites, PET, and nanocomposites PET/organoclay

The temperature of the destruction, T_D, increases with the organoclay content up to 350 °C in case of the composite PBT/organoclay. The thermogravimetrical curves for pure and composite PBTs have similar shapes below 350 °C. The values of temperature T_D depends on the amount of organoclay above 350 °C. The organoclay added becomes a barrier for volatile products being formed during the destruction [61, 62]. Such example of the improvement of the thermal stability was studied in papers [63, 64]. The mass of the remnant at 600 °C increases with organoclay content.

Following obtained data authors draw the conclusion that the optimal results for thermal properties are being obtained if 2weight% of the organoclay is added [11, 12, 19].

The great number of studies on the polymeric composite organoclay—based materials show [11, 12, 13, 19] that the inclusion of the inorganic component into the organic polymer improves the thermal stability of the latter, see Tables 12.3 and 12.4.

TABLE 12.4 The basic properties of the nanocomposite based on PBT with varying organoclay content

Organoclay content, weight %	I.V.[a]	T_g	T_m (°C)	T_D^{i} [b] (°C)	Wt_R^{600} [c] (%)
0 (pure PBT)	0,84	27	222	371	1
2	1,16	33	230	390	6
3	0,77	34	230	388	7
4	0,88	33	229	390	7
5	0,86	33	231	389	9

a—viscosities were measured at 30 °C using 0.1 gram of polymer on 100 milliliters of solution in mix phenol/tetrachlorineethane (50/50);
b—initial temperature of the weight loss;
c—weight percentage of the coke remnant at 600 °C.

The values of the melting temperature increase from 222 to 230°C if the amount of the organoclay added reaches 2weight% and then stay constant. This effect can be explained by both thermal isolation of the clay and interaction between the polymeric chain and organoclay [43, 64]. The increase in the glass transition temperature also occurs what can be a consequence of several reasons [55, 56, 58]. One of the main among them is the limited motion of the segments of the polymeric chain in the galleries within the organoclay.

If the organoclay content in the polymeric matrix of the PBT reaches 2weight% then both the temperature of the destruction increases and the amount of the coke remnant increase at 660 °C and then both stay practically unchanged with the further increase of the organoclay content up to 5weight%. The loss of the weight due to the destruction of the polymer in pure PBT and its composites looks familiar in all cases below 350 °C. The amount of the organoclay added becomes important above that temperature because the very clays possess good thermal stability and make thermal protection by their layers and form a barrier preventing the volatile products of the decomposition to fly off [43, 60]. Such instance of the improvement in thermal properties was observed in many polymeric composites [64–68]. The weight of the coke remnant increases with the rising organoclay content up to 2weight% and stays constant after that. The increase of the remnant may be linked with the high thermal stability of the organoclay itself. Also it is worth noticing that the polymeric chain closed in interlayer space of the organoclay has fewer degrees of oscillatory motion at heating due to the limited interlayer space and the formation of the abundant intermolecular bonds between the polymeric chain and the clay surface. And the best result is obtained at 2–3weight% content of the organoclay added to the polymer.

If one considers the influence of the organoclay added to the polytheleneterephtalate, briefly PET, [69] then the temperature of the destruction increases on 16°C at optimal amount of organoclay of 3weight%. The coke remnant at 600°C again increases with the rising organoclay content, see Table 12.3.

Regarding the change in the temperature of the destruction in the cases of PET and PBT versus the organoclay content one can note that both trends look similar. However the coke remnant considerably increases would the tripheyldodecylphosphonium cation be present within the clay. The melting temperature does not increase in case of the organoclay added into the PET in contrary to the case of PBT. Apparently this may be explained by the more crystallinity of the PBT and the growth of the degree of crystallinity with the organoclay content.

It becomes obvious after analyzing the above results that the introduction of the organoclay into the polymer increases the thermal stability of the latter according the (1) thermal isolating effect from the clay layers and (2) barrier effect in relation to the volatile products of destruction.

The studying of the mechanical properties of the nanocomposites, see Table 12.2, have shown that the limit of the tensile strength increases with the organoclay added up to 3weight% for the majority of the composites. Further addition of the organoclay, up to 5weight%, results in the decreasing limit of the tensile strength. We explain this by the fact that agglomerates appear in the nanocomposite when the organoclay content exceeds the 3weight% value [61, 70, 71]. The proof for the formation of the agglomerates have been obtained from the X-ray study and using the data from electron microscopes.

Nevertheless the rigidity modulus increases with the amount of the organoclay added into the polymeric matrix, the resistance of the clay itself being the explanation for that. The oriented polymeric chains in the clay layers also participate in the increase of the rigidity modulus [72]. The percentage of enlargement at breaking became 6–7weight % for all mixes.

Using data of the Table 12.2 we explain the improvement in the mechanical properties of the nanocomposites with added organoclay up to 3weight% by the good degree of distribution of the organoclay within the polymeric matrix. The degree of the improvement also depends on the interaction between the polymeric chain and clay layers.

The study of the influence of the degree of the extract of fibers on the mechanical properties has shown that the limit of strength and the rigidity modulus both increase in PBT whereas they decrease in nanocomposites, Table 12.5. This can be explained by the breaking of the bonds between the organoclay and PBT at greater degree of extract. Such phenomena have been observed in numerous polymeric composites [73–75].

TABLE 12.5 The ability to stretch of the nanocomposites PBT/organoclay at varying degrees of extract

Organoclay content, weight %	Limit of strength, MPa			Rigidity modulus, GPa		
	DR = 1	DR = 3	DR = 6	DR = 1	DR = 3	DR = 6
0 (pure PBT)	41	50	52	1.37	1.49	1.52
3	60	35	29	1.76	1.46	1.39

The first notions on the lowered flammability of the polymeric nanocomposites on the organoclay basis appeared in 1976 in the patent on the composite based on the nylon-6 [5]. The serious papers in the field were absent till the 1995 [76].

The use of the calorimeter is very effective for studying the refractoriness of the polymers. It can help at measuring the heat release, the carbon monoxide depletion and others. The speed of the heat release is one of the most important parameters defining the refractoriness [77]. The data on the flame resistance in various polymer / organoclay systems such as layered nanocomposite nylon-6/organoclay, intercalated nanocomposites polystyrene (or polypropylene)/organoclay were given in paper [78] in where the lowered flammability was reported, see Table 12.6. And the lowered flammability have been observed in systems with low organoclay content, namely in range from 2 to 5weight%.

TABLE 12.6 Calorimetric data

Sample	remnant (%)±0.5	Peak of the HRR (D%) (kW/m²)	Middle of the HRR (D%) (kW/m²)	Average value H_c (MJ/kg)	Average value SEA (m²/kg)	Average CO left (kg/kg)
Nylon-6	1	1010	603	27	197	0.01
Nylon-6 / organoclay, 2%, delaminated	3	686 (32%)	390 (35%)	27	271	0.01
Nylon-6 / organoclay, 5%, delaminated	6	378 (63%)	304 (50%)	27	296	0.02
Polystyrene	0	1120	703	29	1460	0.09
PS / organoclay, 3%, bad mixing	3	1080	715	29	1840	0.09
PS / organoclay, 3%, intercalated / delaminated	4	567 (48%)	444 (38%)	27	1730	0.08
PS w/ DBDPO/ Sb_2O_3, 30%	3	491 (56%)	318 (54%)	11	2580	0.14

| Polypropylene | 0 | 1525 | 536 | 39 | 704 | 0.02 |
| PP / organoclay, 2%, intercalated | 5 | 450 (70%) | 322 (40%) | 44 | 1028 | 0.02 |

H_c—heat of combustion;
SEA—specific extinguishing area;
DBDPO—dekabrominediphenyloxide;
HRR—speed of the heat release

The curve of the heat release for the polypropylene and the nanocomposite on its basis (organoclay content varying from 2 to 4weight%) is given in the Figure 12.16 from which one can see that the speed of the heat release for the nanocomposite enriched with the 4weight% organoclay (the interlayer distance 3.5 nanometers) is 75 percent less than for pure polypropylene.

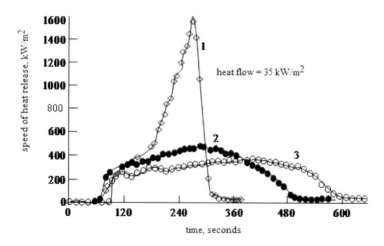

FIGURE 12.16 The speed of the heat release for:
1—pure polypropylene;
2—nanocomposite with 2weight% of organoclay;
3—nanocomposite with 4weight% of organoclay.

The comparison of the experimental data for the nanocomposites on the basis of the nylon-6, polypropylene and polystyrene gathered in Table 12.7 show that the heat of combustion, the smoke release and the amount of the carbon monoxide are almost constant at varying organoclay content. So we conclude that the source for the increased refractoriness of these materials is the stability of the solid phase and not the influence of the vapor phase. The data for the polystyrene with the 30 percent of the dekabrominediphenyloxide and Sb_2O_3 are given in Table 12.6 as the proof of the influence of the vapor phase of bromine. The incomplete combustion of the polymeric material in the latter case results in low value of the heat of the combustion and high quantity of the carbon monoxide released [79].

One should note that the mechanism for the increased fire resistance of the polymeric nanocomposites on the basis of the organoclays is not, in fact, clear at all. The formation of the barrier from the clay layers during the combustion at their collapse is supposed to be the main mechanism. That barrier slows down the combustion [80]. In our paper we study the influence of the nanocomposite structure on the refractoriness. The layered structure of the nanocomposite expresses higher refractoriness comparing to that in intercalated nanocomposite, see Figure 12.17.

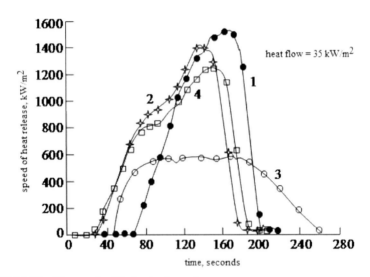

FIGURE 12.17 The speed of the heat release for:
1—pure polystyrene (PS);
2—PS mixed with 3weight% of Na⁺ MMT;
3—intercalated / delaminated PS (3weight% 2C18-MMT) extruded at 170°C;
4—intercalated PS (3weight% C14-FH) extruded at 170°C.

The data on the polymeric polystyrene—based nanocomposites presented in Figure 12.17 are for (1) initial ammoniumfluorine hectorite and (2) quaternary ammonium montmorillonite. The intercalated nanocomposite was produced in first case whereas the layered-intercalated nanocomposite was produced in the second one. But because the chemical nature and the morphology of the organoclay used was quite different it is very difficult to draw a unique conclusion about the flame resistance in polymeric nanocomposites produced. Nonetheless, one should point out that good results of the same quality were obtained for both layered and intercalated structures when studying the aliphatic groupings of the polyimide nanocomposites based on these clays. The better refractoriness is observed in case of polystyrene embedded in layered nanocomposite while intercalated polystyrene—based nanocomposite (with MMT) also exhibits increased refractoriness.

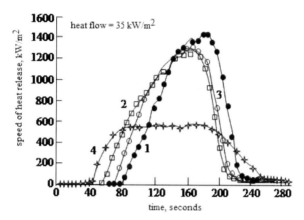

FIGURE 12.18 The speed of the heat release for:
1—pure polysterene (PS);
2—polystyrene with Na-MMT;
3—intercalated PS with organomontmorillonite obtained in extruder at 185 °C;
4—intercalated / layered PS with organoclay obtained in extruder at 170 °C in nitrogen atmosphere or in vacuum.

As one can see the from the Figure 12.18, the speed of the heat release for the nanocomposite produced in nitrogen atmosphere at 170 °C is much lower than for other samples. Probably, the reason for the low refractoriness of the nanocomposite produced in extruder without the vacuuming at 180 °C is the influence of the high temperature and of the oxygen from the air what can lead to the destruction of the polymer in such conditions of the synthesis.

It is impossible to give an exact answer on the question about how the refractoriness of organoclay—based nanocomposites increases basing on only the upper experimental data but the obvious fact is that the increased thermal stability and refractoriness are due to the presence of the clays existing in the polymeric matrix as nanoparticles and playing the role of the heat isolators and elements preventing the flammable products of the decomposition to fly off.

There are still many problems unresolved in the field but indisputably polymeric nanocomposites will take the leading position in the chemistry of the advanced materials with high heat and flame resistance. Such materials can be used either as itself or in combination with other agents reducing the flammability of the substances.

The processes of the combustion are studied for the number of a polymeric nanocomposites based on the layered silicates such as nylon-6.6 with 5weight% of Cloisite 15A—montmorillonite being modified with the dimethyldialkylammonium (alkyls studied C_{18}, C_{16}, C_{14}), maleinated polypropylene and polyethylene, both (1.5%) with 10weight% Cloisite 15A. The general trend is two times reduction of the speed of the heat release. The decrease in the period of the flame induction is reported for all nanocomposites in comparison with the initial polymers [54].

The influence of the nanocomposite structure on its flammability is reflected in the Table 12.7. One can see that the least flammability is observed in delaminated nanocomposite based on the polystyrene whereas the flammability of the intercalated composite is much higher [54].

TABLE 12.7 Flammability of several polymers and composites

Sample	Coke Remnant, Weight%	Max Speed of Heat Release, kW/m²	Average Value of Heat Release, kW/m²	Average Heat of Combustion, MJ/kg	Specific Smoke Release, m²/kg	CO Release, kg/kg
Nylon-6	1	1010	603	27	197	0.01
Nylon-6 + 2% of silicate (delaminated)	3	686	390	27	271	0.01
Nylon-6 + 5% of silicate (delaminated)	6	378	304	27	296	0.02
Nylon-12	0	1710	846	40	387	0.02
Nylon-12 + 2% of silicate (delaminated)	2	1060	719	40	435	0.02
Polystyrene	0	1562	803	29	1460	0.09
PS + 3% of silicate Na-MMT	3	1404	765	29	1840	0.09
PS + 3% of silicate C14-FH (intercalated)	4	1186	705	28	1790	0.09
PS + 3% of silicate 2C18-MMT (delaminated)	4	567	444	28	1730	0.08
Polypropylene	0	1525	536	39	704	0.02
PP + 2% of silicate (intercalated)	3	450	322	40	1028	0.02

Nanostructured Polymeric Composites

The optical properties of the nanocomposites are of much interest too. The same materials could be either transparent or opaque depending on certain conditions. For example in Figure 12.19 we see transparency, plot (a), and turbidity, plot (c), of the material in dependence of the frequency of the current applied.

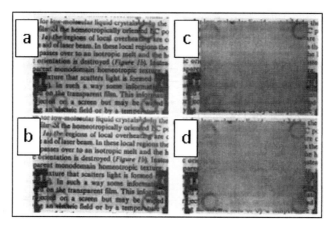

FIGURE 12.19 The optical properties of the clay—based nanocomposites in dependence of the applied electric current:
(a)—low frequency, switched on;
(b)—low frequency, switched off;
(c)—high frequency, switched on;
(d)—high frequency, switched off.

The effect in Figure 12.19 is reversible and can be innumerately repeated. The transparent and opaque states exhibit the memory effect after the applied current switched off, plots (b) and (d) in Figure 12.19. The study of the intercalated nanocomposites based on the smectite clays reveals that the optical and elecrooptical properties depend on the degree of intercalation [81].

12.7 CONCLUSIONS

The quantity of the papers in the field of the nanocomposite polymeric materials has grown multiple times in recent years. The possibility to use almost all polymeric and polycondensated materials as a matrix is shown. The nanocomposites from various organoclays and polymers have been synthesized. Here is just a small part of the compounds for being the matrix referenced in literature: polyacrylate [83], polyamides [82, 84, 85], polybenzoxasene [86], polybutyleneterephtalate [11, 82, 87], polyimides [88], polycarbonate [89], polymethylmetacrylate [90], polypropylene [91,92], polystyrene [90], polysulphones [93], polyurethane [94], polybuthyleneterephtalate and polyethyleneterephtalate [10, 65, 68, 79, 95,99–107], polyethylene [96], epoxies [97].

The organomodified montmorillonite is of the special interest because it can be an element of the nanotechnology and it can also be a carrier of the nanostructure and of asymmetry of length and width in layered structures. The organic modification is being usually performed using the ion-inducing surface-active agents. The nonionic hydrophobization of the surface of the layered structures have been reported either. The general knowledge about the methods of the study is being formed and the understanding of the structure of the nanocomposite polymeric materials is becoming clear. Also scientists come closer to the realizing of the relations between the deformational and strength properties and the specifics of the nanocomposite structure. The growth of researches and their direction into the nanoarea forecasts the fast broadening of the industrial involvement to the novel and attractive branch of the materials science.

KEYWORDS

- **Organoclays**
- **Polymeric matrix**
- **Polymeric nanocomposites**
- **Thermogravimetric analysis**

REFERENCES

1. Romanovsky, B.V.; and Makshina, E. V.; *Sorosovskii obrazovatelniu zhurnal.* **2013**, *8*(2), 50–55.in Russian
2. Golovin, Yu. I.; Priroda, 1 **2009**. in Russian
3. Carter, L. W.; Hendrics, J. G.; and Bolley, D. S.; United States Patent №2, 531.396; **1990**
4. Blumstain, A.; *Bull. Chem. Soc.* **1991**, 899–905.
5. Fujiwara, S.; and Sakamoto, T.; Japanese Application № 109.998; **1996**.
6. Usuki, A.; Kojima, Y.; Kawasumi, M.; Okada, A.; Fukushima, Y.; Kurauchi T.; Kamigatio, O.; *J. Appl. Polym. Sci.* **1995**, *55*, 119.
7. Usuki, A.; Koiwai, A.; Kojima, Y.; Kawasumi, M.; Okada, A.; Kurauchi, T.; and Kamigaito, O.; *J. Appl. Polym. Sci.* **1995**, *55*, 119.
8. Okada, A.; and Usuki, A.; *Mater. Sci. Eng.* **1995**, *3*, 109.
9. Okada, A.; Fukushima, Y.; Kawasumi, M.; Inagaki, S.; and Usuki, A.; Sugiyama, S.; Kurauchi, T.; and Kamigaito, O; United States Patent №4,739,007; **1988**.
10. Mikitaev, M. A.; Lednev, O. B.; Kaladjian, A. A.; Beshtoev, B. Z.; Bedanokov, A. Yu.; and Mikitaev, A. K. Second International Conference (Nalchik 2005).
11. Chang, J.-H.; An, Y. U.; Kim, S. J.; Im, S.; *Polymer*, **2003**, *44*, 5655–5661.
12. Mikitaev, A. K.; Bedanokov, A. Y.; Lednev, O. B.; and Mikitaev, M. A; Polymer/silicate nanocomposites based on organomodified clays. Polymers, Polymer Blends, Polymer Composites

and Filled Polymers. Synthesis, Properties, Application. Nova Science Publishers: New York; **2006**.
13. Delozier, D. M.; Orwoll, R. A.; Cahoon, J. F.; Johnston, N. J.; Smith, J. G.; and Connell, J. W.; *Polymer*, **2002**;*43*, 813–822.
14. Kelly, P.; Akelah, A.; and Moet, A. J.; *Mater. Sci.* **1994**. *29*, 2274–2280.
15. Chang, J.-H.; An Y. U.; Cho, D. E., and Giannelis, P.; *Polymer*, **2003**, *44*, 3715–3720.
16. Yano, K.; Usuki, A; Okada, A; *J. Poly. Sci. Part A: Polym Chem* **1997**, *35*, 2289.
17. Garcia-Martinez, J. M.; Laguna, O.; Areso, S.; and Collar, E. P.; *J. Polym Sci, Part B: Polym Phys.* **2000**, *38*, 1564.
18. Giannelis, E. P.; Krishnamoorti, R.; and Manias, E.; Advances in Polymer Science, Vol. 138, Springer-Verlag: Berlin Heidelberg; **1999**.
19. Delozier, D. M.; Orwoll, R. A.; Cahoon, J. F.; Ladislaw, J. S.; Smith, J. G and Connell, J. W. *Polymer*, **2003**, *44*, 2231–2241.
20. Shen, Y.-H.; *Chemosphere*, **2001**, *44*, 989–995.
21. Giannelis, E. P.; *Adv. Mater.* **1996**, *8*, 29–35.
22. Dennis, H. R.; Hunter, D. L.; Chang, D.; Kim, S.; White, J. L.; Cho, J. W.; and Paul D. R.; *Polymer*, **2001**, *42*, 9513–9522.
23. Kornmann, X.; Lindberg, H.; and Berglund, L. A.; *Polymer.* **2001**, *42*, 1303–1310.
24. Fornes, T. D.; and Paul, D. R.; Formation and properties of nylon 6 nanocomposites. *Polímeros 13*(4) São Carlos Oct/Dec. **2003**.
25. Voulgaris, D.; and Petridis, D. *Polymer*, **2002**, *43*, 2213–2218.
26. Tyan, H.-L.; Liu, Y.-C.; and Wei, K.-H.; *Polymer*, **1999**, *40*, 4877–4886.
27. Davis, C. H., Mathias. L. J.; Gilman, J. W.; Schiraldi, D. A.; Shields, J. R.; Trulove, P.; Sutto, T. E.; and Delong, H. C.; *J. Polym. Sci. Part. B: Polym. Phys.* **2002**, *40*, 2661.
28. Morgan, A. B.; and Gilman, J. W.; *J. Appl. Polym. Sci.* **2003**, *87*, 1329.
29. Hay, John N.; and Steve, J.; Shaw. Organic-inorganic hybridss the best of both worlds? *Europhysics News.* **2003**, *34*, 3.
30. Chang, J. H; An, Y. U.; Sur, G. S.; *J. Polym. Sci. Part. B: Polym. Phys.* **2003**, *41*, 94.
31. Chang, J. H.; Park, D. K.; and Ihn, K. J.; *J. Appl. Polym. Sci.* **2002**, *84*, 2294.
32. Pinnavaia, T. J.; *Science.* **1983**, *220*, 365.
33. Messersmith, P. B.; and Giannelis, E. P; *Chem. Mater.* **1993**, *5*, 1064.
34. Vaia, R. A.; Ishii, H.; and Giannelis, E. P.; *Adv. Mater.* **1996**, *8*, 29.
35. Gilman, J. W.; *Appl. Clay. Sci.* **1999**, *15*, 31.
36. Fukushima, Y.; Okada, A; Kawasumi, M.; Kurauchi, T.; and Kamigaito, O; *Clay Min.* **1988**, *23*, 27.
37. Akelah, A.; and Moet, A.; *J. Mater. Sci.* **1996**, *31*, 3589.
38. Vaia, R. A.; Ishii, H.; and Giannelis, E. P; *Adv Mater.* **1996**, *8*, 29.
39. Vaia, R. A.; Jandt, K. D., Kramer, E. J., Giannelis, E. P.; *Macromolecules* **1995**, *28*, 8080.
40. Greenland, D. G.; *J. Colloid. Sci.* **1963**, *18*, 647.
41. Chang, J. H.; and Park, K. M.; *Polym. Eng. Sci.* **2001**, *41*, 2226.
42. Greenland, D. G.; *J Colloid Sci.* **1963**, *18*, 647.
43. Chang, J. H.; Seo, B. S.; and Hwang, D. H.; *Polymer.* **2002**, *43*, 2969.
44. Vaia, R. A.; Jandt, K. D.; Kramer, E. J.; Giannelis, E. P.; *Macromolecules* **1995**, *28*, 8080.
45. Fukushima, Y.; Okada, A.; Kawasumi, M.; Kurauchi, T.; and Kamigaito, O; *Clay. Min.* **1988**, *23*, 27.
46. Chvalun, S. N. Priroda **2000**; 7 in Russian
47. Brinker, C. J.; and Scherer, G. W.; *Sol-Gel Sci*. Boston; **1990**.
48. Mascia, L.; Tang, T; *Polymer* **1998**, *39*, 3045.
49. Tamaki, R.; and Chujo, Y.; *Chem. Mater.* **1999**, *11*, 1719.

50. Serge Bourbigot e.a. Investigation of Nanodispersion in Polystyrene–Montmorillonite Nanocomposites by Solid-State NMR. *J. Polym. Sci.: Part B: Polym. Phys.* **2003**, *41*, 3188–3213.
51. Lednev, O. B., Kaladjian, A. A.; Mikitaev, M. A.; Tlenkopatchev, M. A.; New polybutylene terephtalate and organoclay nanocomposite materials. Abstracts of the International Conference on Polymer materials (México, 2005)
52. Tretiakov, A. O.: Oborudovanie I instrument dlia professionalov №02(37) 2003 in Russian.
53. Sergeev, G. B. *Ros. Chem. J.* The journal of the D. I. Mendeleev Russian chemical society, **2002**, *XLVI*, №5 in Russian
54. Lomakin, S. M.; and Zaikov, G. E.; *Visokomol. Soed. B.* **2005**, *47*(1), 104–120 in Russian
55. Xu, H.; Kuo, S. W.; Lee, J. S.: and Chang, F. C.; *Macromolecules.* **2002**, *35*, 8788.
56. Haddad, T. S.; and Lichtenhan, J. D.; *Macromolecules,* **1996**, *29*, 7302.
57. Mather, P. T.; Jeon, H. G.; Romo-Uribe, A.; Haddad, T. S; Lichtenhan, J. D;. *Macromolecules,* **1996**, *29*, 7302.
58. Hsu, S. L. C.; and Chang, K. C.; *Polymer,* **2002**, *43*, 4097.
59. Chang, J. H.; Seo, B. S.; Hwang, D. H.; *Polymer,* **2002**, *43*, 2969.
60. Fornes, T. D. Yoon, P. J. Hunter, D. L. Keskkula, H.; and Paul, D. R.; *Polymer.* **2002**, *43*, 5915.
61. Chang, J. H.; Seo, B. S.; and Hwang, D. H.; *Polymer,* **2002**, *43*, 2969.
62. Fornes, T. D.; Yoon, P. J.; Hunter, D. L; Keskkula, H.; Paul, D. R.; *Polymer* **2002**, *43*, 5915.
63. Wen, J., Wikes, G. L.; *Chem. Mater.* **1996**, *8*, 1667.
64. Zhu, Z. K.; Yang, Y., Yin, J.; Wang, X.; Ke, Y.; Qi, Z.; *J. Appl. Polym. Sci.* **1999**, *3*, 2063.
65. Mikitaev, M. A.; Lednev, O. B.; Beshtoev, B. Z.; Bedanokov, A. Yu, Mikitaev, A. K.; Second International conference. Polymeric composite materials and covers (Yaroslavl 2005, may) in Russian
66. Fischer, H. R.; Gielgens, L. H.;, Koster, T. P. M.; *Acta. Polym.* **1999**, *50*, 122.
67. Petrovic, X. S.; Javni, L., Waddong, A., and Banhegyi, G. J.; *J. Appl. Polym. Sci.* **2000**, *76*, 133.
68. Lednev, O. B.; Beshtoev, B. Z.; Bedanokov, A. Yu.; Alarhanova, Z. Z.; Mikitaev A. K.; Second International Conference (Nalchik 2005) in Russian
69. Chang, J.-H.; Kim, S. J.; Joo, Y. L. Im, S.; *Polymer,* **2004**, *45*, 919–926.
70. Lan, T., and Pinnavaia, T. J.; *Chem. Mater.* **1994**, *6*, 2216.
71. Masenelli-Varlot, K.; Reynaud, E.; Vigier, G.; Varlet, J.; *J. Polym. Sci. Part. B: Polym. Phys.* **2002**, *40*, 272.
72. Yano, K.;Usuki, A.; and Okada, A.; *J. Polym. Sci. Part A: Polym. Chem.* **1997**, *35*, 2289.
73. Shia, D., Hui, Y.; Burnside, S. D.; and Giannelis, E. P.; *Polym. Eng. Sci.* **1987**, *27*, 887.
74. Curtin, W. A.; *J. Am. Ceram. Soc.* **1991**, *74*, 2837.
75. Chawla, K. K.; Composite Materials Science and Engineering. Springer New York: **1987**.
76. Burnside, S. D.; and Giannelis, E. P.; *Chem. Mater.* **7**, **1995**, 4597.
77. Babrauskas, V., and Peacock, R. D.; *Fire. Saf. J, 18,* **1992**, 225.
78. Gilman, J.; Kashiwagi, T.; Lomakin, S.; Giannelis, E.; Manias, E.; Lichtenhan, J.; and Jones, P.; In Fire Retardancy of Polymers: The Use of Intumescence. The Royal Society of Chemistry, Cambridge, **1998**, 203–221.
79. Mikitaev, A. K.; Kaladjian, A. A.; Lednev, O. B.; Mikitaev, M. A.; *Plastic Masses.* **2004**, *12*, 45–50 in Russian
80. Gilman, J.; and Morgan, A.; 10th Annual BCC Conference, May 24–26, **1999**.
81. Hay, John N.; and Shaw, Steve J.; Organic-inorganic hybrids: The best of both worlds?. *Europhysics News,* **2003**, *34*(3).
82. Delozier, D. M.; Orwoll, R. A.; Cahoon, J. F.; Johnston, N. J.; Smith, J. G., and Connell, J. W. *Polymer,* **2002**, *43*, 813–822.
83. Chen, Z.; Huang, C.; Liu, S.; Zhang, Y.; Gong, K. J.; *Apply. Poly. Sci.* **2000**, *75*, 796–801.

84. Okado, A.; Kawasumi, M.; and Kojima, Y.; Kurauchi, T.; and Kamigato, O.; *Mater. Res. Soc. Symp. Proc.* **1990**, 171, 45.
85. Leszek, A.; Utracki, Jorgen Lyngaae-Jorgensen. *Rheologica Acta*, **2002**, *41*, 394–407.
86. Wagener, R., and Reisinger, T. J. G.; *Polymer.* **2003**; *44*, 7513–7518.
87. Li, X., Kang, T., Cho, W. J., Lee, J. K.; and Ha, C. S.; *Macromol. Rapid. Commun.*
88. Tyan, H.- L.; Liu, Y.- C.; and Wei, K.- H; *Polymer*, **1999**, *40*, 4877–4886.
89. Vaia, R.; Huang, X.; Lewis, S., and Brittain, W.; *Macromolecules* **2000**, *33*, 2000–2004.
90. Okamoto, M., Morita, S., Taguchi. H., Kim, Y., Kotaka, T., and Tateyama H., *Polymer.* **2000**, *41*, 3887–3990.
91. Chow, W. S.; Mohd Ishak, Z. A.; Karger-Kocsis, J.; Apostolov, A. A.; and Ishiaku, U. S.; *Polymer.* **2003**, *44*, 7427–7440.
92. Antipov, E. M.; Guseva, M. A.; Gerasin, V. A.; Korolev, Yu. M.; Rebrov A. V.; Fisher, H. R.; Razumovskaya, I. V.; *Visokomol soed. A.* **2003**, *45*(11), 1885–1899 in Russian.
93. Sur, G.; Sun, H.; Lyu, S.; and Mark, J.; *Polymer.* **2001**, *42*, 9783–9789.
94. Wang, Z.; Pinnavaia, T.; *Chem Mater.* **1998**, *10*, 3769–3771.
95. Bedanokov, A.Yu.; Beshtoev, B. Z.; Malij polimernij congress (Moscow 2005) in Russian.
96. Antipov, E. M.; Guseva, M. A.; Gerasin, V. A.; Korolev, Yu. M.; Rebrov A. V.; Fisher, H. R.; and Razumovskaya, I. V. *Visokomol. Soed. A.* **2003**, *45*(11), 1874–1884 in Russian.
97. Lan, T.; Kaviartna, P.; Pinnavaia, T.; Proceedings of the ACS PMSE **1994**, *71*, 527–528.
98. Kawasumi et al.; Nematic liquid crystal/clay mineral composites. *Science and Engineering C*, **1998**, *6*, 135–143,.
99. Lednev, O. B.; Kaladjian, A. A.; Mikitaev, M. A.; Second International Conference (Nalchik 2005) in Russian
100. Mikitaev, A. K.; Kaladjian, A. A.; Lednev, O. B.; Mikitaev, M. A.; and Davidov E. M. *Plastic Masses.* **2005**, *4*, 26–31in Russian
101. Eid, A.; Mikitaev, M. A.; Bedanokov, A.Y.; Mikitaev, A. K.; Recycled Polyethylene Terephthalate/Organo-Montmorillanite Nanocomposites, Formation And Properties. The first Afro-Asian Conference on Advanced Materials Science and Technology (AMSAT 06), Egypt, **2006**.
102. Mikitaev, A. K.; Bedanokov, A. Y.; Lednev, O. B.; Mikitaev, M. A.; Polymer/silicate nanocomposites based on organomodified clays/Polymers, Polymer Blends, Polymer Composites and Filled Polymers. Synthesis, Properties, Application. Nova Science Publishers: New York, **2006**.
103. Malamatov, A. H.; Kozlov, G. V.; and Mikitaev, A. M; Mechanismi uprochnenenia polimernih nanokompozitov, (Moscow, RUChT **2013**) 240 p. in Russian
104. Eid, A.; Doctor Thesis, (Moscow, RUChT **2013**) 121 p. in Russian
105. Lednev, O. B.; Doctor Thesis (Moscow, RUChT **2013**) 128 p. in Russian
106. Malamatov, A. H.; Professor Thesis (Nalchik, KBSU **2006**) 296 p. in Russian
107. Borisov, V. A.; Bedanokov, A. Yu.; Karmokov, A. M.; Mikitaev, A. K.; and Mikitaev, M. A.; Turaev, E. R.; Plastic masses **2013**; 5 in Russian

CHAPTER 13

STRUCTURE, PROPERTIES, AND APPLICATION OF DENDRITIC MACROMOLECULES IN VARIOUS FIELDS: MOLECULAR SIMULATION TECHNIQUES IN HYPERBRANCHED POLYMER AND DENDRIMERS

M. HASANZADEH and B. HADAVI MOGHADAM

13.1 INTRODUCTION

Dendritic architectures, as highly branched and three-dimensional macromolecules that have unique chemical and physical properties, offer potential as the next great technological revolution. This review gives a brief introduction to some of the structural properties and application of dendritic polymer in various fields. The focus of the paper is a survey of multiscale modeling and simulation techniques in hyperbranched polymer and dendrimers. Results of modeling and simulation calculations on dendritic architecture are reviewed.

The field of dendritic architectures, as a general class of macromolecules, has found widespread interest in the past decades. Much has been achieved in the preparation of three-dimensional structures such as comb- and star-shaped polymers and dendrimers. These materials have comparable physical and chemical properties to their linear analogous that make them very attractive for numerous applications [1–4].

Intensive studies in the area of dendritic macromolecules, which include applied research and are generally interdisciplinary, have created a need for a more systematic approach to dendritic architectures development that employs a multiscale modeling and simulation approach. A possible way is to determine the atomic-scale characteristics of dendritic molecules using computer simulation and computational approaches. Computer simulation, as a powerful and modern tool for solving scientific problems, can be performed for dendritic architectures without synthesizing them. Computer simulation not only used to reproduce experiment to elucidate the invisible microscopic details and further explain experiments, but also can be used as a useful predictive tool. Currently, Monte Carlo, Brownian dynamics, and molecular dynamics are the most widely used simulation methods for molecular systems [5].

The objective of this paper is to address recent advances in molecular simulation methodologies and computational power. In this paper, we will first briefly review the structure, properties, and application of dendritic macromolecules in various fields. Next, molecular simulation techniques in hyperbranched polymer and dendrimers will be reviewed. Lastly, we will survey the most characteristic and important recent examples in molecular simulation of dendritic architectures. The paper ends with a conclusion.

13.2 DENDRITIC ARCHITECTURES

13.2.1 BASIC PRINCIPLES

Dendritic architectures are highly branched polymers with tree like branching having an overall spherical or ellipsoidal shape and are known as additives having peripheral functional groups. These macromolecules consist of three subsets namely dendrimers, dendrigraft polymers, and hyperbranched polymers (Figure 13.1).

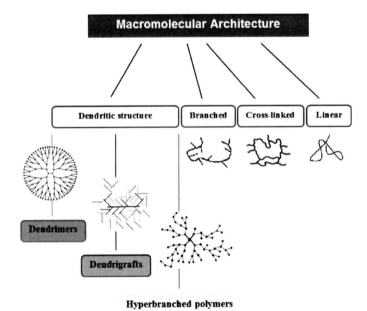

FIGURE 13.1 Classification of macromolecular architecture.

Dendrimers resemble star polymers except that each leg of the star exhibit repetitive branching in the manner of a tree. There are two general routes for the synthesis of the dendrimers: divergent methodology and convergent methodology (Figure 13.2). *Dendrigraft (arborescent) polymers* are prepared by linking macromolecular building blocks. In dendrigraft polymers branching sites are usually distributed randomly along the dendritic chains of the dendrigraft interior whereas in dendrimers the branching sites conducing to the next generation occur only at the chain end. *Hyperbranched polymers* (HBPs) are highly branched, polydisperse, and three-dimensional macromolecules; and it is synthesized from a multifunctional monomer to produce a molecule with dendritic structure [6–16]. Table 13.1 shows typical characteristics of dendritic architectures.

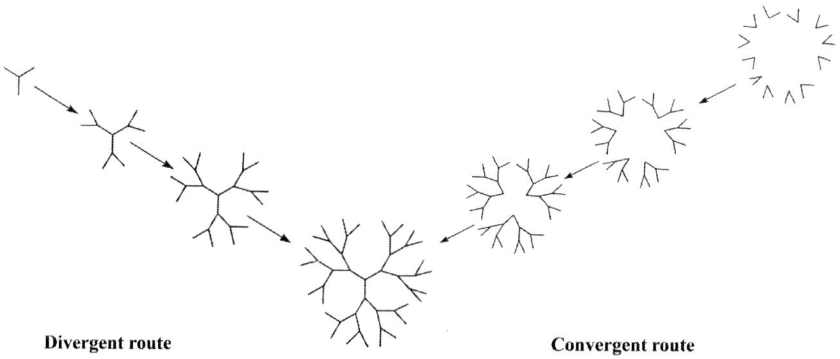

Divergent route   **Convergent route**

FIGURE 13.2 Two general routes for synthesis of dendrimers.

Dendrimers are well-defined and need a stepwise route to construct the perfectly symmetrical structure. Hence, synthesis of dendrimers is time-consuming and expensive procedures. Although hyperbranched polymers are irregularly shaped and not perfectly symmetrical like dendrimers, hyperbranched polymers rapidly prepared and generally synthesized by one-step process via polyaddition, polycondensation, radical polymerization, and so on, of AB_x (mostly AB_2, equal reactivity of all Bs) type monomers and no purification steps are needed for their preparation (Figure 13.3). Therefore HBPs are attractive materials for industrial applications due to their simple production process [6–9, 17–25]. In general, according to molecular structures and properties, hyperbranched polymers represent a transition between linear polymers and perfect dendrimers [18]. Comparison of hyperbranched polymers with their linear analogs indicated that HBPs have remarkable properties, such as low melt and solution viscosity, low chain entanglement, and high solubility, as a result of the large amount of functional end groups and globular structure [16–22].

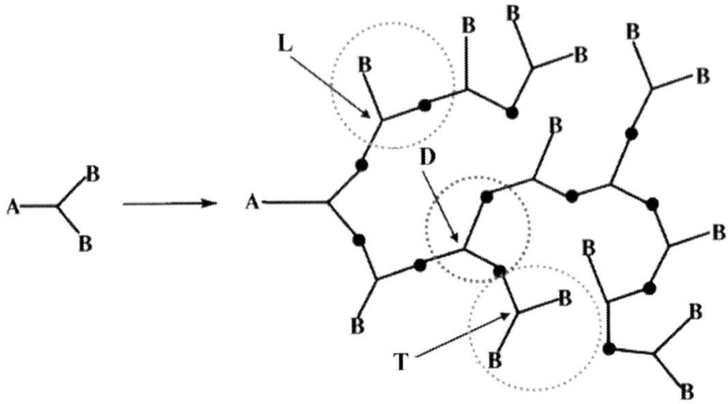

FIGURE 13.3 Schematic representation of a hyperbranched polymer construction and its structural units include terminal (T), dendritic (D) and linear (L) units.

TABLE 13.1 Comparison of different dendritic polymers

Properties	Dendritic architectures		
	Dendrimers	Dendrigraft	Hyperbranched
Terminal units	Small	Linear chains	Small
Molecular mass distribution	Narrow	Narrow	Broad
Synthetic steps	4-20	2-5	1
Purification steps	4-20	2-5	0
Cost	Very high	Moderate	Low

13.2.2 GENERAL STRUCTURAL CONSIDERATIONS

13.2.2.1 DEGREE OF BRANCHING

Different structural parameters such as degree of polymerization, degree of branching and Wiener index can be used to characterize the topologies of hyperbranched polymers. The degree of branching (DB) is defined as follows:

$$DB = \frac{2D}{(2D+L)} \tag{13.1}$$

Structure, Properties, and Application of Dendritic 245

where D is the number of dendritic units and L is the number of linear units. This value varies from 0 for linear polymers to 1 for dendrimers or fully branched hyperbranched polymers [26–28].

13.2.2.2 WIENER INDEX

In addition to the degree of branching, the Wiener index is also used to distinguish polymers of different topologies and defined as:

$$W = \frac{1}{2} \sum_{j=1}^{N_s} \sum_{i=1}^{N_s} d_{ij} \qquad (13.2)$$

where N_s is the number of beads per molecule and d_{ij} is the number of bonds separating site i and j of the molecule. This parameter only describes the connectivity and is not a direct measure of the size of the molecules. Larger Wiener index numbers indicate higher numbers of bonds separating beads in molecules and hence more open structures of polymer molecules [28]. Table 13.2 shows the DB of polymers with different architecture and the same degree of polymerization.

TABLE 13.2 Degree of branching for different polymer architectures of the same molecular weight (white beads representing linear units and gray beads representing branching units)

Polymer type	Polymers architecture	Degree of Branching
Linear polymer		DB=0
Hyperbranched polymer		0<DB<1

Dendrimers

DB=1

Fully branched
hyperbranched polymer

13.2.2.3 RADIUS OF GYRATION

The long chain branching on the polymer s can have a major effect on the rheological properties. Comparison of a long branching polymer with a linear polymer with the same molecular weight shows a drastic decrease in the mean-square dimensions of the polymer in solution. The radius of gyration, which is the trace of the tensor of gyration, can describe the size of a polymer molecule. The mean-square radius of gyration tensor of molecules can be calculated according to the formula:

$$\langle R_g R_g \rangle \equiv \left\langle \frac{\sum_{\alpha=1}^{N_s} m_\alpha (r_\alpha - r_{CM})(r_\alpha - r_{CM})}{\sum_{\alpha=1}^{N_s} m_\alpha} \right\rangle \quad (13.3)$$

where r_α is the position of site α, r_{CM} is the position of the molecular center of mass, m_α is the mass of site α and the angle brackets denote an ensemble average. The value of the squared radius of gyration is defined as the trace of the tensor $R_g^2 = Tr(\langle R_g R_g \rangle)$. By studying the tensor of gyration, the shape of hyperbranched polymers can be investigated. Polymers with higher degree of polymerization normally have larger radius of gyration. Furthermore for a given value of molecular weight, the value of the radius of gyration for different branched polymers can vary depending on the branching topology [28].

13.2.2.4 DISTRIBUTION OF MASS

The radial distribution function is a useful tool to describe the structure of dendritic structure. The distribution of sites from the molecular center of mass is given as:

$$g_{CM} = \frac{\left\{\sum_{i=1}^{N}\sum_{\alpha=1}^{N_s} \delta(|r-(r_{i\alpha}-r_{CM})|)\right\}}{N} \tag{13.4}$$

where N is the total number of molecules, r_{CM} is the position of the center of mass and α runs over all other sites belonging to the same molecule. Similarly, the distribution from the central site (the core) can be defined as:

$$g_{core}(r) = \frac{\left\{\sum_{i=1}^{N}\sum_{\alpha=1}^{N_s} \delta(|r-(r_{i\alpha}-r_{i1})|)\right\}}{N} \tag{13.5}$$

where r_{i1} is the position of the core. Another useful function for characterization of internal structure and spatial ordering of sites composing the materials is the atomic radial distribution.

$$g_A(r) = \frac{\left\{\frac{1}{2}\sum_{i=1}^{N_{total}}\sum_{j \neq i}^{N_{total}} \delta(|r-r_{ij}|)\right\}}{4\pi r^2 N_{total}\rho} \tag{13.6}$$

where r_{ij} is the distance between the sites i and j, $N_{total} = NNs$ is the total number of sites in the studied system, and ρ is the density [28].

13.2.3 APPLICATIONS

A number of excellent reviews have been published on synthesis, functionalization, and applications of dendritic polymer [7, 17–18, 29]. Many application fields based on dendritic polymer have been steadily extended especially in recent years. A schematic diagram illustrating perspective application areas is summarized in Figure 13.4. Although some of these applications have not reached their industry level, their promising potential is believed to be attracting attentions and investments from academia, governments, and industry all over the world.

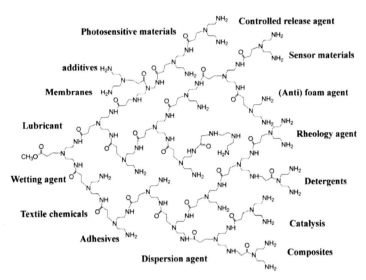

FIGURE 13.4 Potential application of dendritic polymers.

Dendritic polymers and their substitutes can be used as nanomaterials for host-guest encapsulation [30], fabrication of organic-inorganic hybrids [31] and nanoreactors [32]. Moreover, due to the low cost and well-defined architecture with multifunctional terminal groups, these types of polymers have recently attracted special interest in biomaterials application, as biocarriers and biodegradable materials [33]. Furthermore, based on the unique chemical and physical properties, hyperbranched polymer have been used as cross-linking or adhesive agents [34], dye-receptive additives [35], rheology modifiers or blend components [36], tougheners for thermosets [37], and so on. Table 13.3 shows dendritic architectures applications in various fields and their main characteristics.

In the case of textile chemistry, studies have demonstrated that the dyeability of textile fabrics, such as cotton [38–41], poly(ethylene terephthalate) (PET) [3, 42–43], polyamide-6 [44], and polypropylene (PP) [35], were significantly improved by hyperbranched polymer. In the most recent investigation in this filed, the dyeability of modified PET fabrics by amine-terminated HBP was investigated by Hasanzadeh et al. [3, 42]. They reported that the dye uptake of HBP-treated PET fabrics is significantly greater than that of untreated PET ones due to the presence of terminal primary amino groups in the molecular structure of the HBP, that will protonate in the liquid phase and give rise to positive charge at lower pH values.

TABLE 13.3 Dendritic architectures applications in various fields and its main characteristics

Application	Main characteristics	Refs.
Textile dyeing	Modification of PET fabrics for improving dyeability of acid dyes	[3]
	Modification of fiber grade PET for dyeing of disperse dyes	[43]
	Modification PA6 fabrics for studying dyeability with acid dyes	[44]
	Dyeing of modified polypropylene fibers	[35]
	Reactive dyeing on silk fabric	[45]
	Antibacterial activity and anti-ultraviolet properties	[46]
	Modification of cotton for improving dyeability	[38–41]
	Modification of cotton for improving dyeability of reactive dyes	[47]
Drug delivery	Synthesis of water-soluble and degradable hyperbranched polyesters for drug delivery	[48]
DNA delivery	Degradability, cytotoxicity, and in vitro DNA transfection efficiency of the second type of hyperbranched poly(amino ester)s with different terminal amine groups	[49]
Gene carrier	Degradability, great ability to form complexes with DNA and suitable physicochemical properties of poly(ester amine)	[50]
Gene delivery	Biodegradability, very low toxicity and the ability to transfect cells	[51]
Metal ion extractant	Synthesis and application of new nitrogen centered hyperbranched polyesters	[52]
Gas separation	Synthesis and application of a series of wholly aromatic hyperbranched polyimides for gas separation application	[53]
Extractive distillation and solvent extraction	Cost-savings of extractive distillation using commercially available hyperbranched polymers	[54]

TABLE 13.3 *(Continued)*

Application	Main characteristics	Refs.
Fuel cells	Ionic conductivity, thermal properties, and fuel cell performance of the cross-linked membrane	[55]
Coating materials	Synthesis of fluorinated hyperbranched polymers holding different surface free energies and their applications as coating materials to afford highly hydrophobic and/or oleophobic cotton fabrics by solution-immersion method	[56]
Cationic UV curing	Effect of the hyperbranched polyol as flexibilizer and chain-transfer agent on cationic UV curing	[57]

13.3 MOLECULAR SIMULATION METHODS

According to the literature, the molecular simulation methods for nanosystems can be mainly classified into two categories: (i) atomistic modeling, and (ii) continuum mechanics approaches [58–70]. In general, the atomistic modeling technique includes three important categories, namely the molecular dynamics (MD), Monte Carlo (MC), and *ab initio* approaches. In general, *ab initio* approaches are constructed on the basis of an accurate solution of the Schrödinger equation to extract the locations of each atom. Meanwhile, the main objective of MD and MC simulation is to solve the governing equations of particle dynamics based on the second Newton's law. In addition to these methods, tight bonding molecular dynamics (TBMD), local density approximation (LDA), and density functional theory (DFT) have also been proposed [71], but they are often computationally expensive. Although atomistic modeling technique provides a valuable insight into complex structures, due to its huge computational effort especially for large-scale systems, its application is limited to the systems with small number of atoms. Therefore, the alternative continuum and multiscale models were proposed for larger systems or larger time.

13.3.1 MONTE CARLO SIMULATION

The Monte Carlo method is a computerized mathematical technique that relies on repeated random sampling to obtain numerical results. This technique generates large numbers of configurations or microstates of equilibrated systems by stepping from one microstate to the next in a particular statistical ensemble. The advantage of the Monte Carlo simulation technique is that the probability distributions within the model can be easily and flexibly used, without the need to approximate them.

Underlying matrix or trial move which must satisfy the principle of microscopic reversibility and timesaving as only the potential energy is required, are some advantage of Monte Carlo simulation [72–73].

13.3.2 MOLECULAR DYNAMICS SIMULATION

MD simulation is one of the most important techniques to study the macroscopic behavior of systems by following the evolution of the system at the molecular scale. This method is in many respects very similar to real experiments. As mentioned above, this technique deals with the case of Newton's equations of classical mechanics. Coupled Newton's equations of motion, which describe the positions and momenta, are solved for a large number of particles in an isolated cluster or in the bulk using periodic boundary conditions. The equations of motion for these particles which interact with each other via intra- and inter-molecular potentials can be solved accurately using various numerical integration methods such as the common predictor-corrector or Verlet methods. Molecular dynamics efficiently evaluates different configurational properties and dynamic quantities which cannot generally be obtained by Monte Carlo [74]. These configurations provide information on the types of structures and also completely describe the detailed dynamics of structural change and energy flow within the classical model. More details can be found in [75–76].

13.3.3 INTERMOLECULAR INTERACTION

It is believed that the particles interact with each other and, moreover, may be subject to external influence. Interatomic forces are represented in the form of the classical potential force (the gradient of the potential energy of the system). The interaction between atoms is described by means of van der Waals forces (intermolecular forces), mathematically expressed by the Lennard-Jones (LJ) potential [75]:

$$U_{ij}^{LJ} = 4\varepsilon\left[\left(\frac{\sigma}{r_{ij}}\right)^{12} - \left(\frac{\sigma}{r_{ij}}\right)^{6}\right] \quad (13.7)$$

where r_{ij} is the distance between ith and jth particles normalized by typical length σ, and ε is the potential well depth, which governs the strength of interaction. It should be noted that the second term corresponds to the attractive force, which is caused by instantaneous induced van der Waals dipole-dipole attraction between atoms, and the first term corresponds to the Pauli repulsion. Generally, in order to shorten the computer time, the pair interactions beyond the distance of r_c are neglected. Therefore the LJ potential actually used in simulations is a truncated version defined as: (typically $r_c = 2.5\sigma$)

$$U_{ij}^{LJ} = 4\varepsilon\left[\left(\frac{\sigma}{r_{ij}}\right)^{12} - \left(\frac{\sigma}{r_{ij}}\right)^{6}\right], \quad for \quad r_{ij} \leq r_c$$
$$U_{ij}^{LJ} = 0 \quad\quad\quad for \quad r_{ij} > r_c \quad (13.8)$$

This potential gives a reasonable representation of intermolecular interactions in noble gases, such as *Ar* and *Kr*, and systems composed of almost spherical molecules, such as methane. To get an idea of the magnitude, for *Ar* values are $\frac{\varepsilon}{k_B} = 120K$ and $\sigma = 0.34nm$. Figure 13.5 shows the chart of the potential energy of intermolecular interaction.

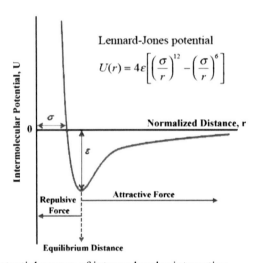

FIGURE 13.5 Potential energy of intermolecular interaction

13.4 MOLECULAR SIMULATION IN DENDRITIC ARCHITECTURES

13.4.1 BACKGROUND

According to the literature, most experimental studies have been provided vital information on the large-scale interactions between dendritic molecules and other molecules, but many atomic-level questions and unsolved problems, which cannot be answered by experiments, are also pointed out. Therefore, theoretical and computational approaches in determining the atomic-scale characteristics of dendritic molecules have attracted a great interest. A tremendous amount of theoretical research has been applied to investigate the structure and dynamics of dendritic polymers. Recent advances in simulation methods and computational power have made it possible to study the physicochemical properties of dendritic polymers and their interaction with other molecules.

13.4.2 HYPERBRANCHED POLYMER

In 1998 the first numerical studies of hyperbranched polymer was reported by Aerts [77] in which configurations of hyperbranched polymers were modeled using the bead model of Lescanec and Muthukumar [78] and intrinsic viscosities were calculated. The application of the algorithm with no configurational relaxation and the questionable deduction of the intrinsic viscosity from the radius of gyration of branched structures were limitations of this work. In the study on structural properties of hyperbranched polymer in the melt under shear using a coarse-grained model and nonequilibrium molecular dynamics (NEMD) techniques by Le et al. [79], it was demonstrated that most of the microscopic structural properties have significant changes induced by the shear flow, which depend on the size and geometry of the molecules. Analysis of the tensor of gyration shows that hyperbranched polymer molecules have a prolate ellipsoid shape under shear which is slightly flatter than the ellipsoid shape of dendrimers. They also found that the conformational behavior of large hyperbranched polymers is similar to that of linear polymers whereas for small hyperbranched polymers, the behavior is similar to that of dendrimers. Moreover, the distribution of terminal beads was investigated and the results show that simulated hyperbranched polymers have similar distribution of mass with terminal groups existing everywhere inside the molecules [79]. Lyulin et al. [80] employed Brownian dynamics simulations of hyperbranched polymers with different degrees of branching (DB) under shear with excluded-volume and hydrodynamic interactions to obtain the intrinsic viscosity data. The results show that hyperbranched polymers with low values of DB reveal sparse structures while those with high values of DB possess very compact structures. Moreover, as the molecular weight of highly branched structures increases, the zero shear rate intrinsic viscosity reaches a maximum and begins to fall similar to the intrinsic viscosity behavior of dendrimers [80]. In another study, theoretical models for the structures of randomly hyperbranched polymers in solution were derived by Konkolewicz et al. [81]. Their models are based on the random assembly of various simple units which can be monomers, linear chains, or larger branched species. They presented a comparison of Monte Carlo and molecular dynamics simulation with theoretical model conformation results including the radii of gyration and full density profiles for randomly hyperbranched polymers in solution [81].

13.4.3 DENDRIMERS

de Gennes and Hervet [82], for the first time used a mean field approach to determine the scaling properties of dendrimers. They reported that the size of dendrimers changes with the number of monomers as $R_g \propto N^{1/5}$, where R_g is the mean molecular radius of gyration. Maiti et al. [83] performed the first atomistic based MD simulations of G4 to G6 polyamidoamine (PAMAM) dendrimers in explicit water.

The radius of gyration calculated from the simulation is roughly in agreement with experimental findings. Also, they found that the extent of ionization significantly affected the R_g of the dendrimer [83]. Cagin et al. [84–85] used the Continuous Configurational Boltzmann Biased (CCBB) direct Monte Carlo method in building the 3D molecular representations of dendrimers. The energetic and structural properties of dendrimers were also studied using molecular dynamics after annealing these molecular representations. Atomistic simulation using a self-avoiding walk and also molecular dynamics simulation were performed by some researchers [78]. Lue et al. [86] studied the behavior of dendrimers in dilute and concentrated solution using Monte Carlo simulation. The equation of state and the second virial coefficients was calculated by them. By comparison with linear chain systems, it is found that for low concentration, the second virial coefficients and the pressure of dendrimers were smaller than for linear chain molecules. Other series of equilibrium molecular simulation studies analyzed and reported the properties of isolated dendrimers [87]. According to the literature, most theoretical and computational studies of dendrimers reported the properties of isolated molecules or molecules in dilute solution. As a first atomistic simulation of dendrimers in the melt, Zacharopoulos et al. [88] reported the morphology of poly(propyleneimine) dendrimers. They found that the redial density profile decrease with the distance from the center of mass. Moreover, the radius of gyration was reported to depend on the number of beads to the power of ~ 0.29. More information on recent advances in simulation of dendritic polymers is summarized in Table 13.4.

TABLE 13.4 Molecular simulation of dendritic architectures in the literature

Dendritic Polymer	Method	Focus of the Research	Refs.
Dendrimers	Atomistic based molecular dynamics	Properties for generations up to the limiting growth size	[83]
	Molecular dynamics	Generation 2 through 6 at several pH values	[89]
	Molecular dynamics	The static properties of the PAMAM dendrimers in aqueous solution	[90]
	Continuous Configurational Boltzmann Biased (CCBB) direct Monte Carlo	Building the 3D molecular representations of dendrimers and study the energetic and structural properties of dendrimers	[80–81]
	Rouze-Zimm hydrodynamic	Dynamic properties of dendrimers in dilute solution	[91]
	Monte Carlo	The behavior of dendrimers in dilute solution	[86]
	Atomistic simulation	Morphology of poly(propyleneimine) dendrimers in melt	[88]
	Metropolis Monte Carlo algorithm	Calculating the quantitative structure–affinity relationship in a dendrimer–drug system	[92]

TABLE 13.4 *(Continued)*

Dendritic Polymer	Method	Focus of the Research	Refs.
Hyper-branched polymers	Reverse Monte Carlo	Generation of randomly branched polymers with different architectures and sizes in solution.	[81]
	Nonequilibrium molecular dynamics	Structural properties of hyperbranched polymers in the melt under shear	[79]
	Monte Carlo	Copolymerization of a self-condensing vinyl monomer and a conventional vinyl monomer in the presence of a multifunctional initiator at equal rate constants	[93]
	Brownian dynamics	Simulations of hyperbranched polymers with different degrees of branching	[80]
	Brownian dynamics	Simulations of hyperbranched polymers up to the sixth generation under elongational flow	[94]
	Brownian dynamics	The structure and transport properties of dendritic polymers in dilute solution	[95]
	Brownian dynamics	Simulations of complexes formed by hyperbranched polymers with linear polyelectrolytes under steady shear flow	[96]
	Monte Carlo and molecular dynamics simulation	Testing theoretical models for randomly hyperbranched polymers in solution.	[97]

13.5 CONCLUDING REMARKS

Dendritic architectures described in this review are highly branched and three-dimensional macromolecules that have new and unusual properties. Compared to their linear analogs, dendritic polymers are expected to have remarkable properties, such as lower melt and solution viscosity, lower chain entanglement, and higher solubility, as a result of the large amount of functional end groups and globular structure. In recent years, great efforts have been devoted to dendritic macromolecules. Although tremendous progress has been made in synthesis and characterization of dendritic polymers as well as their application during past decade, we have only vague answer concerning the atomic-scale characteristics of dendritic molecules. Much useful information about the atomic-scale characteristics of dendritic polymers has been gained in recent years using molecular simulation techniques, such as Monte Carlo, Brownian dynamics simulations and molecular dynamics. Molecular simulations provide atomic-scale insights into structure and properties of dendritic polymers and their interactions with other molecules. This information from simulations has successfully matched experimentally measured properties, and can help in the rational design of dendritic architectures for application in many areas.

KEYWORDS

- Dendritic architectures
- Hyperbranched polymers
- molecular dynamics (MD)
- tight bonding molecular dynamics (TBMD)

REFERENCES

1. Yan, D; Gao, C.; and Frey, H.; Hyperbranched Polymers Synthesis, Properties, and Applications. John Wiley & Sons: New Jersey; **2011**.
2. Frechet, J. M. J.; and Tomalia, D. A.; Dendrimers and Other Dendritic Polymers. John Wiley & Sons: UK; **2011**.
3. Hasanzadeh, M.; Moieni, T.; Hadavi Moghadam, B.; Synthesis and characterization of an amine terminated AB_2-type hyperbranched polymer and its application in dyeing of poly(ethylene terephthalate) fabric with acid dye. *Adv. Poly. Technol.* **2013**, *32*, 792–799.
4. Mishra, M. K.; and Kobayashi, S.; Star and Hyperbranched Polymers. Marcel Dekker: New York; **1999**.
5. Gates, T. S.; Odegard, G. M.; Frankland, S. J. V.; Clancy, T. C.; Computational materials: multi-scale modeling and simulation of nanostructured materials. *Composites. Sci. Technol.* **2005**, *65*, 2416–2434.
6. Jikei, M.; and Kakimoto, M;. Hyperbranched polymers: a promising new class of materials. *Prog. Polym. Sci.* **2001**, *26*, 1233–1285.
7. Gao, C.; and Yan, D.; Hyperbranched polymers: from synthesis to applications. *Prog. Polym. Sci.* **2004**, *29*, 183–275.
8. Voit, B. I.; and Lederer, A.; Hyperbranched and highly branched polymer architectures synthetic strategies and major characterization aspects. *Chem. Rev.* **2009**, *109*, 5924–5973.
9. Kumar, A., and Meijer, E. W.; Novel hyperbranched polymer based on urea linkages. *Chem. Commun.* **1998**, 1629–1630.
10. Grabchev, I., Petkov, C., and Bojinov, V.; Infrared spectral characterization of poly(amidoamine) dendrimers peripherally modified with 1,8-naphthalimides. *Dyes. Pigmen.* **2004**, *62*, 229–234.
11. Qing-Hua, C., Rong-Guo, C., Li-Ren, X., Qing-Rong, Q., and Wen-Gong, Z.; Hyperbranched poly(amide-ester) mildly synthesized and its characterization. *Chinese. J. Struct. Chem.* **2008**, *27*, 877–883.
12. Kou, Y., Wan, A., Tong, S., Wang, L., and Tang, J.; Preparation, characterization and modification of hyperbranched polyester-amide with core molecules. *React. Funct. Polym.* **2007**, *67*, 955–965.
13. Schmaljohann, D, P.; Pötschke, P.; Hässler, R.; Voit, B. I.; Froehling, P. E.; Mostert, B.; and Loontjens J. A.; Blends of amphiphilic, hyperbranched polyesters and different polyolefins. *Macromolecules.* **1999**, *32*, 6333–6339.
14. Kim, Y. H.; Hyperbranched polymers 10 years after. *J. Polym. Sci. Part A: Polym. Chem.* **1998**, *36*, 1685–1698.
15. Liu, G., and Zhao, M.; Non-isothermal crystallization kinetics of AB3 hyper-branched polymer/polypropylene blends. *Iran. Polym. J.* **2009**, *18*, 329–338.
16. Inoue, K.; Functional dendrimers, hyperbranched and star polymers. *Prog. Polym. Sci.* **2000**, *25*, 453–571.

17. Seiler, M.; Hyperbranched polymers: Phase behavior and new applications in the field of chemical engineering. *Fluid. Phase. Equilibria.* **2006**, *241*, 155–174.
18. Yates, C. R., and Hayes, W.; Synthesis and applications of hyperbranched polymers. *Eur. Polym. J.* **2004**, *40*, 1257–1281.
19. Voit, B.; New developments in hyperbranched polymers. *J. Polym. Sci.: Part A: Polym. Chem.* **2000**, *38*, 2505–2525.
20. Nasar, A. S.; Jikei, M.; and Kakimoto, M.; Synthesis and properties of polyurethane elastomers crosslinked with amine-terminated AB2-type hyperbranched polyamides. *Eur. Polym. J.* **2003**, *39*, 1201–1208.
21. Froehling, P. E.; Dendrimers and dyes-a review. *Dyes. Pigm.* **2001**, *48*, 187–195.
22. Jikei, M.; Fujii, K.; to Kakimoto, M.; Synthesis and characterization of hyperbranched aromatic polyamide copolymers prepared from AB_2 and AB monomers. *Macromol. Symp.* **2003**, *199*, 223-232.
23. Radke, W.; Litvinenko, G.; and Müller, A. H. E.; Effect of core-forming molecules on molecular weight distribution and degree of branching in the synthesis of hyperbranched polymers. *Macromolecules.* **1998**, *31*, 239–248.
24. Maier, G.; Zech, C.; Voit, B.; and Komber, H.; An approach to hyperbranched polymers with a degree of branching of 100%. *Macromol. Chem. Phys.* **1998**, *199*, 2655–2664.
25. Voit, B.; Beyerlein, D.; Eichhorn, K.; Grundke, K.; Schmaljohann, D.; and Loontjens, T.; Functional hyper-branched polyesters for application in blends, coatings, and thin films. *Chem. Eng. Technol.* **2002**, *25*, 704–707.
26. Frey, H.; and Hölter, D.; Degree of branching in hyperbranched polymers. 3 Copolymerization of ABm-monomers with AB and ABn-monomers. *Acta Polym.* **1999**, *50*, 67–76.
27. Hölter, D.; Burgath, A.; Frey, H.; Degree of branching in hyperbranched polymers. *Acta Polymerica.* **1997**, *48*, 30–35.
28. Tu, C.; Le, B.; Todd, D.; Daivis, P. J.; and Uhlherr, A.; Structural properties of hyperbranched polymers in the melt under shear via nonequilibrium molecular dynamics simulation. *J. Chem. Phys.* **2009**, *130*, 074901.
29. Mohammad Haj-Abed, Engineering of Hyperbranched Polyethylene and its Future Applications, MSc thesis, McMaster University, **2008**.
30. Stiriba, S. E. Kautz; H.; Frey, H.; Hyperbranched molecular nanocapsules: comparison of the hyperbranched architecture with the perfect linear analogue. *J. Am. Chem, * **2002**, 210–224.
31. Zou J.; Zhao, Y.; Shi, W.; Shen, X.; and Nie, K.; Preparation and characters of hyperbranched polyester-based organic-inorganic hybrid material compared with linear polyester. *Polym. Adv. Technol.* **2005**, *16*, 55–60.
32. Zhu, L.; Shi, Y.; Tu, C.; Wang, R.; and Pang, Y.; Qiu, F.; Zhu, X.; Yan, D.; He, L.; Jin, C.; and Zhu, B.; Construction and application of a pH-sensitive nanoreactor via a double-hydrophilic multiarm hyperbranched polymer. *Langmuir.* **2010**, *26*(11), 8875–8881.
33. Zhou, Y.; Huang, W.; Liu, J.; Zhu, X.; and Yan, D.; Self-assembly of hyperbranched polymers and its biomedical applications. *Adv. Mater.* **2010**, *22*, 4567–4590.
34. Dodiuk-Kenig, H., and Lizenboim, K. … The effect of hyper-branched polymers on the properties of dental composites and adhesives. *J. Adhesion.* **2004**, 601–613.
35. Burkinshaw, S. M.; Froehling, P. E.; Mignanelli, M.; The effect of hyperbranched polymers on the dyeing of polypropylene fibres , Dyes and Pigments,
36. Mulkern, T. J. and Tan, N. C.; Processing and characterization of reactive polystyrene/hyperbranched polyester blends. *Polymer,* **2000**.
37. Boogh, L.; Pettersson, B.; and Månson, J. A. E.; Dendritic hyperbranched polymers as tougheners for epoxy resins. *Polymer,* **1999**, 201–212.

38. Zhang, F.; Chen, Y.; Lin, H.; and Lu, Y.; Synthesis of an amino-terminated hyperbranched polymer and its application in reactive dyeing on cotton as a salt-free dyeing auxiliary. *Coloration Technol.* **2007**, *123*, 351–357.
39. Zhang, F.; Chen, Y.; Lin, H.; Wang, H.; and Zhao, B.; HBP-NH2 grafted cotton fiber: preparation and salt-free dyeing properties. *Carbohyd. Polym.* **2008**, *74*, 250–256.
40. Zhang, F.; Chen, Y. Y.; Lin, H.; and Zhang, D. S.; Performance of cotton fabric treated with an amino-terminated hyperbranched polymer. *Fibers. Polym.* **2008**, *9*, 515–520.
41. Zhang, F.; Chen, Y.; Ling, H.; and Zhang, D.; Synthesis of HBP-HTC and its application to cotton fabric as an antimicrobial auxiliary. *Fibers. Polym.* **2009**, *10*, 141–147.
42. Hasanzadeh, M.; Moieni, T.; Hadavi Moghadam, B.; 'Modification of PET fabrics by hyperbranched polymer: A comparative study of artificial neural networks (ANN) and statistical approach. *J. Polym. Eng.* **2013**, 413–420.
43. Khatibzadeh, M.; Mohseni, M.; and Moradian, S.; Compounding fibre grade polyethylene terephthalate with a hyperbranched additive and studying its dyeability with a disperse dye. *Coloration Technol.* **2010**, *126*, 269–274.
44. Mahmoodi, R.; Dodel, T.; Moieni, T.; and Hasanzadeh, M.; Synthesis and Application of Amine-Terminated AB2-Type Hyperbranched Polymer to Polyamide-6 Fabrics. *J. Polym. Res.* **2013**, *7*, 14–19.
45. Hua, Y.; Zhang, F.; Lin, H.; Chen, Y.; Application of Amino-terminated Hyperbranched Polymers in Reactive Dyeing on Silk Fabric, Silk, **2008–2009**
46. Zhang, F.; Zhang, D.; Chen, Y.; and Lin, H.; The antimicrobial activity of the cotton fabric grafted with an amino-terminated hyperbranched polymer. *Cellulose.* **2009**, *16*, 281–288.
47. Burkinshaw, S. M.; Mignanelli, M., Froehling, P. E.; and Bide, M. J.; The use of dendrimers to modify the dyeing behaviour of reactive dyes on cotton. *Dyes. Pigm. 47,* **2000**, 259±267.
48. Gao, C.; Xu, Y.; Yan, D.; and Chen, W.; Water-Soluble Degradable Hyperbranched Polyesters: Novel Candidates for Drug Delivery?. *Biomacromolecules.* **2003**, *4*, 704–712.
49. Wu, D.; Liu, Y.; Jiang, X.; He, C.; Goh, S. H.; and Leong, K. W.; Hyperbranched Poly(amino ester)s with Different Terminal Amine Groups for DNA Delivery. *Biomacromolecules.* **2006**, *7*, 1879–1883.
50. Kima, T. H.; Cooka, S. E.; Arote, R. B.; Cho, M. H.; Nah, J. W.; Choi, Y. J.; and Cho, C. S.; A Degradable Hyperbranched Poly(ester amine) Based on Poloxamer Diacrylate and Polyethylenimine as a Gene Carrier. *Macromol. Biosci.* **2007**, *7*, 611–619.
51. Reul, R.; Nguyen, J.; and Kissel, T.; Amine-modified hyperbranched polyesters as non-toxic, biodegradable gene delivery systems. *Biomaterials.* **2009**, *30*, 5815–5824.
52. Goswami, A.; and Singh, A. K.; Hyperbranched polyester having nitrogen core: synthesis and applications as metal ion extractant. *React. Funct. Polym.* **2004**, *61*, 255–263.
53. Fang, J.; Kita, H.; and Okamoto, K.; Hyperbranched Polyimides for Gas Separation Applications. 1. Synthesis and Characterization. *Macromolecules.* **2000**, *33*, 4639–4646.
54. Seiler, M.; Köhler, D.; and Arlt, W.; Hyperbranched polymers: new selective solvents for extractive distillation and solvent extraction. *Sep. Purif. Technol.* **2003**, *30*, 179–197.
55. Itoh, T.; Hirai, K.; Tamura, M.; Uno, T.; Kubo, M.; and Aihara, Y.; Synthesis and characteristics of hyperbranched polymer with phosphonic acid groups for high-temperature fuel cells. J. Solid. State. Electrochem.
56. Tang, W.; Huang, Y.; Meng, W.; and Qing, F. L.; Synthesis of fluorinated hyperbranched polymers capable as highly hydrophobic and oleophobic coating materials. *Eur. Polym. J.* **2010**, *46*, 506–518.
57. Hong, X.; Chen, Q.; Zhang, Y.; and Liu, G.; Synthesis and Characterization of a Hyperbranched Polyol with Long Flexible Chains and Its Application in Cationic UV Curing. *J. Appl. Poly. Sci.* **2000**. *77*, 1353–1356.

58. Shokrieh, M. M. and Rafiee, R.; Prediction of Young's modulus of graphene sheets and carbon nanotubes using nanoscale continuum mechanics approach. *Mater. Des.* **2010**. *31*, 790–795.
59. Tserpes, K. I.; and Papanikos, P.; Finite element modeling of single-walled carbon nanotubes. *Composites: Part B*. **2005**, 36, 468–477.
60. Arani, A. G.; Rahmani, R.; and Arefmanesh, A.; Elastic buckling analysis of single-walled carbon nanotube under combined loading by using the ANSYS software. *Physica E*. **2008**, *40*, 2390–2395.
61. Ruoff, R. S.; Qian, D.; Liu, W. K.; Mechanical properties of carbon nanotubes: theoretical predictions and experimental measurements. *C. R. Physique.* **2003**, *4*, 993–1008.
62. Guo, X.; Leung, A. Y. T.; He, X. Q.; Jiang, H.; and Huang, Y.; Bending buckling of single-walled carbon nanotubes by atomic-scale finite element. *Composites: Part B*. **2008**, *39*, 202–208.
63. Xiao, J. R.; Gama, B. A.; Gillespie Jr, J. W.; An analytical molecular structural mechanics model for the mechanical properties of carbon nanotubes. *Int. J. Solid. Struct.* **2005**, *42* 3075–3092.
64. Ansari, R.; and Motevalli, B.; The effects of geometrical parameters on force distributions and mechanics of carbon nanotubes: A critical study, Commun. Nonlinear. *Sci. Numer. Simulat.* **2009**. *14*, 4246–4263.
65. Li, C.; and Chou, T. W.; Modeling of elastic buckling of carbon nanotubes by molecular structural mechanics approach. *Mech. Mater.* **2004**. *36*, 1047–1055.
66. Natsuki, T.; and Endo, M.; Stress simulation of carbon nanotubes in tension and compression. *Carbon*. **2004**, *42.*, 2147–2151,.
67. Ansari, R.; Sadeghi, F.; and Motevalli, B.; A comprehensive study on the oscillation frequency of spherical fullerenes in carbon nanotubes under different system parameters. *Commun. Nonlinear. Sci. Numer. Simulat.* **2013**, *18*, 769–784,.
68. Alisafaei, F.; Ansari, R.; Mechanics of concentric carbon nanotubes: Interaction force and suction energy. *Comput. Mater. Sci.* **2011**, *50*, 1406–1413.
69. natsuki, T.; Ni, Q. Q.; and Endo, M.; Vibrational analysis of fluid-filled carbon nanotubes using the wave propagation approach. *Appl. Phys. A.* **2008**, *90*, 441–445.
70. Joshi, U. A.; Sharma, S. C.; and Harsha, S. P.; Modelling and analysis of mechanical behavior of carbon nanotube reinforced composites. *Proc. IMechE, Part N: J. Nanoeng. Nanosyst.* **2011**, 225, 23–31.
71. Shokrieh, M. M.; and Rafiee, R.; A review of the mechanical properties of isolated carbon nanotubes and carbon nanotube composites. *Mech. Compos. Mater*. **2010**, *46*, 155–172.
72. Sadus, R. J.; Molecular Simulation of Fluids: Algorithms and Object- Orientation. Elsevier: Amsterdam; **1999**.
73. Allen, M. P.; and Tildesley, D. J.; Computer Simulation in Chemical Physics. Kluwer Academic Publishers Dordrecht; **1993**.
74. Haile, J. M.; Molecular Dynamics Simulation: Elementary Methods. Wiley: New York, **1997**.
75. Liu, W. K.; Karpov E. G.; and Park, H. S.; Nano Mechanics and Materials: Theory, Multiscale Methods and Applications. John Wiley & Sons; **2006**.
76. Rapaport, D. C.; The art of molecular Dynamics Simulation. Cambridge University Press: New York; **1995**.
77. Aerts, J.; Prediction of intrinsic viscosities of dendritic, hyperbranched and branched polymers. *Comput. Theor. Polym. Sci*. **1998**, *8*, 49–54.
78. Lescanec, R. L. and Muthukumar, M. () Configurational Characteristics and Scaling Behavior of Starburst Molecules: A Computational Study. *Macromolecules*, 23, 2280–2288., **1990**
79. Tu, C.; Le, B. D.; Todd, P. J.; Daivis, and Uhlherr, A.; Structural properties of hyperbranched polymers in the melt under shear via nonequilibrium molecular dynamics simulation. *J. Chem. Phys. 130*, **2009**, 074901.

80. Lyulin, Alexey V.; Adolf, David B.; and Davies, Geoffrey R.; Computer Simulations of Hyperbranched Polymers in Shear Flows. *Macromolecules.* **2001**, *34*, 3783–3789.
81. Konkolewicz, D.; Thorn-Seshold, O.; and Gray-Weale, A.; Models for randomly hyperbranched polymers: Theory and simulation. *J. Chem. Phys.* **2008**. *129*, 054901.
82. de Gennes, P. G.; and Hervet, H.; *J Phys Lett (Paris).* **1983**, *44*, L351.
83. Maiti, P. K.; Cagın, T.; Wang, G. F.; and Goddard, W. A.; *Macromolecules.* **2004**, *37*, 6236–6254.
84. Cagin, T.; Wang, G.; Martin, R.; Breen, N.; and Goddard III, W. A; Molecular modelling of dendrimers for nanoscale applications. *Nanotechnology.* **2000**, *11*, 77–84.
85. Cagin, T.; Wang, G.; Martin, R.; Zamanakos, G.; Vaidehi, N.; Mainz, D. T.; Goddard III, W. A.; Multiscale modeling and simulation methods with applications to dendritic polymers, *Comput. Theor. Poly. Sci.* 00 (2001) 000±000.
86. Lue, L.; Volumetric behavior of athermal dendritic polymers: Monte Carlo simulation. *Macromolecules.* **2000**, *33*, 2266–2272.
87. Cam Le, T.; Computational Simulation Of Hyperbranched Polymer Melts Under Shear, Ph.D. thesis, Swinburne University of Technology; **2010**.
88. Zacharopoulos, N.; Economou, I. G.; Morphology and organization of poly (propylene imine) dendrimers in the melt from molecular dynamics simulation. *Macromolecules*, **2002**.
89. Lee I.; Athey, B. D.; Wetzel, A. W.; Meixner, W.; and Baker, J. R.; *Macromolecules.* **2002**, *35*, 4510–4520.
90. Hana, M.; Chen, P.; and Yang, X.; Molecular dynamics simulation of PAMAM dendrimer in aqueous solution, Polymer, **2005**, *46,* 3481–3488.
91. Laferla, R.; Conformations and dynamics of dendrimers and cascade macromolecules. *J. Chem. Phys.* **1997**, *106*, 688–700.
92. Avila-Salas, F.; Sandoval, C.; Caballero, J.; Guiñez-Molinos, S.; Santos, L. S.; Cachau, R. E.; and González-Nilo, F. D.; Study of Interaction Energies between the PAMAM Dendrimer and Nonsteroidal Anti-Inflammatory Drug Using a Distributed Computational Strategy and Experimental Analysis by ESI-MS/MS. *J. Phys. Chem. B.* **2012**, *116*, 2031–2039.
93. He, X.; Liang, H.; and Pan, C.; Monte Carlo Simulation of Hyperbranched Copolymerizations in the Presence of a Multifunctional Initiator. *Macromol. Theory. Simul.* **2001**, *10*, 196–203.
94. Neelov, I. M. and Adolf, D. B. Brownian dynamics simulation of hyperbranched polymers under elongational flow. *J. Phys. Chem.B.* **2004**, *108*, 7627–7636.
95. Bosko, J. T.; and Prakash, J. R.; Effect of molecular topology on the transport properties of dendrimers in dilute solution at Theta temperature: A Brownian dynamics study. *J Chem. Phys.* **2008**. *128*, 034902.
96. Dalakoglou, G. K.; Karatasos, K.; Lyulin, S. V.; and Lyulin, A. V.; Shear-induced effects in hyperbranched-linear polyelectrolyte complexes. *J. Chem. Phys.* **2008**, *129*, 034901.
97. Konkolewicz, D.; Gilbert, G. R.; and Gray-Weale, A.; Randomly Hyperbranched Polymers. Physical Review Letters, *98*, 238301. **2007**.

CHAPTER 14

A STUDY ON INFLUENCE OF ELECTROSPINNING PARAMETERS ON THE CONTACT ANGLE OF THE ELECTROSPUN PAN NANOFIBER MAT USING RESPONSE SURFACE METHODOLOGY (RSM) AND ARTIFICIAL NEURAL NETWORK (ANN)

B. HADAVI MOGHADAM and M. HASANZADEH

14.1 INTRODUCTION

The wettability of solid surfaces is a very important property of surface chemistry, which is controlled by both the chemical composition and the geometrical microstructure of a rough surface [1–3]. When a liquid droplet contacts a rough surface, it will spread or remain as droplet with the formation of angle between the liquid and solid phases. Contact angle (CA) measurements are widely used to characterize the wettability of rough surface [3–5]. There are various methods to make a rough surface, such as electrospinning, electrochemical deposition, evaporation, chemical vapor deposition (CVD), plasma, and so on.

Electrospinning as a simple and effective method for preparation of nanofibrous materials have attracted increasing attention during the last two decade [6]. Electrospinning process, unlike the conventional fiber spinning systems (melt spinning, wet spinning, etc.), uses electric field force instead of mechanical force to draw and stretch a polymer jet [7]. This process involves three main components including syringe filled with a polymer solution, a high voltage supplier to provide the required electric force for stretching the liquid jet, and a grounded collection plate to hold the nanofiber mat. The charged polymer solution forms a liquid jet that is drawn toward a grounded collection plate. During the jet movement to the collector, the solvent evaporates and dry fibers deposited as randomly oriented structure on the surface of a collector [8–13]. The electrospun nanofiber mat possesses high specific surface area, high porosity, and small pore size. Therefore, they have been suggested as excellent candidate for many applications including filtration [14], multifunctional membranes [15], biomedical agents [16], tissue engineering scaffolds [17–18], wound dressings [19], full cell [20] and protective clothing [21].

The morphology and the CA of the electrospun nanofibers can be affected by many electrospinning parameters including solution properties (the concentration, liquid viscosity, surface tension, and dielectric properties of the polymer solution), processing conditions (applied voltage, volume flow rate, tip to collector distance, and the strength of the applied electric field), and ambient conditions (temperature, atmospheric pressure and humidity) [9–12].

In this work, the influence of four electrospinning parameters, comprising solution concentration, applied voltage, tip to collector distance, and volume flow rate, on the CA of the electrospun PAN nanofiber mat was carried out using response surface methodology (RSM) and artificial neural network (ANN). First, a central composite design (CCD) was used to evaluate main and combined effects of above parameters. Then, these independent parameters were fed as inputs to an ANN while the output of the network was the CA of electrospun fiber mat. Finally, the importance of each electrospinning parameters on the variation of CA of electrospun fiber mat was determined and comparison of predicted CA value using RSM and ANN are discussed.

14.2 EXPERIMENTAL

14.2.1 MATERIALS

PAN powder was purchased from Polyacryle Co. (Iran). The weight average molecular weight (M_w) of PAN was approximately 100,000 g/mol. N-N, dimethylformamide (DMF) was obtained from Merck Co. (Germany) and was used as a solvent. These chemicals were used as received.

14.2.2 ELECTROSPINNING

The PAN powder was dissolved in DMF and gently stirred for 24 h at 50 °C. Therefore, homogenous PAN/DMF solution was prepared in different concentration ranged from 10 wt.% to 14 wt.%. Electrospinning was set up in a horizontal configuration as shown in Figure 14.1. The electrospinning apparatus consisted of 5 ml plastic syringe connected to a syringe pump and a rectangular grounded collector (aluminum sheet). A high voltage power supply (capable to produce 0–40 kV) was used to apply a proper potential to the metal needle. It should be noted that all electrospinnings were carried out at room temperature.

A Study on Influence of Electrospinning Parameters

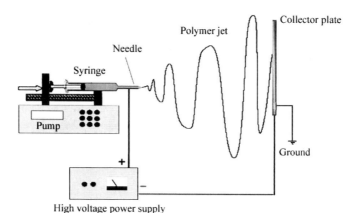

FIGURE 14.1 Schematic of electrospinning set up.

14.2.3 MEASUREMENT AND CHARACTERIZATION

The morphology of the gold-sputtered electrospun fibers were observed by scanning electron microscope (SEM, Philips XL-30). The average fiber diameter and distribution was determined from selected SEM image by measuring at least 50 random fibers. The wettability of electrospun fiber mat was determined by CA measurement. The CA measurements were carried out using specially arranged microscope equipped with camera and PCTV vision software as shown in Figure 14.2. The droplet used was distilled water and was 1 μl in volume. The CA experiments were carried out at room temperature and were repeated five times. All CAs measured within 20 s of placement of the water droplet on the electrospun fiber mat. A typical SEM image of electrospun fiber mat, its corresponding diameter distribution and CA image are shown in Figure 14.3.

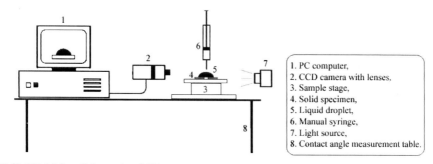

FIGURE 14.2 Schematic of CA measurement set up.

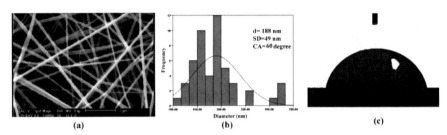

FIGURE 14.3 A typical (a) SEM image, (b) fiber diameter distribution, and (c) CA of electrospun fiber mat.

14.2.4 EXPERIMENTAL DESIGN

14.2.4.1 RESPONSE SURFACE METHODOLOGY

RSM is a combination of mathematical and statistical techniques used to evaluate the relationship between a set of controllable experimental factors and observed results. This optimization process is used in situations where several input variables influence some output variables (responses) of the system [22–23].

In the present study, CCD was employed to establish relationships between four electrospinning parameters and the CA of electrospun fiber mat. The experiment was performed for at least three levels of each factor to fit a quadratic model. Based on preliminary experiments, polymer solution concentration (wt.%), applied voltage (kV), tip to collector distance (cm), and volume flow rate (ml/h) were determined as critical factors with significance effect on CA of electrospun fiber mat. These factors were four independent variables and chosen equally spaced, while the CA of electrospun fiber mat was dependent variable. The values of -1, 0, and 1 are coded variables corresponding to low, intermediate, and high levels of each factor respectively. The experimental parameters and their levels for four independent variables are shown in Table 14.1. The regression analysis of the experimental data was carried out to obtain an empirical model between processing variables. The contour surface plots were obtained using Design-Expert software.

TABLE 14.1 Design of experiment (factors and levels)

Factor	Variable	Unit	Factor level −1	Factor level 0	Factor level 1
X_1	Solution concentration	(wt.%)	10	12	14
X_2	Applied voltage	(kV)	14	18	22
X_3	Tip to collector distance	(cm)	10	15	20
X_4	Volume flow rate	(ml/h)	2	2.5	3

The quadratic model, Eq. (14.1) including the linear terms, was fitted to the data.

$$Y = \beta_0 + \sum_{i=1}^{4} \beta_i x_i + \sum_{i=1}^{4} \beta_{ii} x_i^2 + \sum_{i=1}^{3} \sum_{j=2}^{4} \beta_{ij} x_i x_j \quad (14.1)$$

where, Y is the predicted response, x_i and x_j are the independent variables, β_0 is a constant, β_i is the linear coefficient, β_{ii} is the squared coefficient, and β_{ij} is the second-order interaction coefficients [22, 23].

The quality of the fitted polynomial model was expressed by the determination coefficient (R^2) and its statistical significance was performed with the Fisher's statistical test for analysis of variance (ANOVA).

14.2.4.2 ARTIFICIAL NEURAL NETWORK

ANN is an information processing technique, which is inspired by biological nervous system, composed of simple unit (neurons) operating in parallel (Figure 14.4). A typical ANN consists of three or more layers, comprising an input layer, one or more hidden layers, and an output layer. Every neuron has connections with every neuron in both the previous and the following layer. The connections between neurons consist of weights and biases. The weights between the neurons play an important role during the training process. Each neuron in hidden layer and output layer has a transfer function to produce an estimate as target. The interconnection weights are adjusted, based on a comparison of the network output (predicted data) and the actual output (target), to minimize the error between the network output and the target [6, 24–25].

In this study, feed forward ANN with one hidden layer composed of four neurons was selected. The ANN was trained using back-propagation algorithm. The same experimental data used for each RSM designs were also used as the input variables of the ANN. There are four neurons in the input layer corresponding to four electrospinning parameters and one neuron in the output layer corresponding to CA of electrospun fiber mat. Figure 14.4 illustrates the topology of ANN used in this investigation.

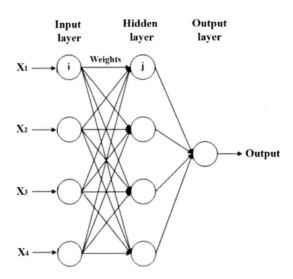

FIGURE 14.4 The topology of artificial neural network used in this study.

14.3 RESULTS AND DISCUSSION

This section discusses in details the wettability behavior of electrospun fiber mat concluded from CA measurements. The results of the proposed RSM and ANN models are also presented followed by a comparison between those models.

13.3.1 THE ANALYSIS OF VARIANCE (ANOVA)

All 30 experimental runs of CCD were performed according to Table 14.2. A significance level of 5 percent was selected; that is, statistical conclusions may be assessed with 95 percent confidence. In this significance level, the factor has significant effect on CA if the p-value is less than 0.05. On the other hand, when p-value is greater than 0.05, it is concluded the factor has no significant effect on CA.

The results of ANOVA for the CA of electrospun fibers are shown in Table 14.3. Equation (14.2) is the calculated regression equation.

TABLE 14.2 The actual design of experiments and response

No.	X_1 Concentration	X_2 Voltage	X_3 Distance	X_4 Flow rate	CA (°)
1	10	14	10	2	44±6
2	10	22	10	2	54±7
3	10	14	20	2	61±6
4	10	22	20	2	65±4
5	10	14	10	3	38±5
6	10	22	10	3	49±4
7	10	14	20	3	51±5
8	10	22	20	3	56±5
9	10	18	15	2.5	48±3
10	12	14	15	2.5	30±3
11	12	22	15	2.5	35±5

No.	X_1 Concentration	X_2 Voltage	X_3 Distance	X_4 Flow rate	CA (°)
13	12	18	20	2.5	30±4
14	12	18	15	2	33±4
15	12	18	15	3	25±3
16	12	18	15	2.5	26±4
17	12	18	15	2.5	29±3
18	12	18	15	2.5	28±5
19	12	18	15	2.5	25±4

20	12	18	15	2.5	24±3
21	12	18	15	2.5	21±3
22	14	14	10	2	31±4
23	14	22	10	2	35±5
24	14	14	20	2	33±6
25	14	22	20	2	37±4
26	14	14	10	3	19±3
27	14	22	10	3	28±3
28	14	14	20	3	39±5
29	14	22	20	3	36±4
30	14	18	15	2.5	20±3

$$CA = 25.80 - 9.89X_1 + 2.17X_2 + 4.33X_3 - 2.33X_4$$
$$-1.63X_1X_2 - 1.63X_1X_3 + 1.63X_1X_4 - 0.88X_2X_3 - 0.63X_2X_4 + 0.37X_3X_4 \quad (14.2)$$
$$+7.90X_1^2 + 6.40X_2^2 - 0.096X_3^2 + 2.90X_4^2$$

TABLE 14.3 Analysis of variance for the CA of electrospun fiber mat

Source	SS	DF	MS	F-value	Probe > F	Remarks
Model	4175.07	14	298.22	32.70	<0.0001	Significant
X_1-Concentration	1760.22	1	1760.22	193.01	<0.0001	Significant
X_2-Voltage	84.50	1	84.50	9.27	0.0082	Significant
X_3-Distance	338.00	1	338.00	37.06	<0.0001	Significant
X_4-Flow rate	98.00	1	98.00	10.75	0.0051	Significant
X_1X_2	42.25	1	42.25	4.63	0.0481	Significant
X_1X_3	42.25	1	42.25	4.63	0.0481	Significant
X_1X_4	42.25	1	42.25	4.63	0.0481	Significant
X_2X_3	12.25	1	12.25	1.34	0.2646	
X_2X_4	6.25	1	6.25	0.69	0.4207	Significant
X_3X_4	2.25	1	2.25	0.25	0.6266	

X_1^2	161.84	1	161.84	17.75	0.0008	Significant
X_2^2	106.24	1	106.24	11.65	0.0039	Significant
X_3^2	0.024	1	0.024	0.0026	0.9597	
X_4^2	21.84	1	21.84	2.40	0.1426	
Residual	136.80	15	9.12			
Lack of Fit	95.30	10	9.53	1.15	0.4668	

From the p-values presented in Table 14.3, it can be concluded that the p-values of terms X_3^2, X_4^2, X_2X_3, X_2X_4 and X_3X_4 is greater than the significance level of 0.05, therefore they have no significant effect on the CA of electrospun fiber mat. Since the above terms had no significant effect on CA of electrospun fiber mat, these terms were removed. The fitted equations in coded unit are given in Eq. (14.3).

$$CA = 26.07 - 9.89X_1 + 2.17X_2 + 4.33X_3 - 2.33X_4$$
$$-1.63X_1X_2 - 1.63X_1X_3 + 1.63X_1X_4$$
$$+9.08X_1^2 + 7.58X_2^2 \qquad (14.3)$$

Now, all the p-values are less than the significance level of 0.05. ANOVA showed that the RSM model was significant (p<0.0001), which indicated that the model has a good agreement with experimental data. The determination coefficient (R^2) obtained from regression equation was 0.958.

14.3.2 ARTIFICIAL NEURAL NETWORK

In this study, the best prediction, based on minimum error, was obtained by ANN with one hidden layer. The suitable number of neurons in the hidden layer was determined by changing the number of neurons. The good prediction and minimum error value were obtained with four neurons in the hidden layer. The weights and bias of ANN for CA of electrospun fiber mat are given in Table 4. The R^2 and mean absolute percentage error were 0.965 and 5.94 percent respectively, which indicates that the model was shows good fitting with experimental data.

TABLE 14.4 Weights and bias obtained in training ANN

Hidden layer	Weights		IW_{11} 1.0610	IW_{12} 1.1064	IW_{13} 21.4500	IW_{14} 3.0700
			IW_{21} −0.3346	IW_{22} 2.0508	IW_{23} 0.2210	IW_{24} −0.2224
			IW_{31} −0.6369	IW_{32} −1.1086	IW_{33} −41.5559	IW_{34} 0.0030
			IW_{41} −0.5038	IW_{42} −0.0354	IW_{43} 0.0521	IW_{44} 0.9560
	Bias		b_{11} −2.5521	b_{21} −2.0885	b_{31} −0.0949	b_{41} 1.5478
Output layer	Weights		LW_{11} 0.5658			
			LW_{21} 0.2580			
			LW_{31} −0.2759			
			LW_{41} −0.6657			
	Bias		b 0.7104			

14.3.3 EFFECTS OF SIGNIFICANT PARAMETERS ON RESPONSE

The morphology and structure of electrospun fiber mat, such as the nanoscale fibers and interfibrillar distance, increases the surface roughness as well as the fraction of contact area of droplet with the air trapped between fibers. It is proved that the CA decrease with increasing the fiber diameter [26], therefore the thinner fibers, due to their high surface roughness, have higher CA than the thicker fibers. Hence, we used this fact for comparing CA of electrospun fiber mat. The interaction contour plot for CA of electrospun PAN fiber mat are shown in Figure 14.5.

As mentioned in the literature, a minimum solution concentration is required to obtain uniform fibers from electrospinning. Below this concentration, polymer chain entanglements are insufficient and a mixture of beads and fibers is obtained. On the other hand, the higher solution concentration would have more polymer chain entanglements and less chain mobility. This causes the hard jet extension and disruption during electrospinning process and producing thicker fibers [27]. Figure 14.5a show the effect of solution concentration and applied voltage at middle level of distance (15 cm) and flow rate (2.5 ml/h) on CA of electrospun fiber mat. It is obvious that at any given voltage, the CA of electrospun fiber mat decrease with increasing the solution concentration.

Figure 14.5b shows the response contour plot of interaction between solution concentration and spinning distance at fixed voltage (18 kV) and flow rate (2.5 ml/h). Increasing the spinning distance causes the CA of electrospun fiber mat to increase. Because of the longer spinning distance could give more time for the solvent to evaporate, increasing the spinning distance will decrease the nanofiber diameter and increase the CA of electrospun fiber mat [28, 29]. As demonstrated in Figure 14.5b, low solution concentration cause the increase in CA of electrospun fiber mat at large spinning distance.

The response contour plot in Figure 14.5c represented the CA of electrospun fiber mat at different solution concentration and volume flow rate. Ideally, the volume flow rate must be compatible with the amount of solution removed from the tip of the needle. At low volume flow rates, solvent would have sufficient time to evaporate and thinner fibers were produced, but at high volume flow rate, excess amount of solution fed to the tip of needle and thicker fibers were resulted [28-30]. Therefore the CA of electrospun fiber mat will be decreased.

As shown by Eq. (14.4), the relative importance (RI) of the various input variables on the output variable can be determined using ANN weight matrix [31].

$$RI_j = \frac{\sum_{m=1}^{N_h}((|IW_{jm}|/\sum_{k=1}^{N_i}|IW_{km}|)\times|LW_{mn}|)}{\sum_{k=1}^{N_i}\left\{\sum_{m=1}^{N_h}((|IW_{km}|/\sum_{k=1}^{N_i}|IW_{km}|)\times|LW_{mn}|)\right\}}\times 100 \qquad (14.4)$$

where RI_j is the relative importance of the jth input variable on the output variable, N_i and N_h are the number of input variables and neurons in hidden layer, respectively ($N_i = 4$, $N_h = 4$ in this study), IW and LW are the connection weights, and subscript "n" refer to output response (n = 1) [31].

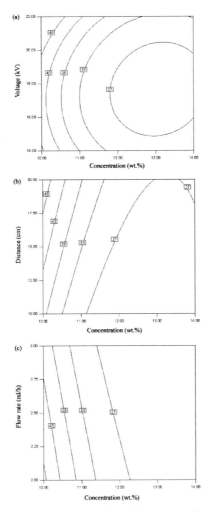

FIGURE 14.5 Contour plots for contact angle of electrospun fiber mat showing the effect of: (a) solution concentration and applied voltage, (b) solution concentration and spinning distance, (c) solution concentration and volume flow rate.

The RI of electrospinning parameters on the value of CA calculated by Eq. (14.4) and is shown in Figure 14.6. It can be seen that, all of the input variables have considerable effects on the CA of electrospun fiber mat. Nevertheless, the solution concentration with RI of 49.69 percent is found to be most important factor affecting the CA of electrospun nanofibers. These results are in close agreement with those obtained with RSM.

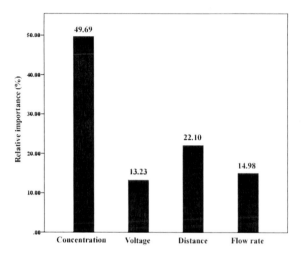

FIGURE 14.6 Relative importance of electrospinning parameters on the CA of electrospun fiber mat.

14.3.4 OPTIMIZING THE CA OF ELECTROSPUN FIBER MAT

The optimal values of the electrospinning parameters were established from the quadratic form of the RSM. Independent variables (solution concentration, applied voltage, spinning distance, and volume flow rate) were set in range and dependent variable (CA) was fixed at minimum. The optimal conditions in the tested range for minimum CA of electrospun fiber mat are shown in Table 14.5. This optimum condition was a predicted value, thus to confirm the predictive ability of the RSM model for response, a further electrospinning and CA measurement was carried out according to the optimized conditions and the agreement between predicted and measured responses was verified. Figure 14.7 shows the SEM, average fiber diameter distribution and corresponding CA image of electrospun fiber mat prepared at optimized conditions.

TABLE 14.5 Optimum values of the process parameters for minimum CA of electrospun fiber mat

Solution concentration (wt.%)	Applied voltage (kV)	Spinning distance (cm)	Volume flow rate (ml/h)	Predicted CA (°)	Observed CA (°)
13.2	16.5	10.6	2.5	20	21

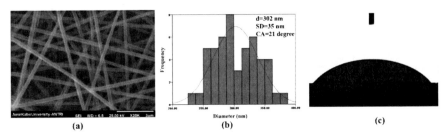

FIGURE 7 (a) SEM image, (b) fiber diameter distribution, and (c) CA of electrospun fiber mat prepared at optimized conditions.

14.3.5 COMPARISON BETWEEN RSM AND ANN MODEL

Table 14.6 gives the experimental and predicted values for the CA of electrospun fiber mat obtained from RSM as well as ANN model. It is demonstrated that both models performed well and a good determination coefficient was obtained for both RSM and ANN. However, the ANN model shows higher determination coefficient ($R^2 = 0.965$) than the RSM model ($R^2 = 0.958$). Moreover, the absolute percentage error in the ANN prediction of CA was found to be around 5.94 percent, while for the RSM model, it was around 7.83 percent. Therefore, it can be suggested that the ANN model shows more accurately result than the RSM model. The plot of actual and predicted CA of electrospun fiber mat for RSM and ANN is shown in Figure 14.8.

TABLE 14.6 Experimental and predicted values by RSM and ANN models

No.	Experimental	Predicted RSM	Predicted ANN	Absolute error (%) RSM	Absolute error (%) ANN
1	44	47	48	6.41	9.97
2	54	54	54	0.78	0.46
3	61	59	61	3.70	0.42
4	65	66	61	2.06	6.06
5	38	39	38	2.37	0.54
6	49	47	49	5.10	0.68
7	51	51	51	0.35	0.45
8	56	58	56	4.32	0.17
9	48	45	60	6.17	24.37
10	30	31	27	4.93	9.35
11	35	36	31	2.34	11.15
12	22	22	21	1.18	4.15
13	30	30	32	1.33	6.04

A Study on Influence of Electrospinning Parameters

14	33	28	33	13.94	0.60
15	25	24	25	5.04	0.87
16	26	26	26	0.27	1.33
17	29	26	26	10.10	9.16
18	28	26	26	6.89	5.91
19	25	26	26	4.28	5.38
20	24	26	26	8.63	9.77
21	21	26	26	24.14	25.45
22	31	30	31	2.26	0.57
23	35	31	35	10.34	0.66
24	33	36	32	8.18	2.18
25	37	37	37	0.59	0.34
26	19	29	21	52.11	10.23
27	28	30	30	7.07	8.20
28	39	34	31	12.05	21.30
29	36	35	36	1.72	0.04
30	20	25	20	26.30	2.27
R^2		0.958	0.965		
Mean absolute error (%)				7.83	5.94

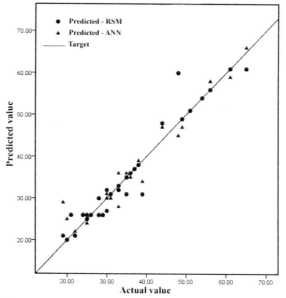

FIGURE 14.8 Comparison between the actual and predicted contact angle of electrospun nanofiber for RSM and ANN model

14.4 CONCLUSIONS

The morphology and properties of electrospun nanofibers depends on many processing parameters. In this work, the effects of four electrospinning parameters namely; solution concentration (wt.%), applied voltage (kV), tip to collector distance (cm), and volume flow rate (ml/h) on CA of PAN nanofiber mat were investigated using two different quantitative models, comprising RSM and ANN. The RSM model confirmed that solution concentration was the most significant parameter in the CA of electrospun fiber mat. Comparison of predicted CA using RSM and ANN were also studied. The obtained results indicated that both RSM and ANN model shows a very good relationship between the experimental and predicted CA values. The ANN model shows higher determination coefficient ($R^2=0.965$) than the RSM model. Moreover, the absolute percentage error of prediction for the ANN model was much lower than that for RSM model, indicating that ANN model had higher modeling performance than RSM model. The minimum CA of electrospun fiber mat estimated by RSM equation obtained at conditions of 13.2 wt.% solution concentration, 16.5 kV of the applied voltage, 10.6 cm of tip to collector distance, and 2.5 ml/h of volume flow rate.

KEYWORDS

- **Artificial neural network (ANN)**
- **Central composite design (CCD)**
- **Electrospinning parameters**
- **Response surface methodology (RSM)**

REFERENCES

1. Miwa, M.; Nakajima, A.; Fujishima, A.; Hashimoto K.; and Watanabe, T.; *Langmuir.* **2000**, *16*, 5754.
2. Öner D.; and McCarthy, T. J.; *Langmuir*, *16*, **2000**, 7777.
3. Abdelsalam, M. E.; Bartlett, P. N.; Kelf T.; and Baumberg, J.; *Langmuir.* **2005**, *21*, 1753.
4. Nakajima, A.; Hashimoto, K.; Watanabe, T.; Takai, K.; Yamauchi G.; and Fujishima, A.; *Langmuir.* **2000**, *16*, 7044.
5. Zhong, W.; Liu, S.; Chen, X.; Wang Y.; and Yang, W.; *Macromolecules* **2006**, *39*, 3224.
6. Shams Nateri A.; and Hasanzadeh, M.; *J. Comput. Theor. Nanosci.* **2009**, *6*, 1542.
7. Kilic, A.; Oruc F.; and Demir, A.; *Text. Res. J.* **2008**, *78*, 532.
8. Reneker D. H.; and Chun, I.; *Nanotechnology.* **1996**, *7*, 216.
9. Shin, Y. M.; Hohman, M. M.; Brenner M. P.; and Rutledge, G. C.; *Polymer.* **2001**, *42*, 9955.
10. Reneker, D. H.; Yarin, A. L.; Fong H.; and Koombhongse, S.; *J. Appl. Phys.* **2000** *87*, 4531.
11. Zhang, S.; Shim W. S.; and Kim, J.; *Mater. Design. 30*, **2009**, 3659.
12. Yördem, O. S.; Papila M.; and Menceloğlu, Y. Z.; *Mater. Design.* **2008**, *29*, 34.

13. Chronakis, I. S.; *J. Mater. Process. Tech.* **2005,** *167*, 283.
14. Dotti, F.; Varesano, A.; Montarsolo, A.; Aluigi, A.; Tonin C.; and Mazzuchetti, G.; *J. Ind. Text.* **2007,** *37*, 151.
15. Lu, Y.; Jiang, H.; Tu K.; and Wang, L.; *Acta Biomater.* **2009, 5, 1562.**
16. Lu, H.; Chen, W.; Xing, Y.; Ying D.; and Jiang, B.; *J. Bioact. Compat. Pol.* **2009**, *24*,158.
17. Nisbet, D. R.; Forsythe, J. S.; Shen, W.; Finkelstein D. I.; and Horne, M. K. ; *J. Biomater. Appl.* **2009,** *24*, 7.
18. Ma, Z.; Kotaki, M.; Inai R.; and Ramakrishna, S.; *Tissue Eng.* **2005,** *11*, 101.
19. Hong, K. H.; *Polym. Eng. Sci.*, **2007**. *47*, 43.
20. Zhang W.; and Pintauro, P. N.; *ChemSusChem.* **2011**, *4*, 1753.
21. Lee S.; and Obendorf, S. K.; *Text. Res. J.* **2007**, *77*, 696.
22. H. R.; Myers, Montgomery D. C.; and Anderson-cook, C. M.; Response Surface Methodology: Process and Product Optimization Using Designed Experiments, 3rd ed.. John Wiley and Sons: USA; **2009.**
23. Gu, S. Y.; Ren J.; and Vancso, G. J.; *Eur. Polym. J.* **2005,** *41*, 2559.
24. Dev, V. R. G.; Venugopal, J. R.; Senthilkumar, M.; Gupta D. ;and Ramakrishna, S.; *J. Appl. Polym. Sci.* **2009,** *113*, 3397.
25. Galushkin, A. L.; Neural Networks Theory. Springer: Moscow Institute of Physics & Technology; **2007.**
26. Ma, M.; Mao, Y.; Gupta, M.; Gleason K. K.; and Rutledge, G. C. : *Macromolecules.* **2005**, *38*, 9742.
27. Haghi A. K.; and Akbari, M.; *Phys. Status. Solidi. A.* **2007,** *204*, 1830.
28. Ziabari, M.; Mottaghitalab V.; and Haghi, A. K.; In W. N. Chang (Ed) Nanofibers: Fabrication, Performance, and Applications. Nova Science Publishers: USA; **2009.**
29. Ramakrishna, S.; Fujihara, K.; Teo, W. E.; Lim T. C.; and Ma, Z.; An Introduction to Electrospinning and Nanofibers. World Scientific Publishing: Singapore; **2005.**
30. Zhang, S.; Shim W. S.; and Kim, J.; *Mater. Design,* **2009,** *30*, 3659.
31. Kasiri, M. B.; Aleboyeh H.; and Aleboyeh, A.; *Environ. Sci. Technol.* **2008** *42*, 7970.

CHAPTER 15

FABRICATION AND CHARACTERIZATION OF THE METAL NANOSIZED BRANCHED STRUCTURES AND THE COMPOSITE NANOSTRUCTURES GROWN ON INSULATOR SUBSTRATES BY THE EBID PROCESS

GUOQIANG XIE, MINGHUI SONG, KAZUO FURUYA, and AKIHISA INOUE

Using an electron-beam-induced deposition (EBID) process in a transmission electron microscope, we fabricated self-standing metal nanosized branched structures including nanowire arrays, nanodendrites, and nanofractal-like trees, as well as their composite nanostructures with controlled size and position on insulator (SiO_2, Al_2O_3) substrates. The fabricated nanostructures were characterized with high-resolution transmission electron microscopy and X-ray energy dispersive spectroscopy. The growth mechanism was discussed. Effect of the electron-beam accelerating voltage on crystallization of the nanostructures was investigated. The nanostructures of the different morphologies were obtained by controlling the intensity of the electron beam during the EBID process. A mechanism for the growth and morphology of the nanostructures was proposed involving charge-up produced on the surface of the substrate, and the movement of the charges to and charges accumulation at the convex surface of the substrate and the tips of the deposits. High-energy electron irradiation enhanced diffusion of the metallic atoms in the nanostructures and hence promoted crystallization. More crystallized metal nanobranched structures were achieved by the EBID process using high energy electron beams.

15.1 INTRODUCTION

Metal nanostructures are of great importance in nanotechnology due to their potential applications as building blocks in optoelectronic devices, catalysis, chemical sensors, and other areas [1–3]. The formation of nanostructures with controlled size and morphology has been the focus of intensive research in recent years. Such nanostructures are important in the development of nanoscale devices and in the exploitation of the properties of nanomaterials [4–7]. Many fabrication methods including chemical vapor decomposition [8], the arc-discharge method [9], evaporation [10], hydrothermal reactions [11], and so on, have been developed for the production of nanomaterials with controlled sizes and shapes. However, up to date, fabrication of position controllable nanostructures at selected positions on a substrate is still a challenge.

Among the methods, electron-beam-induced deposition (EBID) process is one of the most promising techniques to fabricate small-sized structures on substrates. An advantage of this process is that the deposited position can be controlled. In this approach, an electron beam in a high vacuum chamber is focused on a substrate surface on which precursor molecules, containing the element to be deposited (e.g., organometallic compound or hydrocarbon), are adsorbed. As a result of complex beam-induced surface reactions, the precursor molecules absorbed in or near to the irradiated area, are dissociated into nonvolatile and volatile parts by energetic electrons. The nonvolatile materials are deposited on surface of the substrate, while the volatile components are pumped away. Due to the controllability of electron beam, fabrication of position controllable nanostructures can be realized. Using the technique, a variety of nanometer-sized structures, such as nanodots, nanowires, nanotubes, nanopatterns, two or three dimensional nano-objects, and so on, have been fabricated [12–22]. Due to easy to receive a stable fabrication condition, conductive substrates are generally used in these fabrications [16]. As a result, compact structures are usually fabricated with this process.

In the case using nonconductive substrates, the nanofabrication with the EBID technique is also very important for the technique to be applied in technology. The accumulation of charges readily occurs on an insulator substrate during the EBID process. The deposition of novel structures may occur under this condition. The growth of carbon fractal-like structures has already been observed on insulator substrates under electron-beam irradiation due to the existence of residual pump oil remaining in the atmosphere of the vacuum chamber [23–25], but the growth of metallic nanostructures on insulator substrates using EBID process was first reported recently by Song et al. [26]. By using insulator substrates such as SiO_2 [27–30] and Al_2O_3 [26, 31–35], characteristic morphologies of nanostructures, such as arrays of nanowhiskers (or nanowires), arrays of nanodendrites, and fractal treelike structures, have been fabricated in transmission electron microscopes (TEMs) by the EBID process [26–35]. The typical size of the diameter of a nanowhisker, the tip of a nanodendrite, and the tip of a nanotree is about 3 nm, and is almost completely independent of the size of the electron beam. In the present review we reported the fabrication and characterization of the metal nanosized branched structures and the composite nanostructures grown on insulator substrates by the EBID process.

15.2 FABRICATION AND CHARACTERIZATION OF NANOSIZED BRANCHED STRUCTURES

Thin films of SiO_2 and Al_2O_3 were irradiated by an electron beam in transmission electron microscope (TEM) with organometallic precursor gases at room temperature. The EBID process was carried out using JEM-2010 or JEM-ARM1000 TEMs made by JEOL Co., Ltd. The energy of the electron beam used was from 200 to 1,000 keV. The pressure in the specimen chambers of the TEMs was on the order of 10^{-5} to 10^{-6} Pa. Insulator SiO_2 and Al_2O_3 substrates suitable for TEM observation were used. An organometallic precursor gas was introduced near the substrate using a special designed system comprising a nozzle with a diameter smaller than 0.1 mm and a reservoir containing the precursor [26]. Tungsten hexacarbonyl ($W(CO)_6$) or (Methylcyclopentadienyl)trimethylplatinum ($Me_3MeCpPt$) powders were used as the precursors for fabricating tungsten (W)-containing, or platinum (Pt)-containing nanostructures, respectively. The vapor pressures of these precursors are several Pa at room temperature. These precursors have been previously typically used to fabricate small objects on conductive substrates by EBID process [16]. The intensity of the electron beam for the EBID process was from 3.2 to 111.7×10^{18} e cm^{-2} s^{-1} (current density: 0.51–17.9 A cm^{-2}), which was estimated by measuring the total intensity and the size of the electron beam under operating conditions. The fabricated structure was characterized *in situ* or after fabrication in a TEM. The experiments were performed at room temperature.

Figure 15.1 shows a series of micrographs of growth process of the W-nanodendrite structures using the TEM at an accelerating voltage of 1,000 kV during the electron-beam irradiation with an electron-beam current density of 1.6 A cm^{-2} [29]. Figure 15.1a presents a micrograph before the electron-beam irradiation. No deposits are observed at the SiO_2 substrate. Figure 15.1b shows a micrograph taken 3 s after the beginning of the irradiation at a fluence of 3.0×10^{19} e cm^{-2}. The electron beam irradiated the specimen from a direction perpendicular to the plane of the micrograph. Whisker-like deposits begin to nucleate and grow on the surface of the substrate. The deposits are about 3 nm in diameter and about 5 nm in length. The structures grow self-standing at positions separated from each other at a distance of several nanometers, and in parallel and nearly perpendicular to the surface of the substrate within the irradiated area. The whisker-like deposits grow longer and also denser under further electron-beam irradiation, and new nucleation and growth deposits are not observed. Figure 15.1c shows a micrograph taken 3 min after the beginning of irradiation at a fluence of 1.8×10^{21} e cm^{-2}. Similarity in length and an almost even thickness of all deposits are observed. Branching is observed at the tips of the deposits. The fabricated deposits have a nanodendrite structure. With further electron-beam irradiation, new branches grew from the tips of the grown branches. This process continued as long as the irradiation continued. Figure 15.1d shows a micrograph taken 10 min after the beginning of irradiation at a fluence of

6.0×10^{21} e cm^{-2}. The nanodendrite structures grow at both the tip and trunk. The diameter of the nanodendrite structures increased with the decrease in distance from the substrate, but is about 3 nm at the tips. The average length of the nanodendrite structures increased to about 51.4 nm. Figure 15.2 shows the length of nanodendrite structures grown at various electron-beam irradiation fluences (namely, electron-beam irradiation time). The length of nanodendrite structures increases approximated linearly with electron-beam irradiation fluence [29].

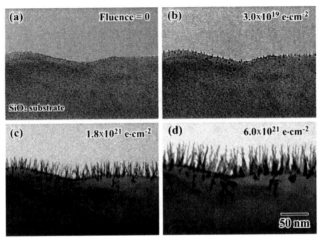

FIGURE 15.1 Bright-field TEM images of W-nanodendrite structures grown on surface of SiO2 substrate by the EBID process at an electron beam accelerating voltage of 1000 kV with a current density of 1.6 A cm^{-2} after beginning of electron beam irradiation. (a) 0 s; (b) 3 s; (c) 3 min; and (d) 10 min [29].

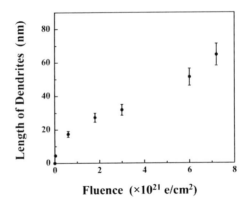

FIGURE 15.2 Relationship between length of the W-nanodendrite structures grown on surface of SiO$_2$ substrate by the EBID process at an electron beam accelerating voltage of 1000 kV with a current density of 1.6 A cm^{-2} and electron beam irradiating fluence [29].

Fabrication and Characterization of the Metal Nanosized 283

Furthermore, the morphology of the nanostructures grown can also be controlled by the intensity of the electron beam [26, 36]. Figure 15.3a shows a micrograph of an array of W nanowhiskers (or nanowires) fabricated on an Al_2O_3 substrate irradiated with a 200 keV electron beam at an intensity of 4.7×10^{18} e cm^{-2} s^{-1} (0.75 A cm^{-2}) to about 120 s [26]. The growth speed was very low. By increasing the current density to a higher value such as 1.6 A cm^{-2}, the growth speed was increased. The nanowhiskers were longer and also denser than those grown at the lower current density with an almost even thickness. Branching is observed to take place at the tips of the nanowhiskers, and the morphology becomes dendritic structures [27, 29]. A further increase in the current density to 17.9 A cm^{-2} resulted in extensive branching and the formation of complicated nanotree structures (Figure 15.3c) [26].

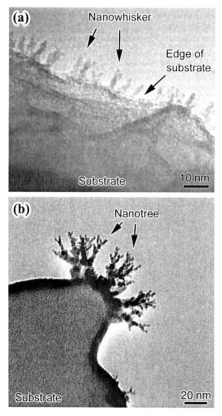

FIGURE 15.3 (a) W nanowhisker arrays grown on the surface of an Al_2O_3 substrate by the EBID process at an accelerating voltage of 200 kV. The current density of the electron beam was 0.75 A cm^{-2} and the irradiation time was 120 s. (b) Fractal-like W nanotrees grown on the surface of an Al_2O_3 substrate by the EBID process. The current density of the electron beam was 17.9 A cm^{-2} and the irradiation time was 10 s [26].

The composition of the fabricated nanodendrites is important, since they are related to the physical and chemical properties of deposits. The chemical composition of the nanodendrites was examined using an X-ray energy dispersive spectroscopy (EDS). Figure 15.4 shows the spectra taken using EDS from an as-deposited nanodendrite fabricated at an electron-beam irradiating fluence of 6.0×10^{21} e cm^{-2} after detaching the precursor source [29]. Figure 15.4a presents a spectrum taken from the tips of a nanodendrite structure, and Figure 15.4b shows that taken from its trunk. The size of the electron beam for the analysis using EDS is about 10 nm. The peaks of tungsten dominate these spectra. No obvious differences in the composition between the tip and the trunk of the nanodendrite structure were observed. On the basis of the analyses using EDS, a relative content of 89.4 percent W compared with 10.6 percent C for the spectrum from the tips is obtained. The relative content of W is higher than the reported values for W-deposits fabricated using the same precursor but at a lower voltage of the electron beam (25 kV) by EBID process, which was 75 percent [17].

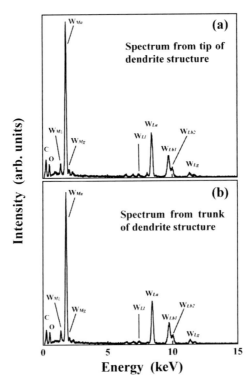

FIGURE 15.4 EDS spectra taken from the nanodendrite grown on surface of SiO$_2$ substrate by the EBID process at an accelerating voltage of 1000 kV with a current density of 1.6 A cm^{-2} to an electron-beam irradiating fluence of 6.0×10^{21} e cm^{-2}. (a) at tip; and (b) at trunk [29].

Fabrication and Characterization of the Metal Nanosized

Using (Methylcyclopentadienyl)trimethylplatinum (Me$_3$MeCpPt) powder as a precursor, one can obtain Pt nanodendritic structures by the EBID process. Figure 15.5 shows TEM micrographs of Pt nanodendrite-like structures grown on an Al$_2$O$_3$ substrate in a 200 kV TEM [34]. The electron beam is defocused to a size of about 600 nm, corresponding to a current density of 0.52 A cm^{-2}. The irradiated time is 1 min. The irradiating fluence of the electron beam is 2.0 × 10^{20} e cm^{-2}. The edge of the electron beam is indicated by arrows in Figure 15.5a. The contrast inside the irradiated area is obviously dark. The size of dendrite structures is much smaller than the diameter of the electron beam. The dendrites show a tendency to grow at the edge of the substrate. The dendrites have branches at the tip, as observed in Figure 15.5b. The diameter of the nanodendrites become thicker near the substrate, which implies that the deposition takes place at both tip and trunk part. The typical thickness of the tips is less than 3 nm, as observed in Figure 15.5b. TEM observation, diffraction pattern analyses, and EDS analyses indicated that a nanodendrite structure with a high Pt content was formed.

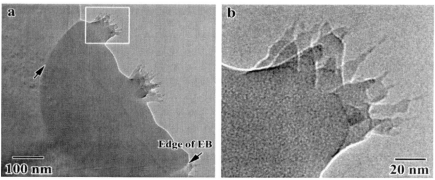

FIGURE 15.5 (a) Bright-field TEM micrographs of Pt-nanodendrite structures grown on the surface of an Al$_2$O$_3$ substrate in a 200 kV TEM with an EBID process 1 min after the start of the electron beam irradiation to a fluence of 2.0×10^{20} e cm^{-2}. (b) Enlargement of the square area in (a), showing the nanodendrites in more detail [34].

Figure 15.6 shows a high-resolution TEM (HRTEM) micrograph of the fabricated Pt nanodendrite [34]. It is confirmed that fcc Pt nanoparticles and an amorphous part are contained in the structure. The morphology of the Pt nanodendrite is considerably different from that of the W nanodendrite fabricated on an Al$_2$O$_3$ substrate, suggesting that the growth of a nanodendrite depends largely on the properties of the precursor.

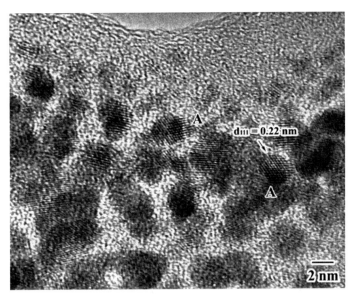

FIGURE 15.6 An HRTEM micrograph of a nanodendrite structure fabricated on an electron beam irradiating fluence of 2.6×10^{21} e cm^{-2} [34].

15.3 GROWTH MECHANISM OF NANOSIZED BRANCHED STRUCTURES

It is well known that the EBID process is caused by the dissociation of molecules adsorbed to a surface by energetic electron beam. In this approach, a molecule of a precursor is first adsorbed on surface of a substrate and then decomposed into volatile and nonvolatile parts by further irradiation of the energetic beam. The nonvolatile fraction accumulates to form a deposit, while the volatile component is pumped away by the vacuum system. The dissociation mechanism is complex and not fully understood until the present time because of the huge number of excitation channels available even for small molecules. The details involving the decomposition have been argued to relate to secondary electrons on surface of a specimen produced by incident electron beam and backscattered electrons in an EBID process [17, 19].

If an EBID process is carried out on the surface of an electric grounded conductive substrate, the molecules absorbed may move or not move, but distribute randomly and be decomposed on the surface. Therefore a compact deposit is usually formed. On the other hand, in a case that the deposition is conducted on an insulator substrate, charge-up may take place on the surface due to emission of secondary electrons under energetic beam irradiation. When specimens are exposed to electron bombardments, the molecules absorbed to a surface of a substrate or near the surface in the irradiated area may be polarized by irradiation of an incident electron beam.

The irradiated area on the insulator is easily charged, forms local electric potential. It is reasonable to consider that the distribution of charges due to charge-up on the surface may be not even in a nanometer scale. This unevenness may be resulted from the nanoscaled unflatness or atomic steps on surface. Charges may accumulate at some places to some extent. Figure 15.7a shows the growth mechanism of a nanowhisker array [26]. The intensity of the electron beam is weak, but the accumulation of charges occurs. The distribution of the charges on the surface is assumed to be uneven at a nanometer scale. Charges may accumulate at some places (charge centers), where an electric field is generated. The dark dots represent the nonvolatile fraction of the precursor molecules of which the deposits are composed, while the brighter dots represent the volatile fraction. Because the intensity of the electron beam is weak, the adsorbed molecules can move around considerably before being decomposed. The precursor molecules adsorbed on the surface may be attracted to the charge centers since the molecules are easily polarized due to the weak bonding between the atoms of the molecules. The precursor molecules then decompose and form a deposit. After a deposit is formed, the charges on the surface of the substrate or the deposit tend to move and accumulate at the tip of the deposit, since W deposits have been reported to be conductive [17, 19]. Thus, molecules are attracted to the tip, and the deposit increases in length upon further electron-beam irradiation. The formation of W nanodot deposits on the surface of a SiO_2 substrate and the growth of W nanowhiskers from these nanodots [30] are consistent with the above mechanism.

Figures 15.7b, c schematically show the growth conditions of nanodendrites and fractal-like nanotrees, respectively [26]. When the current density of the electron beam is increased to a definite extent, an adsorbed molecule on the surface may be decomposed in a very short period before it can move. Hence, a compact layer is deposited on the surface. This results in the entire irradiated area becoming conductive. After this stage, the charges on the surface can move a long distance to the substrate edge, particularly when the surface is convex. A strong electric field is thus generated near the convex surface. Once a deposit grows away from the surface at a point, charges then accumulate there. This deposit thus grows preferentially because the charges attract precursor molecules. When the current density is moderate, the electric field generated near a tip may not strongly affect the trajectory of a molecule. Therefore, the deposit grows at both its tip and its main body because the molecules arrive at the tip and the main body at the same time. Therefore, the main body of the deposit increases in thickness with time and develops a dendritic morphology. On the other hand, if the current density of the electron beam is very strong, the generated electric field will become sufficiently strong to affect the trajectory of a molecule near the tip of a deposit. The molecules are thus likely to be attracted to the tip and are not easily adsorbed on the main body of the deposit. Therefore, a fractal-like tree morphology is formed.

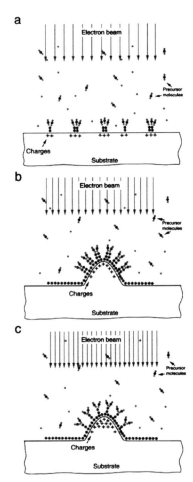

FIGURE 15.7 Schematic drawings showing the growth process of the nanostructures with different morphologies on insulator substrates by the EBID process. The dark dots represent the nonvolatile fraction of the precursor and the deposit, while the brighter dots represent the volatile fraction of the precursor. (a) The growth of nanowhiskers (or nanowires) inside the area of irradiation with a weak electron beam; (b) the growth of nanodendrites on a convex surface under irradiation with a moderate electron beam; and (c) the growth of fractal-like nanotrees on a convex surface under irradiation with a strong electron beam [26].

15.4 EFFECT OF ACCELERATING VOLTAGE ON CRYSTALLIZATION OF THE NANOSTRUCTURES

The fabricated nanostructure morphologies at various electron-beam accelerating voltages have been investigated with a conventional TEM. Figure 15.8 shows a set

Fabrication and Characterization of the Metal Nanosized 289

of TEM micrographs of dendrite-like structures grown on SiO_2 substrate at different electron-beam accelerating voltages to an electron-beam fluence of 6.0×10^{21} e cm^{-2} [27]. The electron beam irradiated the specimen from a direction perpendicular to the plane of the micrograph. The obvious difference in dendrite morphology and their composites for the deposits fabricated by various accelerating voltages is not observed. The length of the nanodendrite structures increases with an increase in electron-beam irradiating fluence, and has not obvious effect with the increase of accelerating voltage, as shown in Figure 15.9.

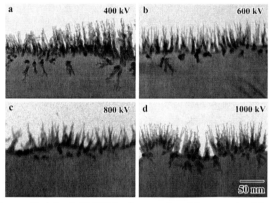

FIGURE 15.8 A set of bright-field TEM micrographs of the nanodendrite structures grown on SiO_2 substrate irradiated to an electron beam fluence of 6.0×10^{21} e cm^{-2} at different electron beam accelerating voltages. (a) 400 kV; (b) 600 kV; (c) 800 kV; (d) 1000 kV. The current density of electron beam for irradiating the specimens was 1.1 A cm^{-2} for 400 kV, and 1.6 A cm^{-2} for 600, 800 and 1000 kV, respectively [27].

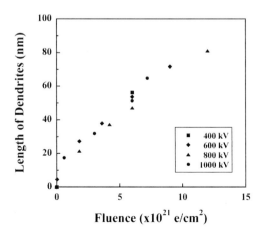

FIGURE 15.9 Dependence of the length of nanodendrites on electron beam irradiating fluence at different electron beam accelerating voltages [27].

The effect of electron-beam accelerating voltage on crystallization of the nanodendrite structures has been investigated with an HRTEM. Figure 15.10 shows a series of HRTEM micrographs of some branches of nanodendrite structures fabricated with various accelerating voltages to a fluence of 6.0 × 10^{21} e cm^{-2} [28]. Figure 15.10a is an HRTEM micrograph of some branches of nanodendrites grown with an accelerating voltage of 400 kV. Lattice fringes are observed at the most places. This indicates that they are crystal in several nanometers. The largest and also the most observed lattice spacing measured from micrographs is 0.22 nm. It is close to the lattice spacing, 0.224 nm, of {110} of bcc W crystals with a deviation smaller than 5 percent. Moreover, lattice fringes with spacing d = 0.22 nm have inter-fringe angle of 60 degrees (grain A), as well as that of 90 degrees (grain B), as indicated in Figure 15.10. They agree to zone axis of [111] and [001] of bcc W structure, respectively. Combined with nanometer-sized area diffraction pattern analyses and EDS analyses, it is clarified that these crystals are W grains in bcc structure. Furthermore, lattice fringes cannot be clearly observed in some places, indicating that a large part of them is in amorphous state, as shown by arrows in Figure 15.10. Figure 15.10b shows a micrograph of some branches of nanodendrites grown with an accelerating voltage of 600 kV. The fraction of amorphous state in the as-fabricated nanodendrites decreases obviously. For the dendrites fabricated with an accelerating voltage of 800 kV, as shown in Figure 15.10c, the amorphous state is observed only in a few places. Figure 15.10d shows an HRTEM micrograph of some branches of nanodendrites grown with an accelerating voltage of 1,000 kV. Lattice fringes are observed clearly in almost all of the grains except in thick region, where small crystals in random orientations overlapped each other, so that their lattice fringes cannot be clearly observed. Therefore, it is indicated that the fraction of amorphous state in the as-fabricated nanodendrites decreases with an increase in electron-beam accelerating voltage.

It is known that the growth of a W nanodendrite during the EBID process is controlled by a random accumulation of nonvolatile elements with thermal energy. Therefore the element does not have enough energy to move and form crystalline grains. The electron beam during the EBID process may transfer some energy to the element, but it is not enough for total deposit to transform into crystalline state with a 400 keV electron beam. Therefore, some amorphous structures are remained in deposits (refer to Figure 15.10a. This also may explain the reason that an amorphous state has often been obtained in the W-deposits fabricated by a lower energy EBID [19, 26]. With an increase in electron-beam accelerating voltage, the fraction transformed into crystalline state in the deposits increases, and the fraction of amorphous structure decreases. The results clearly indicate that high-energy electron irradiation enhances crystallization of an amorphous structure. Song et al. [37] has reported that the effect of 1 MeV electron-beam irradiation on crystallization of nanometer-

Fabrication and Characterization of the Metal Nanosized   291

sized W-dendritic structure fabricated on Al_2O_3 substrates with the EBID process at an accelerating voltage of 200 kV. After irradiation at room temperature for 100 minutes at a current density of 6.4 A cm^{-2}, almost all the grains crystallized. The nanodendrite structure also changes morphology and shrinks its size. In general, there are two important factors in the occurrence of electron irradiation induced crystallization of an amorphous state: (1) the promotion of atomic diffusion by electron irradiation in an amorphous state; and (2) the high stability of crystalline phase against electron irradiation [38]. When the two factors are satisfied simultaneously, electron irradiation induced crystallization of the amorphous structure. The change in morphology and shrinkage in size may be resulted from crystallization and sputtering during the irradiation.

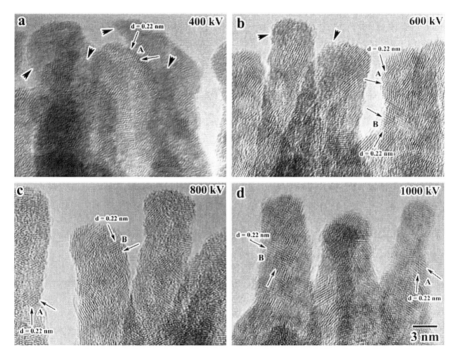

FIGURE 15.10 A series of HRTEM micrographs of some branches of nanodendrites grown on SiO_2 substrate irradiated to a fluence of 6.0×10^{21} e cm^{-2} with different electron beam accelerating voltages. (a) 400 kV; (b) 600 kV; (c) 800 kV; (d) 1000 kV. Arrows indicate lattice fringes in grains A and B, which have inter-fringe angles of 60 (A) and 90 (B) degrees, respectively. The current density of electron beam for irradiating the specimens was 1.1 A cm^{-2} for 400 kV, and 1.6 A cm^{-2} for 600, 800 and 1000 kV, respectively [28].

15.5 COMPOSITE NANOSTRUCTURES OF NANOPARTICLES/NANODENDRITES

Complex-shape nanostructures have attracted great interest because advanced functional materials might be emerged if they can be formed with well-defined three-dimensional (3D) architectures [39–41]. Furthermore, the impact is greater for multielement systems, as in the case of advanced nanostructured systems like Au/Cu, Pt/C composite materials [42, 43] and so on, which are used in catalysis, sensors, energy sources [44, 45] and in many other applications. It has been demonstrated that bimetallic bonding can induce significant changes in the properties of the surface, producing in many cases catalysts that have superior activity and/or selectivity [42, 43, 46, 47].

The nanostructures fabricated on insulator substrates by the EBID process are good base materials for the fabrication of complex-shape nanostructures because of their superior features. Pt nanoparticles or Au nanoparticles were deposited on W nanodendrites fabricated on insulator Al_2O_3 or SiO_2 substrates, and a composite nanostructure consisting of Pt nanoparticles and W nanodendrites (Pt-nanoparticle-decorated W nanodendrites, or Pt nanoparticle/W nanodendrite), or Au nanoparticle and W nanodendrites were fabricated by ion-sputtering [33]. Figure 15.11 shows the Pt nanoparticle/W nanodendrite composite structures. Figure 15.11a is a bright-field TEM image of the as-fabricated composite structure produced with an electron-beam irradiation fluence of 5.0×10^{21} cm^{-2} and an ion sputtering time of 40 s. The Pt-nanoparticles are nearly uniformly distributed on the W-nanodendrites. Figure 15.11b shows an HRTEM image of a tip of the composite nanostructure. Lattice fringes are observed in the image. By measuring the lattice spacing and the inter-fringe angle from this image and other images, one can demonstrate that the as-fabricated composite nanostructures consist of nanocrystals of Pt and W. By using selected-area diffraction (SAD) pattern (Figure 15.11c), these nanocrystals are identified to be equilibrium phases of fcc Pt and bcc W. Figure 15.11d is an EDS spectrum obtained at the tip corresponding to Figure 15.11b. Pt and W peaks dominate in the spectrum although traces of C and O are also observed. Thus, it is confirmed that Pt has been effectively grown on the nanodendritic W structures. From the HRTEM micrographs, the average nanoparticle size at the present ion sputtering conditions was 2.3 nm. The particle size can be easily controlled by variation of the ion sputtering time.

Using this process, various other metal composite nanoparticle/nanodendrite structures, such as Au/W, Mo/W, Au/Pt, and so on, can also be fabricated. Figure 15.12 shows an example of Au-nanoparticle/W-nanodendrite compound nanostructures grown on an insulator SiO_2 substrate. Therefore, the technique may be easily employed to fabricate metal nanoparticle/nanodendrite composite nanostructures of a wide range of materials. Because the nanodendrite structure possesses a compara-

Fabrication and Characterization of the Metal Nanosized 293

tively large specific surface area, it has potential applications in catalysts, sensors, gas storages and so on.

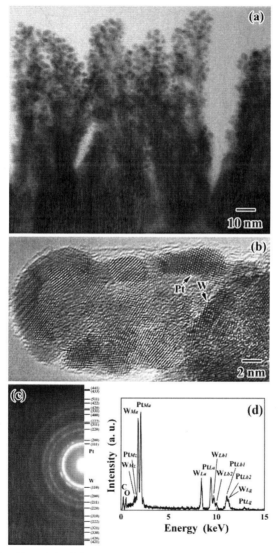

FIGURE 15.11 (a) Bright-field TEM image of the as-fabricated Pt-nanoparticle/W-nanodendrite composite structures on an Al2O3 substrate. (b) an HRTEM image, (c) SAD pattern and (d) EDS spectrum taken from a tip of the composite nanostructures [33].

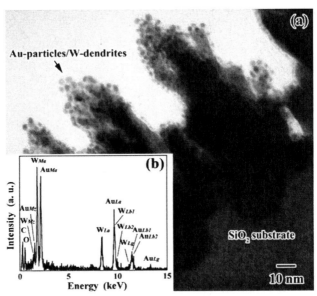

FIGURE 15.12 (a) Bright-field TEM image of the as-fabricated Au-nanoparticle/W-nanodendrite composite structures on a SiO2 substrate and (b) an EDS spectrum taken from a tip of the composite nanostructures. The W nanodendrites were fabricated with an electron-beam irradiation fluence of 4.4×10^{21} cm^{-2}. Au nanoparticles were deposited by a quick auto coater system (JEOL JFC-1500). The anodic voltage used during sputtering was 1 kV, and anodic current was 7 mA. The ion sputtering time was 7 s. The average Au nanoparticle size was measured to be 2.1 nm [33].

15.6 SUMMARY

Self-standing metal nanosized branched structures including nanowire arrays, nanodendrites, and nanofractal-like trees with controlled size and position were grown on insulator (SiO$_2$, Al$_2$O$_3$) substrates by the EBID process. The nanostructures of the different morphologies can be obtained by controlling the intensity of the electron beam during the EBID process. The nucleation and growth of the nanosized branched structures are proposed to be related to a mechanism involving charge-up produced on the surface of the substrate, and the movement of the charges to and charges accumulation at the convex surface of the substrate and the tips of the deposits. High-energy electron irradiation enhances diffusion of the metallic atoms in the nanostructures and hence promotes crystallization. More crystallized metal nanobranched structures are achieved by the EBID process using high energy electron beams. The nanobranched structures can be easily decorated by metallic nanoparticles to form composite nanostructures such as the Pt nanoparticles/nanodendrites. Therefore the nanosized branched structures and their fabrication process may be applied in technology to realize various functional nanomaterials such as catalysts, sensor materials, and emitters.

KEYWORDS

- **Crystallization**
- **Electron-beam-induced deposition**
- **Insulator substrate**
- **Nanofabrication**
- **Nanostructures**

REFERENCES

1. Shi, J.; Gider, S.; Babcock, K.; and Awschalom, D. D.; *Science.* **1996**, *271*, 937.
2. Favier, F.; Walter, E.; Zach, M.; Benter, T.; and Penner, R. M.; *Science.* **2001**, *293*, 2227.
3. Xia, Y. N.; Yang, P. D.; Sun, Y. G.; Wu, Y. Y.; Mayer, B.; Gates, B.; Yin, Y. D.; Kim, F.; and Yan, H. Q.; *Adv. Mater.* **2003**, *15*, 353.
4. Lao, Y. L.; Wen, J. G.; and Ren, Z. F.; *Nano. Lett.* **2002**, *2*, 1287.
5. Manna, L.; Milliron, D. J.; Meisel, A.; Scher, E. C.; and Alivisatos, A. P.; *Nat. Mater.* **2003**, *2*, 382.
6. Yan, H.; He, R.; Pham, J.; and Yang, P.; *Adv. Mater.* **2003**, *15*, 402.
7. Dick, K. A.; Deppert, K.; Larsson, M. W.; Martensson, T.; Seifert, W.; Wallenberg, L. R.; and Samuelson, L.; *Nat. Mater. 3*, **2004**, 380.
8. Zhang, H. F.; Wang, C. M.; Buck, E. C.; and Wang, L. S.; *Nano. Lett.* **2003**, *3*, 577.
9. Shi, Z. J.; Lian, Y. F.; Liao, F. H.; Zhou, X. H.; Gu, Z. N.; Zhang, T.; Iijima, S.; Li, H. D.; Yue, K. T.; and Zhang, S. L.; *J. Phys. Chem. Solids.* **2000**, *61*, 1031.
10. Dai, Z. R.; Pan, Z. W.; and Wang, Z. L.; *Adv. Funct. Mater.* **2003**, *13*, 9.
11. Jin, Y.; Tang, K. B.; An, C. H.; Huang, L. Y.; *J. Cryst. Growth.* **2003**, *253*, 429.
12. Mitsuishi, K.; Shimojo, M.; Han, M.; and Furuya, K.; *Appl. Phys. Lett.* **2003**, *83*, 2064.
13. Dong, L. X.; Arai, F.; and Fukuda, T.; *Appl. Phys. Lett.* **2002**, *81*, 1919.
14. Utke, I.; Hoffmann, P.; Dwir, B.; Leifer, K.; Kapon, E.; and Doppelt, P.; *J. Vac. Sci. Technol. B.* **2000**, *18*, 3168.
15. Brückl, H.; Kretz, J.; Koops, H. W.; and Reiss, G.; *J. Vac. Sci. Technol. B.* **1999**, *17*, 1350.
16. Koops, H. W. P.; Kretz, J.; Rudolph, M.; Weber, M.; Dahm, G.; and Lee, K. L.; *Jpn. J. Appl. Phys. Part 1.* **1994**, *33*, 7099.
17. Koops, H. W. P.; Weiel, R.; Kern, D. P.; and Baum, T. H.; *J. Vac. Sci. Technol. B.* **1988**, *6*, 477.
18. Hiroshima, H.; Suzuki, N.; Ogawa, N.; and Komuro, M.; *Jpn. J. Appl. Phys. Part 1.* **1999**, *38*, 7135.
19. Hoyle, P. C.; Cleaver, J. R. A.; and Ahmed, H.; *J. Vac. Sci. Technol. B.* **1996** *14*, 662.
20. Kohlmann-von Platen, K. T.; Chlebek, J.; Weiss, M.; Reimer, K.; Oertel, H.; and Brünger, W. H.; *J. Vac. Sci. Technol. B. 11*, **1993**, 2219.
21. Matsui, S.; Kaito, T.; Fujita, J.; Komura, M.; Kanda, K.; and Haruyama, Y.; *J. Vac. Sci. Technol. B.* **2000**, *18*, 3181.

22. Liu, Z. Q.; Mitsuishi, K.; and Furuya, K.; *Nanotechnology.* **2004**, *15*, S414.
23. Banhart, F.; *Phys. Rev. E,* **1995**, *52*, 5156.
24. Zhang, J. Z.; Ye, X. Y.; Yang, X. J.; and Liu, D.; *Phys. Rev. E.* **1997**, *55*, 5796 ().
25. Wang, H. Z.; Liu, X. H.; Yang, X. J.; and Wang, X.; *Mat. Sci. Eng. A.* **2001**, *311*, 180.
26. Song, M.; Mitsuishi, K.; Tanaka, M.; Takeguchi, M.; Shimojo, M.; and Furuya, K.; *Appl. Phys. A.* **2004**, *80*, 1431.
27. Xie, G. Q.; Song, M.; Mitsuishi, K.; and Furuya, K.; *J. Nanosci. Nanotechnol.* **2005**, *5*, 615.
28. Xie, G. Q.; Song, M.; Mitsuishi, K.; and Furuya, K.; *Physica E.* **2005**, *29*, 564.
29. Xie, G. Q.; Song, M.; Mitsuishi, K.; and Furuya, K.; *Jpn. J. Appl. Phys.* **2005**, *44*, 5654.
30. Song, M.; Mitsuishi, K.; and Furuya, K.; *Mater. Trans.* **2007**, *48*, 2551.
31. Song, M.; Mitsuishi, K.; Takeguchi, M.; and Furuya, K.; *Appl. Surf. Sci.* **2005**, *241*, 107.
32. Song, M.; Mitsuishi, K.; and Furuya, K.; *Physica. E.* **2005**, *29*, 575.
33. Xie, G. Q.; Song, M.; Furuya, K.; Louzguine, D. V.; and Inoue, A.; *Appl. Phys. Lett.* **2006**, *88*, 263120.
34. Xie, G. Q.; Song, M.; Mitsuishi, K.; and Furuya, K.; *J. Mater. Sci.* **2006**, *41*, 2567.
35. Xie, G. Q.; Song, M.; and Furuya, K.; *J. Mater. Sci.* **2006**, *41*, 4537.
36. Furuya, K.; Takeguchi, M.; Song, M.; Mitsuishi, K.; and Tanaka, M.; *J. Phys. Conf. Ser.* **2008**, *126*, 012024.
37. Song, M.; Mitsuishi, K.; and Furuya, K.; *Mater. Sci. Forum.* **2005**, *475–479*, 4035.
38. Nagase T.; and Umakoshi, Y.; *Mater. Sci. Eng. A.* **2003**, *352*, 251.
39. Gao, P. X.; Ding, Y.; Mai, W.; Hughes, W. L.; Lao, C.; and Wang, Z. L.; *Science.* **2005**, *309*, 1700.
40. Dick, K. A.; Deppert, K.; Larsson, M. W.; Martensson, T.; Seifert, W.; Wallenberg, L. R.; and Samuelson, L.; *Nat. Mater.* **2004**, *3*, 380.
41. Li, M.; Schnablegger, H.; Mann, S.; *Nature,* **1999**, *402*, 393.
42. Pal, U.; Ramirez, J. F. S.; Liu, H. B.; Medina, A.; and Ascencio, J. A.; *Appl. Phys. A.* **2004**, *79*, 79.
43. Joo, S. H.; Choi, S. J.; Oh, I.; Kwak, J.; Liu, Z.; Terasaki, O.; and Ryoo, R.; *Nature.* **2001**, *412*, 169.
44. Ruiz, A.; Arbiol, J.; Cirera, A.; Cornet, A.; and Morante, J. R.; *Mater. Sci. Eng. C.* **2005**, *19*, 105.
45. De Meijer, R. J.; Stapel, C.; Jones, D. G.; Roberts, P. D.; Rozendaal, A.; Macdonald, W. G.; Chen, K. Z.; Zhang, Z. K.; Cui, Z. L.; Zuo, D. H.; and Yang, D. Z.; *Nanostruct. Mater.* **1997**, *8*, 205.
46. Wang, A.; Liu, J.; Lin, S.; Lin, T.; and Mou, C.; *J. Catal.* **2005**, *233*, 1486.
47. Liu, P.; Rodriguez, J. A.; Muckerman, J. T.; and Hrbek, J.; *Surf. Sci.* **2003**, *530*, L313.

CHAPTER 16

A CASE STUDY ON HYPERBRANCHED POLYMERS

RAMIN MAHMOODI, TAHEREH DODEL, TAHEREH MOIENI, and MAHDI HASANZADEH

16.1 INTRODUCTION

In the current chapter, a hyperbranched, functional, water-soluble and amine-terminated polymer is synthesized by melt-polycondensation reaction of methyl acrylate and diethylene triamine. The polymer is then characterized by FTIR spectroscopy. The color characteristics of the dyed samples are evaluated using CIELAB method. The result showed that the color strength of HBP-treated PA6 fabrics is more than pure PA6 fabric due to the presence of terminal primary amino groups in the molecular structure of the HBP. Moreover, the washing fastness of the dyed treated PA6 fabrics were also good compared with that obtained by conventional dyeing.

Dendritic polymers are highly branched polymers with tree like branching having an overall spherical or ellipsoidal shape. These macromolecules consist of three subsets namely dendrimers, dendrigraft polymers and hyperbranched polymers (HBPs) [4]. HBPs are highly branched, polydisperse, and three-dimensional macromolecules synthesized from a multifunctional monomer to produce a molecule with dendritic structure [2]. These polymers have remarkable properties, such as low melt and solution viscosity, low chain entanglement, and high solubility, as a result of the large amount of functional end groups and globular structure, so they are excellent candidates for use in random applications, particularly for modifying fibers [7, 8].

Recently, the application of HBPs in textile industry has been developed especially for improvement of dye uptake of several fibers. For instance, Burkinshaw et al. [1] investigated the dyeability of polypropylene (PP) fibers modified by HBP. The results showed that the incorporation of HBP prior to fiber spinning considerably improved the color strength of PP fiber with C.I. Disperse Blue 56 and has no significant effect on physical properties of the PP fibers. In the study on applying HBP to cotton fabric by Zhang et al. [9, 10, 11, 12], it was demonstrated that HBP treatment on cotton fabrics has no undesirable effect on mechanical properties of fabrics. Dyeing of treated cotton fabrics with direct and reactive dyes in the absence of electrolyte showed that the color strength of treated samples was better than untreated cotton fabrics. Furthermore, it has shown that the application of HBP to cotton fabrics can reduce UV transmission and impart antibacterial properties. In another study, the dyeability of modified poly(ethylene terephthalate) (PET) fabrics by amine-terminated HBP was investigated by Hasanzadeh et al. [3]. They reported that the dye uptake of HBP-treated PET fabrics is significantly greater than that of untreated PET ones due to the presence of terminal primary amino groups in the molecular structure of the HBP, that will protonate in the liquid phase and give rise to positive charge at lower pH values.

Literature review showed that there has not been a previous report regarding the application of HBP on PA6 fabric and dyeing with acid dyes. In this work, new synthesized HBP with an amine terminal group was used to improve the dyeability of PA6 fabrics with acid dye. For this purpose, amine-terminated AB2-type HBP was synthesized from methyl acrylate and diethylene triamine by melt polycondensation. The obtained HBP was characterized using Fourier transform infrared spectroscopy (FTIR). Then the HBP was applied to PA6 fabric and the dyeability of treated samples and untreated one with C.I. Acid Red 131 was investigated.

16.2 EXPERIMENTAL

16.2.1 MATERIALS

Methyl acrylate (molecular weight = 86.09 g/mol) and diethylene triamine (molecular weight = 103.17 g/mol) were purchased from Merck Co. and used as received. PA6 knitted fabrics was used throughout this work and before using, the samples were scoured by a 2 g/L of anionic detergent at 70°C for 30 min and the procedure was followed by washing in cold distilled water and ambient drying. The selected acid dye for this work was C.I. Acid Red 131 provided by the Ciba Ltd. (Iran).

16.2.2 MEASUREMENTS

The FTIR spectrum of AB_2-type monomer and corresponding hyperbranched polymer were recorded by Nicolet 670 FTIR spectrophotometer in the wave number range of 500–4,000 cm^{-1} which nominal resolution for all spectra was 4cm^{-1}. The spectral reflectance of the dyed samples was determined using a spectrophotometer. As shown by Eq. (16.1), the Kubelka–Munk single-constant theory was employed to calculate the color strength as reflectance function (K/S value) (McDonald, 1997).

$$(K/S)_ë = \frac{(1-R_ë)^2}{2R_ë} \quad (16.1)$$

where R_λ is the reflectance value of dyed sample at the wavelength of maximum absorbance (λ_{max}), K is the absorption coefficient, and S is the scattering coefficient. Washing fastness test was performed according to ISO/R 105/IV, Part 8.

16.2.3 SYNTHESIS OF AMINE-TERMINATED HYPERBRANCHED POLYMER

For synthesis of amine-terminated hyperbranched polymer two steps were employed. The first step was the synthesis of AB_2-type monomers, and the second one was the preparation of HBP using a melt-polycondensation reaction. Diethylene triamine (0.5 mol, 52 ml) was added in a three-necked flask equipped with a constant-voltage dropping funnel, condenser, and a nitrogen inlet. The flask was placed in an ice bath and solution of methyl acrylate (0.5 mol, 43 ml) in methanol (100 ml) was added dropwise into the flask. The reaction mixture was stirred with a magnetic stirrer. Then the mixture was removed from the ice bath and left to react with a flow of nitrogen at room temperature (Scheme 16.1). After stirring for 14 hrs, the nitrogen flow was removed and AB_2 type monomer was obtained.

The obtained light yellow and viscous mixture was transferred to an eggplant-shaped flask for an automatic rotary vacuum evaporator to remove the methanol under low pressure. Then the temperature was raised to 150°C using an oil bath and condensation reaction was carried out for 6 hrs. A pale yellow viscous amine-terminated hyperbranched polymer was obtained.

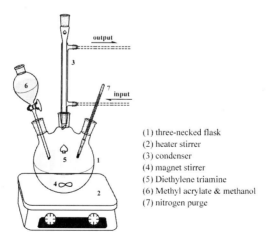

SCHEME 16.1 Schematic of reactor for monomer synthesis.

16.2.4 MODIFICATION OF PA FABRICS USING HBP

PA6 fabrics were immersed into baths containing 10 percent HBP and 0.5 percent nonionic surfactant (Irgasol NA) with the liquor to goods ratio of 60:1 at the temperature 40°C for 30 min. The pH was adjusted to 5.5 with acetic acid prior to adding the fabric. After exhaustion, the samples (PA6-HBP) were allowed to dyeing process.

16.2.5 DYEING PROCEDURE

The dyeing process of PA6-HBP samples was carried out in the same bath. Acid dye (0.5, 0.75, 1% owf) was added to the bath at the temperature 40°C. Then the temperature was raised up to boiling point at a rate of 2°C/min. Dyeing was then continued for 60 min with occasional stirring. Dyeing process was carried out using a liquor ratio of 60:1, and the procedure was followed by washing in warm water at about 50°C and then with cold water. The same procedure was also carried out for pure PA6 fabrics. Figure 16.1 show the dyeing profile.

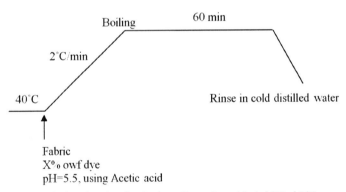

FIGURE 16.1 Method and recipe for dyeing of samples with Acid Red 131.

16.3 RESULTS AND DISCUSSION

16.3.1 SYNTHESIS AND CHARACTERIZATION OF HBP

Amine-terminated hyperbranched polymer were synthesized in two step reaction comprise preparation of AB_2 type monomers (1 and 2) by Michael addition reaction of methyl acrylate and diethylene triamine and synthesis of hyperbranched polymer by polycondensation reaction respectively (Scheme 16.2).

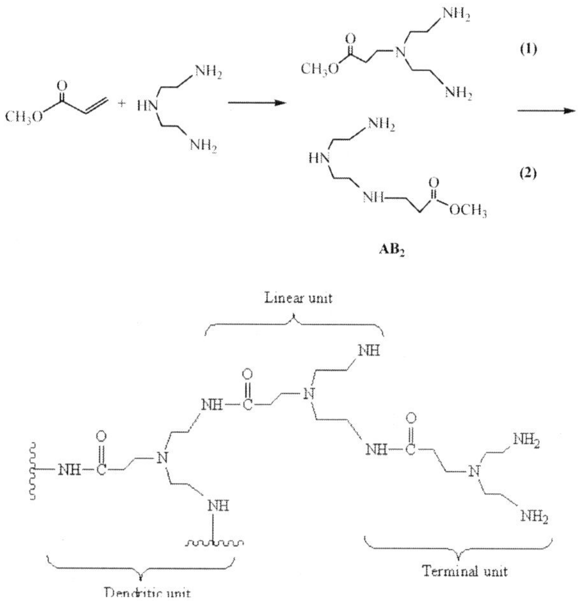

A Case Study on Hyperbranched Polymers

SCHEME 16.2 Chemical structure of amine-terminated hyperbranched polymer.

In order to confirm the polycondensation reaction progress, FTIR of AB_2-type monomer and hyperbranched polymer were studied (Figure 16.2). Spectral differences between AB_2 type monomer and hyperbranched polymer are observed in the fingerprint region between 1,800 and 900 cm^{-1}. The absorption bands in FTIR spectra are assigned according to literature (Pavia et al., 2001). The FTIR spectra of both monomer and polymer show that the peak positions are at 3,384, 2,950, 2,880, and 2,819 cm^{-1}. The band at above 3,300 cm^{-1} is due to N–H stretching vibration, while the band at 1,463 cm^{-1} reflects the CH_2 group bending vibration. In Figure 16.2 a, the absorption at 1,727 cm^{-1} is attributed to C=O stretching vibration of esters. This peak is generally weak in FTIR spectrum of HBP (Figure 16.2b). Moreover, the band at 1,637 cm^{-1} corresponds to C = O stretching vibration of amides. Melt-polycondensation reaction and synthesis can be responsible for this change in the band intensity [9].

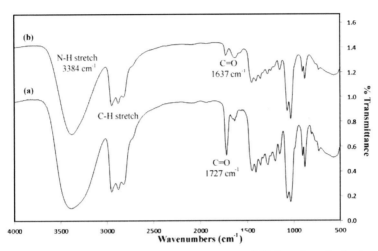

FIGURE 16.2 FTIR spectra of (a) AB$_2$ type monomer and (b) hyperbranched polymer.

16.3.2 DYEING PROPERTIES OF PA6-HBP FABRICS

The effect of HBP treatment on dye absorbance behavior (dyeability) of PA6 fabrics was evaluated by measuring its optical properties. Figure 16.3 shows the reflectance spectra of dyed PA6 and PA6-HBP fabrics. It can be seen that the reflectance spectrum of PA6-HBP samples is generally lower than PA6 samples. Therefore, the treatment of PA6 fabrics by HBP significantly reduced the reflectance.

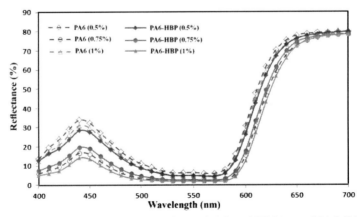

FIGURE 16.3 Reflectance spectra of dyed pure (PA6) and HBP-treated PA6 (PA6-HBP) fabrics.

The color strength (K/S value) results of the PA6 and PA6-HBP fabrics dyed with Acid Red 131 using a competitive dyeing method are shown in Figure 16.4. A higher K/S value of PA6-HBP samples indicated greater dye uptake than that of PA6 sample. The increase in the color strength is due to the positively charged amino groups on the PA6-HBP fabrics. Moreover, the CIELAB color coordinates (L*, a*, b*, C* and h°) of samples are shown in Table 16.1. It can be concluded that the HBP treatment has significant effects on the color coordinates of PA6 fabrics.

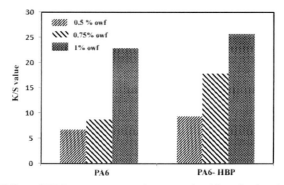

FIGURE 16.4 Effect of HBP treatment on color strength achieved using Acid Red 131.

TABLE 16.1 Color coordinates of dyed samples.

Fabrics	Dye (% owf)	L*	a*	b*	C*	h°
	0.5	48.140	55.798	−17.867	58.58879	−0.30989
PA6	0.75	36.914	60.330	−7.471	60.79083	−0.12321
	1	45.958	57.430	−17.393	60.00601	−0.29407
	0.5	44.995	55.988	−15.714	58.15141	−0.27363
PA6-HBP	0.75	38.949	59.685	−10.187	60.54811	−0.16905
	1	33.099	55.476	−4.346	55.64597	−0.07818

Table 16.2 shows the washing fastness of PA6 and PA6-HBP fabric dyed with acid dye. Compared with the pure PA6 fabrics, the treated ones have good washing fastness. This result showed that, the washing fastness of acid dye was largely unaffected by HBP treatment. Furthermore, the color strength of samples before and

after washing was shown in Figure 16.5. It is clear that the extent of color strength change was relatively small for treated and pure PA6 fabrics.

TABLE 16.2 Color fastness of pure (PA6) and HBP-treated PA6 (PA6-HBP) fabrics.

Fabrics	Dye (% owf)	K/S	Washing fastness Fading	Staining
PA6	0.5	6.78	4–5	4–5
	0.75	8.75	4–5	4–5
	1	22.93	4	4
PA6-HBP	0.5	9.31	5	5
	0.75	17.81	5	5
	1	25.75	4–5	4–5

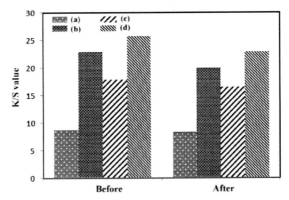

FIGURE 16.5 Effect of washing fastness test on color strength of (a), (b) PA6 fabric dyed with 0.75% and 1% owf, respectively, (c) and (d) PA6-HBP fabric dyed with 0.75% and 1% owf, respectively.

16.4 CONCLUSION

In this study, a hyperbranched, functional, and water-soluble polymer was successfully synthesized by melt-polycondensation reaction of methyl acrylate and diethylene triamine. The FTIR spectroscopy measurement of HBP indicated that this polymer comprised terminal amine group. The HBP was applied to PA6 fabric using the exhaustion process. The study of dyeability of treated samples with C.I. Acid Red 131 indicated that the color strength of HBP-treated PA6 fabrics is more than pure PA6 fabric due to the presence of terminal primary amino groups in the molecular structure of the HBP, that will protonate in the liquid phase and give rise to positive charge at lower pH values. Moreover, the washing fastness of acid dye was largely unaffected by HBP treatment.

ACKNOWLEDGMENT

The author would like to express their grateful thanks to Dr. M. Ghanbar Afjeh from Amirkabir University of Technology for the color measurements.

KEYWORDS

- Dyeability
- Melt-polycondensation reaction
- Polycondensation reaction
- Reflectance

REFERENCES

1. Burkinshaw, S.M.; Froehling, P. E.; and Mignanelli, M.; The effect of hyperbranched polymers on the dyeing of polypropylene fibres. *Dyes. Pigm.* **2002**, *53*, 229–235. doi: 10.1016/S0143-7208(02)00006-2.
2. Gao, C.; and Yan, D.; Hyperbranched polymers: from synthesis to applications. *Progr. Polym. Sci.* **2004**, *29*, 183–275. doi: 10.1016/j.progpolymsci.2003.12.002.
3. Hasanzadeh, M.; Moieni, T.; Hadavi Moghadam, B.; Synthesis and characterization of an amine terminated AB$_2$-type hyperbranched polymer and its application in dyeing of poly(ethylene terephthalate) fabric with acid dye. *Adv. Polym. Technol.* **2013**, *32*, xx. doi: 10.1002/adv.21345
4. Jikei, M.; and Kakimoto, M.; Hyperbranched polymers: a promising new class of materials. *Prog. Polym. Sci.* **2001**, *26*, 1233–1285. doi: 10.1016/S0079-6700(01)00018-1
5. McDonald, R.; Color Physics for Industry. England: Society of dyers and colorists; **1997**.
6. Pavia, D.L.; Lampman, G. M.; Kriz, G. S.; and Vyvyan, J. R.; Introduction to Spectroscopy. Brooks/Cole: US; **2001.**
7. Schmaljohann, D.; Pötschke, P.; Hässler, R.; Voit, B. I.; Froehling, P. E.; Mostert, B.; and Loontjens, J. A.; Blends of amphiphilic, hyperbranched polyesters and different polyolefins. *Macromolecules.* **1999**, *32*, 6333–6339. doi: 10.1021/ma9902504.

8. Seiler, M.; Hyperbranched polymers: Phase behavior and new applications in the field of chemical engineering. *Fluid. Phase. Equilibria.* **2006**, *241*, 155–174. doi: 10.1016/j.fluid.2005.12.042
9. Zhang, F.; Chen, Y.; Lin, H.; and Lu, Y.; Synthesis of an amino-terminated hyperbranched polymer and its application in reactive dyeing on cotton as a salt-free dyeing auxiliary. *Color. Technol.* **2007**, *123*, 351–357. doi: 10.1111/j.1478-4408.2007.00108.x.
10. Zhang, F.; Chen, Y.; Lin, H., Wang, H., and Zhao, B. (2008). HBP-NH$_2$ grafted cotton fiber: Preparation and salt-free dyeing properties. *Carbohydrate Polymer*, 74, 250-256. doi: 10.1016/j.carbpol.2008.02.006
11. Zhang, F.; Chen, Y. Y.; Lin, H.; and Zhang, D. S.; Performance of cotton fabric treated with an amino-terminated hyperbranched polymer. *Fib. Polym.* **2008**, *9*, 515–520.
12. Zhang, F.; Chen, Y.; Ling, H.; and Zhang, D.; Synthesis of HBP-HTC and its application to cotton fabric as an antimicrobial auxiliary. *Fib. Polym.* **2009**, *10*, 141–147. doi: 10.1007/s12221-009-0141-6

CHAPTER 17

A STUDY ON NETWORK OF SODIUM HYALURONATE WITH NANO-KNOTS JUNCTIONS

SHIN-ICHI HAMAGUCHI and TOYOKO IMAE

17.1 INTRODUCTION

Amine-terminated poly(amido amine) (PAMAM) dendrimers have been attached to sodium hyaluronates (NaHAs) by a coupling reaction, and the complexes have been characterized. By the complexation with dendrimers, the morphology of NaHAs was varied from the commom network to the bead and string network, which gave rise to the decrease in viscosity of NaHAs. The bead and string network was more abundant for the covalent network complex than the noncovalent one. The beads, that is, the nano-knots of the network consist of the covalent-bonded NaHA/dendrimer composites, and the strings are NaHA chains. Beads became small and strings decreased in number with decreasing a molecular weight of NaHA. The complexation of sodium poly-L-glutamates (NaPGAs) with PAMAM dendrimers was different in the manner from that of NaHAs with dendrimers. Flexible NaPGAs produced globular composites with dendrimers.

    A hyaluronic acid (HA) is a linear polysaccharide including D-glucuronic acid and N-acetyl-D-glucosamine in a repeating unit [1]. Importantly this polysaccharide is a major biomacromolecular component of the intercellular matrix in most connective tissues (e.g. eye vitreous humor, synovial fluid, and cartilage) [2]. Moreover, its unique physicochemical properties such as remarkable viscoelastic property and biocompatibility led the utilization of HA in the pharmaceutical and medical fields [3–7]. Especially, recent pharmaceutical and medical researchers paid attention to the use of HA in drug delivery systems in addition to the therapeutic products in ophthalmology, dermatology, and osteoarthritis [8, 9]. Incidentally, it has been reported that HAs formed the complexes with another biomaterials like liposomes, proteins, other sugars, graft polymers and DNA [10–13].

On the other hand, a dendrimer is a highly branched synthetic architecture with many functional terminals. Especially, poly(amido amine) (PAMAM) dendrimer, [14–17] one of typical dendrimers, has been focused on the interaction with biomacromolecules, such as DNA, [18–24] polypeptides, [25–28] and polysaccharides [29–31]. Such researches are valuable, because the PAMAM dendrimer is nontoxic and biocompatible. A complexation of sodium hyaluronate (NaHA) with PAMAM dendrimer has been reported by one of authors and her coworkers [32]. Then the motive force of the complexation was the electrostatic interaction between negatively charged carboxylates in NaHA and positively charged amino terminal groups in PAMAM dendrimer. In addition, the formation of the complex was played an important role by the hydrogen bonding between hydroxyl groups in NaHA and amide and amino groups in dendrimer so as to overcome the repulsive interaction between charged amino groups in dendrimers. However, on the utilization of the complex, the complex must be chemically stable so that it does not break apart under the variation of conditions (e.g. pH, temperature, ionic strength, etc.)

In the present work, amine-terminated dendrimers were immobilized on NaHAs by means of the chemical reaction, which is well-known to form an amide bond [33]. By use of this procedure, PAMAM dendrimers were covalent-bonded. The covalent bond increases the chemical stability of the complex against the variation of conditions, unlike the non-covalent-bonded complexation by only the electrostatic interaction. The composites were investigated by a viscometry and an atomic force microscopy, which enabled us to understand the behavior of the complex both in aqueous solution and on substrate. Moreover, the difference between covalent-bonded and electrostatic (non-covalent-bonded) complexation and the dependence on molecular weight of NaHA were discussed, and the composites were also compared with those of sodium poly-L-glutamate (NaPGA) with dendrimer.

17.2 EXPERIMENTAL SECTION

17.2.1 MATERIALS

NaHA (repeating unit mass = 401.3) was donated from Shiseido Co., Japan. NaPGA (repeating unit mass = 151.3) was a product from Peptide Institute Inc., Japan. N-hydroxysuccinimide (NHS), 1-ethyl-3-(3-dimethylaminopropyl)-carbodiimide (EDC), and second generation PAMAM dendrimer with NH_2 terminals (molecular weight = 3256; 16 terminal groups) were purchased from Aldrich Chemical Co. Water was distilled, further purified using a Millipore Milli-Q apparatus, and used throughout the experiments.

17.2.1 COUPLING REACTION BETWEEN NAHA AND DENDRIMER

The procedure of a covalent cross-linkage between NaHA and dendrimer is schematically illustrated in Chart 17.1. The cross-linkage was prepared in accordance

A Study on Network of Sodium Hyaluronate

with the well-known coupling reaction: [34, 35]. Equimolar coupling reagents, EDC and NHS, were added to an aqueous NaHA solution. Then an EDC-activated carboxylate group in NaHA reacted with NHS to form a stable NHS carboxylate. After that, an aqueous dendrimer solution was mixed with a reaction solution, and then the NHS carboxylate was combined with the terminal amino group of dendrimer. The molar concentration of dendrimer was equal to that of coupling reagents. A concentration of NaHA in the reaction solution was varied like 0.004 wt% (concentration of the repeating unit = 0.1 mM), 0.008 wt% or 0.04 wt%. Then a molar ratio of NaHA residue and dendrimer [NaHA residue/dendrimer] changed.

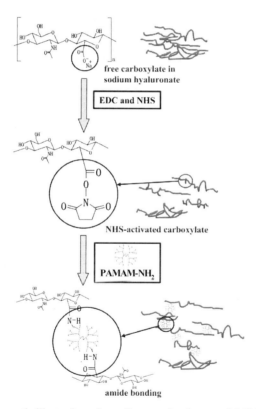

CHART 17.1 Schematic illustration of coupling reaction between NaHA and dendrimer.

17.2.2 ANALYSES

Dynamic light scattering was measured at 25°C on an Otsuka Electronics DLS-700 spectrophotometer equipped with an Ar ion laser at a wavelength of 488 nm. The apparent diffusion coefficients of NaHA were determined. The viscosity measurements were carried out using an Ubbelohde viscometer in a temperature-regulated water bath (25°C). A Digital Instruments NanoScope III microscope was used to take atomic force microscopic (AFM) images at room temperature (~25°C). Spin-coated films on mica substrates were prepared by spinning solutions for 1 min at 2,000 rpm using a spinner Model K-359SD-1 (Kyowa Riken Co., Japan). Dialysis of an aqueous NaHA solution was carried out by using a Visking seamless cellulose tube in excess water (four times over 48 hr with each 1,000 cm^3 of water).

17.3 RESULTS AND DISCUSSION

17.3.1 EFFECT OF COMPLEXATION ON VISCOSITIES OF AQUEOUS NAHA SOLUTIONS

NaHA molecules are estimated to overlap each other at high concentrations. Then an overlapping threshold concentration (c*) of an aqueous NaHA solution is evaluated from a plot of apparent diffusion coefficient D as a function of NaHA concentration c_{NaHA}. As seen in Figure 17.1, the apparent diffusion coefficients decreased initially and increased through a minimum with increasing NaHA concentration. Thus a concentration (around 0.06 wt%) at minimum D value was assigned to an overlapping threshold concentration, which is a threshold from "dilute solutions" to "semi-dilute solutions" [36, 37].

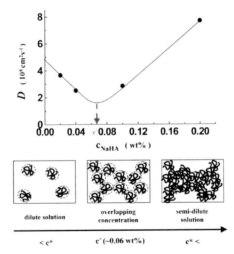

FIGURE 17.1 Apparent diffusion coefficients D of aqueous NaHA solutions as a function of NaHA concentration c_{NaHA}. c* is an overlap threshold concentration.

A Study on Network of Sodium Hyaluronate

Figure 17.2 plots the reduced viscosity η_{red} against NaHA concentration. The viscosity of NaHA solutions (NaHA residue/dendrimer) = [1/0]) was proportional to NaHA concentration below 0.04 wt% but deviated from a linearity above it. This indicates that the NaHA solutions below 0.04 wt% are in the category of "diluted solutions", which was determined from the apparent diffusion coefficients (Figure 17.1). Therefore, the NaHA concentrations below 0.04wt % were chosen for the preparation of composites with dendrimers unless otherwise noted.

The reduced viscosities of aqueous NaHA solutions with and without coupling reaction with dendrimers are included in Figure 17.2. While there was no change in the viscosity behavior of dendrimer-immobilized NaHA at a ratio of [1/0.001] in comparison with that at [1/0], drastic decrease occurred at [1/0.01]. This implies that NaHA molecules shrink by the reaction with much amount of dendrimers, namely, the linkage with dendrimer affects the morphology of NaHA. On the other hand, the viscosities of mixture at [1/0.01] without coupling reaction were same as those with coupling reaction. It can be noted that the decrease of viscosity is relative to the ratio of dendrimer to NaHA but independent of the fact whether the composite is networked by the covalent bond or not.

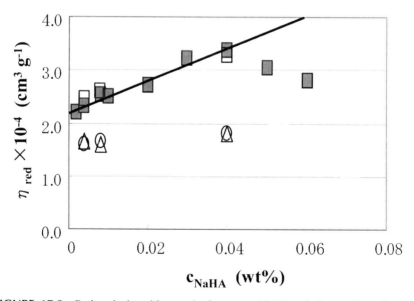

FIGURE 17.2 Reduced viscosities η_{red} of aqueous NaHA solutions with and without coupling reaction with dendrimers as a function of NaHA concentration c_{NaHA}. [NaHA residue/dendrimer]: closed square, [1/0]; open square, [1/0.001] with coupling reaction; open circle, [1/0.01] with coupling reaction; open triangle, [1/0.01] without coupling reaction.

17.3.2 MORPHOLOGY OF NAHA BEFORE AND AFTER COUPLING REACTION

The effect of coupling reaction can be confirmed by means of the morphological observation by AFM. Figures 17.3 and 17.4 show AFM images of spin-coated films prepared from aqueous NaHA solutions at 0.004 and 0.04 wt%, respectively, by comparison between complexes without and with coupling reaction. NaHA molecules formed network domains with monomolecular and multimolecular thicknesses for solutions at 0.004 and 0.04 wt%, respectively, according to their section analyses. Because of intermolecular hydrogen bonding, NaHA molecules attract each other to form domains, which were dispersed as network films on a substrate, if the solution is dilute. Larger amount of residues in NaHA molecules were intermolecularly hydrogen-bonded in a concentrated solution, producing larger network domains in comparison with those in a dilute solution.

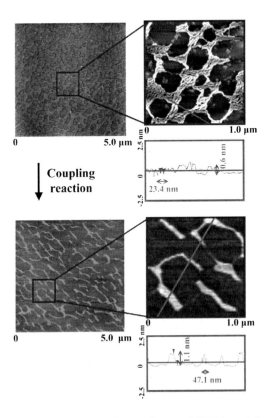

FIGURE 17.3 AFM images and section analyses of NaHA and its coupling-reacted composite with dendrimer at a ratio of [1/0.01]. NaHA concentration = 0.004 wt%.

A Study on Network of Sodium Hyaluronate 313

A drastic change happened, when NaHA molecules in solutions at both low and high NaHA concentrations suffered the coupling reaction with dendrimers at [1/0.01]. As seen in Figure 17.3, NaHA domains in a spin-coated film prepared from a 0.004 wt% solution changed to zones with two-fold thickness of monolayer. It was assumed that the zones became dense by strong intermolecular linkages between NaHAs through dendrimers, since the separation between NaHA strings in the zone was not found any more, different from a case of NaHAs free from the cooperation of dendrimers. On the other hand, on a film from a 0.04 wt% solution (Figure 17.4), large spherical beads (with about 10 nm height and submicron width) were linked by strings. It can be mentioned that a lot of dendrimers and NaHA molecules form large globules or aggregates after the coupling reaction, and the remaining (free) NaHA chains connects globules (or aggregates).

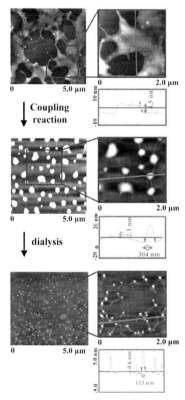

FIGURE 17.4 AFM images and section analyses of NaHA, its coupling-reacted composite with dendrimer at a ratio of [1/0.01], and the dialyzed composite. NaHA concentration = 0.04 wt%.

17.3.3 EFFECTS OF COVALENT NETWORK AND DIALYSIS ON MORPHOLOGY OF NAHA-DENDRIMER COMPOSITES

Figure 17.5 shows an AFM image of a NaNA-dendimer composite without the coupling reaction (NaHA concentration = 0.04 wt%, a ratio = [1/0.01]). Large globular beads and the strings between them were confirmed. The texture is apparently similar to that of the composite with the coupling reaction (Figure 17.4) but the numbers of beads and strings are apparently less, although this difference is not sensitive in viscosity behavior as described above. These results suggest that the electrostatic binding between dendrimers and NaHAs is less abundant than the covalent-bonding between them, indicating the high effectivity of the coupling reaction. Moreover, the formation of globular beads consisting of NaHA-dendrimer composites both with and without coupling reaction gives rise to the decrease in viscosity of NaHA, as seen in Figure 17.2.

An AFM image of a spin-coated film from a NaHA-dendrimer composite (NaHA concentration = 0.04 wt%, a ratio = [1/0.01], with the coupling reaction) after dialysis is compared to that before dialysis in Figure 17.4. The large beads, which were found after the coupling reaction, became smaller after dialysis, although the network texture through strings remained. Instead, small aggregates appeared on the network as nano-knots of the network. This indicates that dendrimers, free from covalent-bonding or bound electrostatically, were removed from the large beads. Then the nano-knots consist of covalent-bonded HaNA-dendrimer composites. The scheme in Figure 17.6 illustrates the changing morphologies of NaHA and its composites on the process of coupling reaction and dialysis.

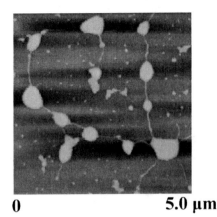

0 5.0 μm

FIGURE 17.5 An AFM image of a composite of HaNA and dendrimer at [1/0.01] without a coupling reaction. NaHA concentration = 0.04 wt%.

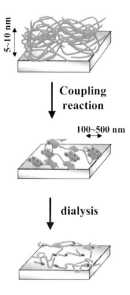

FIGURE 17.6 Schematic illustration of NaHA, its coupling-reacted composite with dendrimer at a ratio of [1/0.01], and the dialyzed composite on substrate. NaHA concentration = 0.04 wt%.

17.3.4 EFFECT OF NAHA SIZE ON COMPLEXATION

For three NaHAs with different molecular weights (molecular weight = 1.0×10^5, 1.0×10^6, 2.0×10^6), which were lower than that used in the examination described above (molecular weight = 3.0×10^6), viscosities and AFM images before and after the coupling reaction were compared in Figure 17.7. The viscosity η_{red} was in same order between NaHAs of molecular weights 3.0×10^6 and 2.0×10^6, but it drastically decreased with further decrease in molecular weight. Especially, the viscosity of NaHA with the lowest molecular weight (1.0×10^5) was as low as that of solvent (water). Although same tendency on molecular weight dependence of viscosity was preserved even after the coupling reaction with dendrimers, the viscosity after the coupling reaction was always smaller than that before the coupling reaction, as well as a case of the highest molecular weight (3.0×10^6) NaHA (see Figure 17.2). It can be noted that regardless of the molecular weight, NaHA in an aqueous solution shrank by the coupling reaction with dendrimer so as to decrease the viscosity.

In AFM images of pristine NaHAs (Figure 17.7), the common network structures were visualized for three NaHAs of different molecular weights as well as a case of the highest molecular weight (3.0×10^6) NaHA (Figure 17.4). With decreasing molecular weight, the texture changed from the network of large scale with multimolecular thickness to the network of small scale with small-size mesh and monomolecular thickness. The network of the lowest molecular weight (1.0×10^6) NaHA consisted of beads (knots) and strings. It is assumed that the intermolecular interaction of high molecular weight NaHAs is too stronger to overcome the centrifugal force of spin-coating than that of low molecular weight NaHAs.

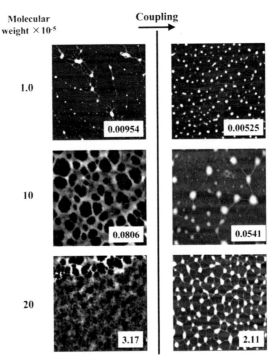

FIGURE 17.7 AFM images of three NaHAs with different molecular weights and their coupling-reacted composites with dendrimer at [1/0.01]. NaHA concentration = 0.04 wt%. Size: 5 μm × 5 μm. Inset: numerical values of reduced viscosities hred.

After the coupling reaction, string-cross-linking beads were observed for all NaHAs of different molecular weights, indicating that the NaHA molecules were coiled by the involvement of dendrimers. The network consisting of beads (aggregates) and strings by a high molecular weight (2.0×10^6) NaHA was in the similar manner to a case of the highest molecular weight (3.0×10^6) NaHA (Figure 17.4), but the former bead size was smaller than the latter one. The size of beads and the number of strings became further small with decreasing molecular weight of NaHA. These results are in consistency with the viscosity behavior. It is inferred that NaHA molecules with molecular weight below 1.0×10^6 are too short to form the network of NaHA-dendrimer beads connected by NaHA strings.

17.3.5 COMPARISON WITH THE COMPLEXATION OF SODIUM POLY-L-GLUTAMATE WITH DENDRIMER

The complexation of biomimetic polymer, NaPGA, with amine-terminated PAMAM dendrimer in water occurs in different manner from a case of the complexation between NaHA and dendrimer [26] Since NaPGA possesses carboxyl group in the repeating unit as well as NaHA, the motive force of the interaction with dendrimer is the electrostatic attraction between carboxyl group of NaPGA and amine group of dendrimer, but the hydrogen bonding is not expected. Another difference from NaHA-dendrimer composites is the fact that flexible NaPGA chains can penetrate into the interior of dendrimer. This suggests the easy production of globular composites.

In the present work, dendrimers were electrostatically combined on NaPGA and then immobilized by the coupling reaction. An AFM image after the coupling reaction was compared to the result for pristine NaPGA at same conditions, as shown in Figure 17.8. The pristine NaPGA molecules formed network domain with many pinholes on a solid substrate. The height (~1 nm) of the flat network indicated that the network sheet consisted of a mono- or bi-molecular layer. Then it is supposed that the network is constructed by the condensed entanglement of NaPGA.

The morphology of a NaHA film remarkably changed after the reaction with dendrimers. In this case, the islands (zones) with various shapes existed, and their heights (~4 nm) were about four times as high as that of pristine NaPGA, indicating that NaPGAs formed complexes with dendrimers. Moreover, it should be noticed that there are many spherical particles besides large polymorphic zones in the NaPGA-dendrimer composite system (Figure 17.8). Particles will be complexes consisting of some NaPGAs and some dendrimers, and large zones will be the aggregates of such complexes or the grown complexes. Similar globules were visualized on a transmission electron microscopic image of a NaPGA-dendrimer mixture without the coupling reaction [26]. These characteristics of NaPGA-dendrimer composites at a 0.04 wt% NaPGA concentration are comparable to those of NaHA-dendrimer composites at the lower NaHA concentration (0.004 wt%) but not at the higher NaHA concentration (0.04 wt%), namely, a concentration corresponding to the examination of NaPGA. This supports that flexible NaPGAs are easy to produce globular composites with dendrimers without bridging of extended NaPGAs.

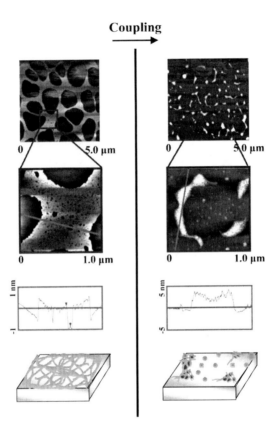

FIGURE 17.8 AFM images, section analyses, and schematic illustration of NaPGA and its coupling-reacted composite with dendrimer at [1/0.01]. NaPGA concentration = 0.04 wt%.

17.4 CONCLUSIONS

In the present work, NaHAs and PAMAM dendrimers have been covalent cross-linked using a coupling reaction and characteristics of the composites have been investigated. Increasing the ratio of PAMAM dendrimer to NaHA caused decreasing viscosity of the composite, which is irrespective of the coupling reaction. An AFM image on a spin-coated film of NaHA-dendrimer composite at a high NaHA concentration showed drastic change from the common network sheet structure of pristine NaHA to the bead-string network structure of the composites, where beads were cross-linked by strings. The bead-string network became more abundant by the coupling reaction between NaHAs and dendrimers than by only the electrostatic binding between them, indicating the high effectivity of the covalent network. Furthermore, it was confirmed by the dialysis that there were nano-knots consisting of covalent-bonding between HaNAs and dendrimers in the network structure. Effect of NaHA size on the morphology of composites was investigated, and it was suggested that the length of NaHA molecule played an important role in the size of beads consisting of NaHA-dendrimer composite as well as in the number of strings connecting the beads. It was inferred from the availment of NaPGA instead of NaHA that the aggregates of NaPGA and dendrimer were globules without cross-linkages by strings. This difference on the morphology of composites is ascribed to the conformational characters of flexible NaPGA and semiflexible NaHA. [32, 38, 39]. It is suggested that the results of the present work can be applied to other linear polymers possessing carboxyl group in the repeating unit, in order to control viscosity and morphology of polymer network. Moreover, the combination of NaHA and PAMAM dendrimer is valuable as biomaterials and applicable as drug delivery systems and biosensors, because both polymers are biocompatible.

ACKNOWLEDGEMENTS

We are grateful to Shiseido Co. for generous donation of NaHAs, which were key materials of the present work. We also acknowledge Mr. Daisuke Onoshima, Nagoya University, for his kind help on the viscosity measurements.

KEYWORDS

- Coupling reaction
- Hyaluronic acid (HA)
- Noncovalent-bonded complexation
- Sodium hyaluronates (NaHAs)

REFERENCES

1. Takayama, K.; Hirata, M.; Machida, Y.; Masada, Y.; Sannan, T.; and Nagai, T.; Chem. Pharm. Bull. **1993,** 1990, 38.
2. Myers, R. R.; Negami, S.; and White, R. H.; Biorheology. **1996**, 3, 197.
3. Tirtaatmadja, V.; Boger, D. V.; and Fraser, J. R. E.; Rheol. Acta **1984**, 23, 311.
4. Nimrod, A.; Ezra, E.; Ezov, N.; Nachum, G.; Parisada, B. J.; Ocular. Pharmacol. **1992**, 8, 161.
5. Peyron, J. G. J.; Rheumatology. **1993**, 20, 10.
6. Lim, S. L.; Martin, G. P.; Berry, D. J.; Brown, M. B.; J. Cont. Release. **2000**, 66, 281.
7. Smeds, K. A.; and Grinstaff, M. K.; J. Biomed. Mater. Res. **2001**, 54, 115.
8. Cho, K. Y.; Chung, T. W.; Kim, B. C.; Kim, M. K.; Lee, J. H.; and Wee, R. W.; Cho, C. S.; Int. J. Pharm. **2003**, 260, 83.

9. Shu, X. Z.; Liu, Y.; Palumbo, F.; and Prestwich, G. D.; Biomaterials. **2003**, 24, 3825.
10. Yerushalmi, N.; Arad, A.; and Margalit, R.; Arch. Biochem. Biophys. **1994**, 313, 267.
11. Burns, J. W.; Burgess, L.; Skinner, K.; Rose, R.; Colt, M. L.; and Diamond, M. P.; Fertil. Steril. **1996**, 66, 814.
12. Bakos, D.; Soldan, M.; and Vanis, M.; Adv. Med. Phys., Biophys. Biomat. **1997**, 54.
13. Asayama, S.; Nogawa, M.; Takei, Y.; Akaike, T.; and Maruyama, A.; Bioconjugate Chem. **1998**, *9*, 476
14. Liu, M.; and Fréchet, J. M. J.; Pharm. Sci. Technol. Today. **1999**, *2*, 393.
15. Esfand, R. D.; and Tomalia, A.; DDT. **2001**, *6*, 427.
16. Beezer, E. A.; King, S. H.; Martin,; I. K.; Mitchel, J. C.; Twyman, L. J.; and Wain C. F.; Tetrahedron. **2003**, 59, 3873.
17. Namazi, H.; and Adeli, M.; Biomaterials, **2005**, 26, 1175.
18. Plank, C.; Mechtler, K.; Szoka Jr, F. C.; and Wagner, E.; Hum. Gene. Therap. **1996**, 7, 1437.
19. Kukowska-Latallo, J. F.; Bielinska, A. U.; Johnson. J.; Spindler, R.; Tomalia, D. A.; and Baker Jr., J. R.; Proc. Natl. Acad. Sci. USA. **1996**, 93, 4897.
20. Bielinska, A. U.; Kukowska-Latallo, J. F.; Johnson, J.; Tomalia, D. A.; and Baker Jr., J. R.; Nucl. Acid. Res. **1996**, 24, 2176.
21. Tang, M. X.; Redemann, C. T.; and Szoka, Jr. F. C.; Bioconjugate. Chem. **1996**, 7, 703.
22. Ottaviani, M. F.; Sacchi, B.; Turro, N. J.; Chen, W.; Jockusch, S.; and Tomalia, D. A.; Macromolecules. **1999**, 32, 2275.
23. Chen, W.; Turro, N.; and Tomalia, J. D. A.; Langmuir. **2000**, 16, 15.
24. Mitra, A.; and Imae, T.; Biomacromolecules. **2004**, 5, 69.
25. Leisner, D.; and Imae, T.; J. Phys. Chem. B. **2003**, 107, 8078.
26. Imae, T.; and Miura, A.; J. Phys. Chem. B **2003**, 107, 8088
27. Leisner, D.; and Imae, T.; J. Phys. Chem. B. **2003**, 107, 13158
28. Leisner, D.; and Imae, T.; J. Phys. Chem. B. **2004**, 108, 1798.
29. Arima, H.; Kihara, F.; Hirayama, F.; and Uekama, K.; Bioconjugate. Chem. **2001**, 12, 476.
30. Kihara, F.; Arima, H.; Tsutsumi, T.; Hirayama, F.; Uekama K., Bioconjugate Chem. 2002, **13**, 1211
31. Kihara, F.; Arima, H.; Tsutsumi, T.; Hirayama, F.; and Uekama, K.; Bioconjugate. Chem. **2003**, 14, 342.
32. Imae, T.; Hirota, T.; Funayama, K.; Aoi, K.; and Okada, M.; J. Coll. Interfa. Sci. **2003**, 263, 306.
33. Sheehan, J. C.; Cruickshank, P. A.; and Boshart, G. L.; J. Org. Chem. **1961**, 26, 2525.
34. Yamazaki, T.; and Imae, T.; J. Nanosci. Nanotech. **2005**, 5, 1066.
35. Yamazaki, T.; Imae, T.; Sugimura, H.; Saito, N.; Hayashi, K.; and Takai, O.; J. Nanosci. Nanotech. **2005**, 5, 1792.
36. de Gennes, P. G.; Scaling Concepts in Polymer Physics. Cornell University Press: Ithaca, London; **1979**.
37. Imae, T.; and Ikeda, S.; J. Phys. Chem. **1986**, 90, 5216.
38. Cleland, R. L.; Biopolymers. **1984**, 23, 647.
39. Ghosh, S.; Li, X.; Reed, C. E.; Reed, W. F.; Biopolymers. **1990**, 30, 1101.

FIGURE CONTENTS

Network of sodium hyaluronate with nano-knots junction of poly(amido amine) dendrimer

Shin-ichi Hamaguchi and Toyoko Imae*

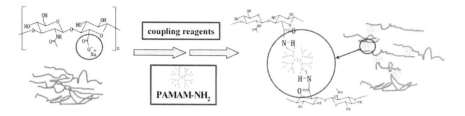

CHAPTER 18

THE MAGNETIC PHOTOCATALYST CONVERSION TO THE MAGNETIC DYE-ADSORBENT CATALYST VIA HYDROTHERMAL FOLLOWED BY TYPICAL WASHING AND THERMAL TREATMENTS

SATYAJIT SHUKLA

18.1 INTRODUCTION

The magnetic dye-adsorbent catalyst has been synthesized via hydrothermal processing of the magnetic photocatalyst followed by typical washing and thermal treatments. The magnetic dye-adsorbent catalyst consists of a core-shell nanocomposite with the core of a magnetic ceramic particle (such as mixed cobalt ferrite and hematite, and pure cobalt ferrite) and the shell of nanotubes of dye-adsorbing material (such as hydrogen titanate). The samples have been characterized for determining the phase structure, morphology, size, and magnetic properties using the X-ray and selected-area electron diffraction, transmission electron microscope, and vibrating sample magnetometer. The photocatalytic activity under the ultraviolet-radiation exposure and the dye-adsorption under the dark-condition have been measured using the methylene blue as a model catalytic dye-agent. It has been demonstrated that, the transformation of the magnetic photocatalyst to the magnetic dye-adsorbent catalyst is accompanied by a change in the mechanism of dye-removal from an aqueous solution from the photocatalytic degradation to the surface-adsorption under the dark-condition. It has been also shown that, due to its magnetic nature, the magnetic dye-adsorbent catalyst can be separated from the treated solution using an external magnetic field and the previously adsorbed dye can be removed from the surface of nanotubes via typical surface-cleaning treatment, which make the recycling of the magnetic dye-adsorbent catalyst possible.

Organic synthetic dyes find applications in various fields including textile, leather tanning, paper production, food technology, agricultural research, light-harvesting arrays, photo-electrochemical cells, and hair-coloring. Due to the large-scale production, extensive use, and subsequent discharge of colored waste-waters containing the toxic and nonbiodegradable pollutants such as organic synthetic dyes, the latter are considered to be environmentally unfriendly and health-hazardous. Moreover, they affect the sunlight penetration and the oxygen solubility in the water-bodies, which in turn affect the under-water photosynthetic activity and life-sustainability. In addition to this, due to their strong color even at lower concentrations, the organic synthetic dyes generate serious aesthetic issues in the waste-water disposal [1–5].

As a consequence, powerful oxidation/reduction methods are required to ensure the complete decolorization and degradation of the organic synthetic dyes and their metabolites present in the waste-water effluents. Over the last two decades, photocatalysis has been the area of rapidly growing interest for the removal organic synthetic dyes from the industrial effluents, which involves the use of semiconductor particles as photocatalyst for the initiation of the redox chemical reactions on their surfaces [6–9]. When the semiconductor oxide particle is illuminated with the radiation having energy comparable to its band-gap energy, it generates highly active oxidizing/reducing sites, which can potentially oxidize/reduce large number of organic-wastes. Metal-oxide and metal-sulfide semiconductors, such as titania (TiO_2) [6–9], zinc oxide (ZnO) [10], tin oxide (SnO_2) [11], zinc sulfide (ZnS) [12], and cadmium sulfide (CdS) [13] have been successfully applied as photocatalyst for the removal of highly toxic and nonbiodegradable pollutants commonly present in air and waste-water. Among them, TiO_2 is believed to be the most promising one since it is cheaper, environmentally friendly, nontoxic, highly photocatalytically active, and stable to chemical and photo-corrosion. However, its effective application as a photocatalyst is hindered due to some of its major limitations. First, TiO_2 nanocrystallites trend to aggregate (or agglomerate) into large-sized nanoparticles, which affects its performance as a photocatalyst due to the decreased specific surface-area. Secondly, it has lower absorption in the visible-region, which makes it less effective in using the readily available solar-energy. Third, the separation of photocatalyst from the treated effluent, via traditional sedimentation and coagulation approaches, has been difficult and time consuming.

The technologies such as adsorption on inorganic or organic matrices and microbiological or enzymatic decomposition have also been developed for the removal of organic synthetic dyes from the waste-water to decrease their impact on the environment [5, 14–16]. However, the treatment of waste-water containing organic synthetic dyes using these techniques is very costly and has lower efficiency in the color removal and mineralization. The adsorption method results in the generation of large amount of sludge, which causes further difficulties in the recovery of the photocatalyst for recycling the product as a catalyst. Therefore, further development of these techniques for the effective waste-water treatment is essential.

In the literature, to overcome the major drawbacks associated with the photocatalytic degradation and adsorption mechanisms, the magnetic photocatalyst has been developed [17–26], which consists of a "core-shell" composite particle with a ceramic magnetic particle as a core and TiO_2-based photocatalyst particles as shell. Such magnetic nanocomposite, which possesses both the photocatalytic and magnetic properties, can be effectively separated from the treated solution using an external magnetic field. However, even the magnetic photocatalyst has been associated with several drawbacks. First, they show limited photocatalytic activity due to the presence of a core magnetic ceramic particle, which reduces the volume fraction of the photocatalyst available for the photo-degradation. Second, the total time of dye-decomposition using the magnetic photocatalyst is substantially higher (few hours). Third, the dye-removal using the magnetic photocatalyst is based predominantly on the photocatalytic degradation mechanism. Forth, being an energy-dependent mechanism (that is, requiring an exposure to the ultraviolet- (UV), visible-, or solar-radiation), the photocatalytic degradation is relatively an expensive process. Fifth, the dye-removal via other mechanism(s) such as the surface-adsorption, which is not an energy-dependent process (that is, it can be carried out in the dark) has never been utilized for the magnetic photocatalyst. This has been mainly due to the non-suitability of the magnetic photocatalyst for the surface-adsorption mechanism as a result of its lower specific surface-area. Sixth, the techniques to enhance the specific surface-area of the magnetic photocatalyst are not yet reported.

From these points of view, we demonstrate here the conversion of a magnetic photocatalyst having lower specific surface-area to a "magnetic dye-adsorbent catalyst", having higher specific surface-area, consisting a composite structure with the core of a magnetic ceramic particle and the shell of nanotubes of a dye-adsorbent material [27–29]. It is demonstrated here that, such conversion is accompanied by a concurrent change in the organic dye-removal mechanism from the photocatalytic degradation under the radiation-exposure to the surface-adsorption under the dark-condition, which offers several advantages over the former.

18.2 EXPERIMENTAL

18.2.1 PROCESSING OF MAGNETIC CERAMIC PARTICLES

A mixed cobalt ferrite ($CoFe_2O_4$) and hematite (Fe_2O_3) (CFH) and pure-$CoFe_2O_4$ magnetic ceramic powders are first processed via polymerized complex technique [28, 30]. 36.94 g of citric acid was dissolved in 40 ml of ethylene glycol as complexing agents and stirred to get a clear solution. 17 g of cobalt(II) nitrate ($Co(NO_3)_2 \cdot 6H_2O$, 98 + %) and 47.35 g of iron(III) nitrate ($Fe(NO_3)_3 \cdot 9H_2O$, 99.99 +

%) (Sigma-Aldrich, India) were added and the solution was stirred for 1 h followed by heating at 80°C for 4 h. The yellowish gel obtained was charred at 300 °C for 1 h in a vacuum furnace. A black colored solid precursor was obtained which, after grinding, was heated at 600 °C for 6 h to obtain a mixed $CoFe_2O_4$-Fe_2O_3 magnetic powder. Further calcination at 900 °C for 4 h resulted in the formation of pure-$CoFe_2O_4$ magnetic powder. The selection of mixed $CoFe_2O_4$-Fe_2O_3 or pure-$CoFe_2O_4$ powder as a core magnetic material for the photocatalytic and dye-adsorption measurements was as per convenience.

18.2.2 PROCESSING OF MAGNETIC PHOTOCATALYST

An insulating layer of silica (SiO_2) was deposited on the surface of magnetic ceramic particles using the Stober process [28, 31]. To 2 g suspension of magnetic powder dispersed in 250 ml of 2-propanol (S.D. Fine-Chem Ltd., India), 1 ml of ammonium hydroxide (NH_4OH, 25 wt.%, Qualigens Fine Chemicals, India) solution was slowly added. This was followed by the drop wise addition of 7.3 ml of tetraethylorthosilicate (TEOS, 98 %, Sigma-Aldrich, India) and the resulting suspension was allowed to settle after stirring for 3 h. The clear top solution was decanted and the powder was washed with 100 ml of 2-propanol and distilled water followed by drying in an oven at 60 °C overnight.

In order to deposit nanocrystalline TiO_2, 2 g of SiO_2-coated $CoFe_2O_4$-Fe_2O_3 magnetic powder was suspended in a clear solution of prehydrolized titanium(IV) iso-propoxide ($Ti(OC_3H_7)_4$, 98 percent, Sigma-Aldrich, India) precursor (4.73 g) dissolved in 125 ml of 2-propanol (Note: The prehydrolized precursor was obtained due to the reaction of pure-$Ti(OC_3H_7)_4$ with the atmospheric moisture over a prolonged period of time). To this suspension, a clear solution consisting 1.5 ml of distilled water (R=5, defined as the ratio of molar concentration of water to that of the precursor), dissolved in 125 ml of 2-propanol, was added drop wise. The suspension was stirred for 10 h, and after settling, the top solution was decanted. The powder was washed with 100 ml of 2-propanol and then dried in an oven at 80 °C overnight followed by calcination at 600 °C for 2 h. The above procedure was utilized for R=10 and 20 using the pure-alkoxide precursor, with a reduced concentration (0.5 g), and pure-$CoFe_2O_4$ as a core magnetic ceramic particle. In this case, the coating process was repeated twice to obtain relatively thicker TiO_2-coating.

18.2.3 HYDROTHERMAL TREATMENT OF MAGNETIC PHOTOCATALYST

The magnetic photocatalyst, as processed above, was subjected to the hydrothermal treatment under highly alkaline condition followed by typical washing and thermal treatments [27, 28]. 0.5 g of conventional magnetic photocatalyst was suspended in a highly alkaline aqueous solution containing 10 M sodium hydroxide (NaOH, 97

%, S.D. Fine-Chem Ltd., India) filled up to 84 vol.% of Teflon-beaker placed in a 200 ml stainless-steel (SS 316) vessel. The process was carried out in an autoclave (Amar Equipment Pvt. Ltd., Mumbai, India) at 120 °C for 30 h under an autogenous pressure. The hydrothermal product was washed once using 100 ml of 1 M hydrochloric acid (HCl, Qualigens Fine Chemicals, India) solution for 1 h followed by washing multiple times with distilled water till the final pH of the filtrate was in between ~5–7. The washed powder was dried in an oven at 110 °C overnight (dried-sample) and then calcined at 400 °C for 1 h (calcined-sample).

18.2.4 CHARACTERIZATION

The morphology of different samples at the nanoscale was examined using the transmission electron microscope (TEM, Tecnai G2, FEI, The Netherlands) operated at 300 kV. The selected-area electron diffraction (SAED) patterns were obtained to confirm the crystallinity and the structure of different samples. The crystalline phases present were determined using the X-ray diffraction (XRD, PW1710, Phillips, The Netherlands). The broad-scan analysis was typically conducted within the $2\text{-}\theta$ range of 10–80° using the Cu $K\alpha$ (λ=1.542 Å) X-radiation. The magnetic properties of different samples were measured using a vibrating sample magnetometer (VSM) attached to a Physical Property Measurement System (PPMS). The pristine samples were subjected to different magnetic field strengths (H) and the induced magnetization (M) was measured at 270 K. The external magnetic field was reversed on saturation and the hysteresis loop was traced [28].

18.2.5 DYE-ADSORPTION AND PHOTOCATALYTIC ACTIVITY MEASUREMENTS

The dye-adsorption measurements in the dark were conducted using the methylene blue (MB) (>96 %, S.D. Fine-Chem Ltd., India) as a model catalytic dye-agent. A 75 ml of aqueous suspension was prepared by dissolving 7.5 µmol•L^{-1} of MB dye and then dispersing 1.0 g•L^{-1} of the catalyst powder in pure distilled water. The suspension was stirred in the dark and 3 ml sample suspension was separated after each 30 min time interval for total 180 min. The catalyst powder was separated using a centrifuge (R23, Remi Instruments India Ltd.) and the solution was used to obtain the absorption spectra using the UV-visible absorption spectrophotometer (UV-2401 PC, Shimadzu, Japan). The normalized concentration of surface-adsorbed MB dye was calculated using the equation of the form,

$$\%MB_{adsorbed} = \left(\frac{C_0 - C_t}{C_0}\right)_{MB} \times 100 \qquad (18.1)$$

which is equivalent of the form,

$$\%MB_{adsorbed} = \left(\frac{A_0 - A_t}{A_0}\right)_{MB} \times 100 \qquad (18.2)$$

where, C_0 and C_t correspond to the MB dye concentration at the start and after string time "t", under the dark-condition, with the corresponding absorbance of A_0 and A_t. In few experiments, the powder was used for the successive cycles of dye-adsorption, under the dark-condition, to demonstrate its reusability as a catalyst. The dye-adsorption experiments were typically conducted under two different solution-pH (6.4 (neutral) and 10).

During the measurement of photocatalytic activity, the aqueous suspension of MB dye and the catalyst powder was stirred in the dark for 1 h to stabilize the surface-adsorption of the former on the surface of latter. The suspension is then exposed to the UV-radiation, having the wavelength within the range of 200–400 nm peaking at 360 nm, in a Rayonet photoreactor (The Netherlands) and 3 ml sample suspension was separated after each 10 min time interval for total 60 min. The powder was separated using a centrifuge and the solution was used to obtain the absorption spectra. The normalized residual MB dye concentration was calculated using the equation of the form,

$$\%MB_{adsorbed} = \left(\frac{C_t}{C_0}\right)_{MB} \times 100 \qquad (18.3)$$

which is equivalent of the form,

$$\%MB_{adsorbed} = \left(\frac{A_t}{A_0}\right)_{MB} \times 100 \qquad (18.4)$$

where, C_0 and C_t correspond to the MB dye concentration just at the beginning of the UV-radiation exposure (that is, after stirring the suspension under the dark-condition for 1 h) and after the UV-radiation exposure time of 't' with the corresponding absorbance of A_0 and A_t.

18.3 RESULTS AND DISCUSSION

The XRD broad-scan spectra, as obtained using the mixed $CoFe_2O_4$-Fe_2O_3 and pure-$CoFe_2O_4$ magnetic ceramic powders, are presented in Figure 18.1a, b. The major peaks corresponding to $CoFe_2O_4$ and Fe_2O_3 are identified after comparison with the JCPDS card numbers 22-1086 and 33-663. The formation of magnetic ceramic powders consisting mixed $CoFe_2O_4$-Fe_2O_3 and pure-$CoFe_2O_4$ is, thus, confirmed via broad-scan XRD analyses.

The TEM image of a magnetic photocatalyst particle, exhibiting a "core-shell" structure, with a core of magnetic ceramic particle and the shell of a sol-gel derived nanocrystalline coating of anatase-TiO$_2$ particles, is shown in Figure 18.2a. The SAED pattern as obtained from the core is shown as an inset in Fig. 2(a), which confirms the crystalline nature of the mixed CoFe$_2$O$_4$-Fe$_2$O$_3$ magnetic ceramic particle. The presence of anatase-TiO$_2$ in the shell has also been confirmed via XRD analysis as reported elsewhere [27, 28]. As observed in Figure 18.2b, the hydrothermal treatment results in the morphological transformation within the shell involving the conversion of nanocrystalline anatase-TiO$_2$ particles into the nanotubes of hydrogen titanate (H$_2$Ti$_3$O$_7$) as identified via high-magnification images of the shell presented in Figure 18.2c, d, and the corresponding SAED pattern shown as an inset in Figure 18.2c. The "core-shell" magnetic nanocomposite, Figure 18.2b, with the core of a magnetic ceramic particle and the shell of nanotubes of H$_2$Ti$_3$O$_7$ is termed as a "magnetic dye-adsorbent catalyst", which possesses the magnetic, dye-adsorption (in the dark), and catalytic properties.

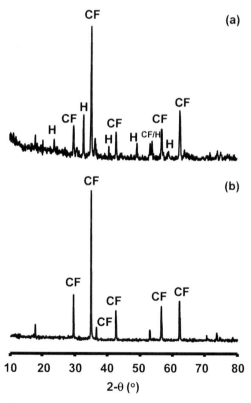

FIGURE 18.1 XRD patterns as obtained for the mixed CoFe$_2$O$_4$-Fe$_2$O$_3$ (CFH) (a) and pure-CoFe$_2$O$_4$ (b) magnetic ceramic particles.

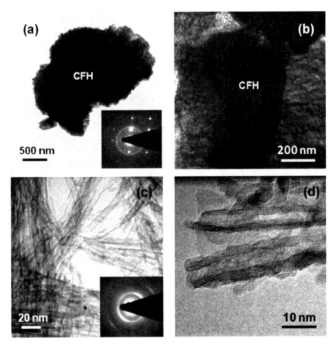

FIGURE 18.2 TEM images of the magnetic photocatalyst (**a**) and the magnetic dye-adsorbent catalyst (**b**). The high-magnification images of the shell of magnetic dye-adsorbent catalyst are presented in (**c**) and (**d**). The SAED pattern corresponding to the core of magnetic photocatalyst is shown as an inset in (**a**); while, that corresponding to the shell of magnetic dye-adsorbent catalyst is shown as an inset in (**c**). The samples are processed with $R=5$.

The comparison of the morphologies of magnetic photocatalyst and the magnetic dye-adsorbent catalyst is schematically shown in Figure 18.3. The formation mechanism of magnetic dye-adsorbent catalyst via hydrothermal treatment of the magnetic photocatalyst, under highly alkaline condition, can be explained using the model originally proposed for the free-standing powder [32–34], which is applied here for the similar conversion in the form of coating. When the anatase-TiO_2 particles, present in the coating form on the surface of magnetic ceramic particle, Figure 18.3a, are subjected to the hydrothermal treatment under highly alkaline condition, an exfoliation of single-layer nanosheets of sodium titanates ($Na_2Ti_3O_7$) results from the bulk anatase-TiO_2 structure, which continuously undergo the dissolution and crystallization processes [34].

$$3TiO_2 + 2NaOH \rightarrow Na_2Ti_3O_7 + H_2O \tag{18.5}$$

$$Na_2Ti_3O_7 \longleftrightarrow 2Na^+ + Ti_3O_7^{2-} \tag{18.6}$$

Due to their higher surface-area to volume ratio and the presence of dangling bonds along the two long-edges, the nanosheets of $Na_2Ti_3O_7$ have a strong drive to rollup. However, this rollup tendency is opposed by the repulsive force produced by the charge on the nanosheets created by the presence of Na^+-ions. These Na^+-ions are easily replaced via ion-exchange mechanism with H^+-ions in the subsequent washing steps in an acidic aqueous solution and pure-water, which reduces the repulsive force to rollup.

$$Na_2Ti_3O_7 + 2HCl \rightarrow H_2Ti_3O_7 + 2NaCl \tag{18.7}$$

$$Na_2Ti_3O_7 + 2H_2O \rightarrow H_2Ti_3O_7 + 2NaOH \tag{18.8}$$

This results in the formation of nanotubes of $H_2Ti_3O_7$, which are normally transformed to those of anatase-TiO_2 following the calcination treatment at higher temperature.

$$H_2Ti_3O_7 \xrightarrow{\Delta} 3TiO_2 + H_2O \tag{18.9}$$

The XRD pattern of the magnetic dye-adsorbent catalyst (calcined-sample), however, did not reveal the presence of anatase-TiO_2 on the surface [27, 28]. Since the $H_2Ti_3O_7$-to-anatase TiO_2 transformation has been observed earlier in the powder form [35], it appears that, such transformation is possibly retarded in the coating form due to the substrate effect.

The *M-H* graphs, as obtained for the magnetic photocatalyst and the magnetic dye-adsorbent catalyst (calcined-sample), are presented in Figure 18.4. The presence of a hysteresis loop is noted, which suggests the ferromagnetic nature of these composite particles. For the magnetic photocatalyst, the obtained values of saturation magnetization, remenance magnetization, and coercivity are 59 emu•g^{-1}, 24 emu•g^{-1}, 1410 Oe; while, those for the magnetic dye-adsorbent catalyst (calcined-sample) are 45 emu•g^{-1}, 15 emu•g^{-1}, 578 Oe. It is noted that, the magnetic dye-adsorbent catalyst (calcined-sample) shows reduced saturation magnetization, remenance magnetization, and coercivity relative to those observed for the magnetic photocatalyst. This has been attributed to the combined effect of decrease in the volume fraction and increase in the average particle size and crystallinity of the core magnetic ceramic particle following the hydrothermal, drying, and calcination treatments [28]. Nevertheless, the hydrothermal product (calcined-sample) also possesses the magnetic property and is suitable for its separation, after the photocatalytic and dye-adsorption experiments, using an external magnetic field.

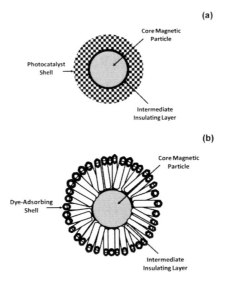

FIGURE 18.3 Schematic representation of the morphologies of magnetic photocatalyst (**a**) and magnetic dye-adsorbent catalyst (**b**).

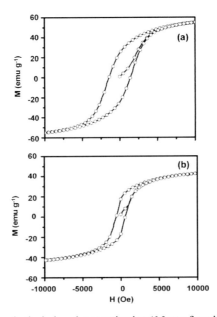

FIGURE 18.4 Variation in the induced magnetization (M) as a function of external magnetic field strength (H) as obtained for the magnetic photocatalyst (**a**) and magnetic dye-adsorbent catalyst (calcined-sample) (**b**). The samples are processed with $R = 5$.

The variation in the normalized residual MB dye concentration as a function of stirring time, under the UV-radiation exposure, as obtained for the pure and Pd-deposited [36] magnetic photocatalyst is shown in Figure 18.5a. It has been demonstrated earlier that, the sol-gel derived pure nanocrystalline anatase-TiO$_2$ particles completely remove the MB dye via photocatalytic degradation mechanism under similar test-conditions within the UV-radiation exposure time of 1 h [37]. It, hence, appears that the magnetic photocatalyst possesses very slow MB dye degradation kinetics. The comparison of the kinetics of MB dye removal via surface-adsorption mechanism, under the dark-condition, using the magnetic photocatalyst and the magnetic dye-adsorbent catalyst is presented in Figure 18.5b. Due to its higher specific surface-area, the magnetic dye-adsorbent catalyst (dried-sample) exhibits >99 percent of dye-adsorption, which is larger than that (~90 %) of the calcined-sample. This is attributed here to some loss in the specific surface-area of the magnetic dye-adsorbent catalyst as a result of the calcination treatment. Nevertheless, the magnetic dye-adsorbent catalysts (dried- and calcined-samples) show significantly higher amount of dye-adsorption on the surface, under the dark-condition, relative to that (~50–55 %) shown by the magnetic photocatalyst. It is, thus, successfully demonstrated that under the dark-condition, the magnetic dye-adsorbent catalyst removes an organic dye from an aqueous solution predominantly via surface-adsorption mechanism. On the other hand, the magnetic photocatalyst cannot completely remove the MB dye neither via photocatalytic degradation mechanism (under the UV-radiation exposure) nor via surface-adsorption mechanism (under the dark-condition). This is further supported by the comparison of the qualitative variation in the color of the MB dye solution as a function of stirring time, as observed for the magnetic photocatalyst and the magnetic dye-adsorbent catalyst (dried-sample), Figure 18.6a. In Figure 18.6b, the qualitative variation in the color of the H$_2$Ti$_3$O$_7$ nanotubes (without the core magnetic ceramic particle) is presented under different conditions. It is noted that, the initial white-color of the nanotubes changes to blue after the surface-adsorption of the MB dye under the dark-condition. After a typical surface-cleaning treatment [27], the surface-adsorbed MB dye can be decomposed, which is suggested by the disappearance of the blue-color. This produces surface-cleaned H$_2$Ti$_3$O$_7$ nanotubes which can be recycled for the next cycle of dye-adsorption under the dark-condition with the dye-adsorption capacity comparable with the original powder. (Note: Since the surface-cleaning treatment is effective with pure-H$_2$Ti$_3$O$_7$ nanotubes, it would also be effective in the presence of the core magnetic ceramic particle). As a result, it seems that, the magnetic dye-adsorbent can be recycled and used as a catalyst for the dye-removal application.

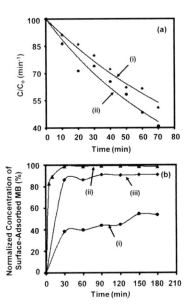

FIGURE 18.5 (a) Variation in the normalized residual MB dye concentration as a function of stirring time under the UV-radiation exposure as obtained for the pure (i) and palladium (Pd)-deposited (ii) magnetic photocatalyst. (b) Variation in the normalized concentration of surface-adsorbed MB as a function of stirring time under the dark-condition as obtained for the magnetic photocatalyst (i) and magnetic dye-adsorbent catalyst, dried (ii) and calcined (iii) samples. The samples are processed with $R = 5$.

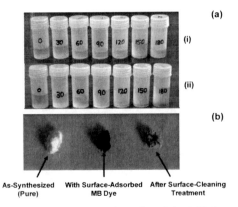

FIGURE 18.6 (a) Qualitative variation in the color of the MB dye solution as a function of stirring time under the dark-condition as obtained for the magnetic photocatalyst (i) and magnetic dye-adsorbent catalyst (dried-sample) (ii). The samples are processed with $R = 5$. (b) Qualitative variation in the color of the $H_2Ti_3O_7$ nanotubes powder (without the core magnetic ceramic particle) under different conditions.

The increased surface-adsorption of the MB dye, under the dark-condition, as observed earlier in Figure 18.5b for $R = 5$ following the hydrothermal treatment of the magnetic photocatalyst, is also shown by those processed with larger R-values (10 and 20), Figure 18.7a and 18.7b. Interestingly, comparison of Figures 18.8 and 18.7b shows that, at higher solution-pH in the basic region (pH = 10), both the magnetic photocatalyst as well as the magnetic dye-adsorbent catalyst exhibit high and comparable surface-adsorption of the MB dye under the dark-condition. However, when the same catalyst is used for the repetitive dye-adsorption cycles, at higher solution-pH in the basic region (pH = 10), the magnetic photocatalyst rapidly loses its maximum dye-adsorption capacity, Figure 18.9a, during each successive cycles. On the other hand, under similar test-conditions, the magnetic dye-adsorbent catalyst (calcined-sample) retains its maximum dye-adsorption capacity, Figure 18.9b.

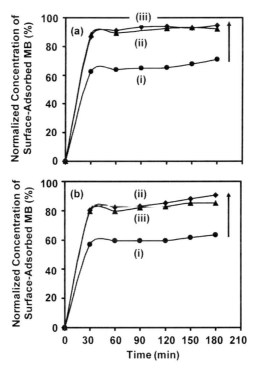

FIGURE 18.7 Variation in the normalized concentration of surface-adsorbed MB as a function of stirring time under the dark-condition as obtained for different samples: magnetic photocatalyst (i) and magnetic dye-adsorbent catalyst, dried (ii) and calcined (iii) samples. The samples are processed with $R = 10$ (a) and $R = 20$ (b), and the dye-adsorption measurements are conducted at the neutral solution-pH (6.4).

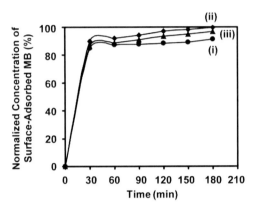

FIGURE 18.8 Variation in the normalized concentration of surface-adsorbed MB as a function of stirring time under the dark-condition as obtained for different samples: magnetic photocatalyst (i) and magnetic dye-adsorbent catalyst, dried (ii) and calcined (iii) samples. The samples are processed with $R = 20$ and the dye-adsorption measurements are conducted at the basic solution-pH (10).

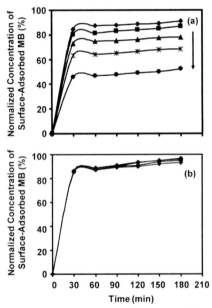

FIGURE 18.9 Variation in the normalized concentration of surface-adsorbed MB as a function of stirring time under the dark-condition, for the successive dye-adsorption cycles, as obtained for different samples—magnetic photocatalyst (**a**) and magnetic dye-adsorbent catalyst (calcined- sample) (**b**). The samples are processed with $R = 20$ and the dye-adsorption measurements are conducted at the basic solution-pH (10).

As a result, overall it appears that, the transformation of magnetic photocatalyst to the magnetic dye-adsorbent catalyst is accompanied by a concurrent change in the dye-removal mechanism from the photocatalytic degradation under the UV-radiation exposure to the surface-adsorption under the dark-condition.

18.4 CONCLUSIONS

The magnetic photocatalyst has been successfully converted to the magnetic dye-adsorbent catalyst via hydrothermal followed by typical washing and thermal treatments. The latter consists of a core-shell nanocomposite with the core of a magnetic ceramic particle, such as mixed $CoFe_2O_4$-Fe_2O_3 and pure-$CoFe_2O_4$, and the shell of nanotubes of dye-adsorbing material such as $H_2Ti_3O_7$. The magnetic dye-adsorbent catalyst has been successfully utilized for the removal of organic dye from an aqueous solution, under the dark-condition, via surface-adsorption mechanism. Due to their magnetic nature, the magnetic dye-adsorbent catalyst can be separated from the treated solution using an external magnetic field. The successful removal of previously adsorbed dye from the surface of nanotubes, via typical surface-treatment, suggests that the magnetic dye-adsorbent catalyst can be recycled.

ACKNOWLEDGEMENTS

Author thanks CSIR, India for funding the photocatalysis and nanotechnology research at NIIST-CSIR, India through the Projects # NWP0010 and P81113.

KEYWORDS

- Dye-adsorbent catalyst
- Magnetic ceramic particle
- Magnetic photocatalyst
- Methylene blue (MB)

REFERENCES

1. Forgacsa, E.; Cserhati, T.; and Oros, G.; *Environ. Int.* **2004**, *30*, 953–971.
2. Gupta, G. S.; Shukla, S.; Prasad, G.; and Singh, V. N.; *Environ. Technol.* **1992**, *13*, 925–936.
3. Shukla, S. P.; and Gupta, G. S.; *Ecotoxicol. Environ. Saf.* **1992**, *24*, 155–163.
4. Sokolowska-Gajda, J.; Freeman, H. S.; and Reife, A.; *Dyes. Pigm.* **1996**, *30*, 1–20.
5. Robinson, T.; McMullan, G.; Marchant, R.; and Nigam, P.; *Bioresource. Technol.* **2001**, *77*, 247–255.
6. Carp, O.; Huisman, C. L.; and Reller, A.; *Prog. Solid State Ch.* **2004**, *32*, 33–177.

7. Tachikawa, T.; Fujitsuka, M.; and Majima, T.; *J. Phys. Chem. C.* **2007**, *111*, 5259–5275.
8. Chen, X.; and Mao, S. S.; *Chem. Rev.* **2007**, *107*, 2891–2959.
9. Fujishima, A.; Zhang, X.; Tryk, D. A.; *Surf. Sci. Rep.* **2008**, *63*, 515–582.
10. Marto, J.; Marcos, P.S.; Trindade, T.; and Labrincha, J. A.; *J. Haz. Mat.* **2009**, *163*, 36–42.
11. Pan, S. S.; Shen, Y. D.; Teng, X. M.; Zhang, Y.X.; Li, L.; Chu, Z. Q.; Zhang, J.P.; Li, G.H.; and Hub, X.; *Mater. Res. Bull.* **2009**, *44*, 2092–2098.
12. S. Feng, J. Zhao, Z. Zhu, *Mater. Sci. Eng. B* 2008, 150, 116-120.
13. A. Datta, A. Priyama, S.N. Bhattacharyya, K.K. Mukherjee, A. Saha, *J. Colloid Interf Sci.* 2008, 322, 128-135.
14. Shaul, G. M.; Holdsworth, T. J.; Dempsey, C. R.; and Dostal, K. A.; *Chemosphere* **1991**, *22*, 107–119.
15. Gupta, V.; and Suhas K.; *J. Environ. Manage.* **2009**, *90*, 2313–2342.
16. Crini, G.; *Bioresource. Technol.* **2006**, *97*, 1061–1085.
17. Beydoun, D.; Amal, R.; Scott, J.; Low, G.; and Evoy, S. M. *Chem. Eng. Technol.* **2001**, *24*, 745–748.
18. Song, X.; Gao, L.; *J. Am. Ceram. Soc.* **2007**, *90*, 4015–4019.
19. Gao, Y.; Chen, B.; Li, H.; and Ma, Y.; *Mater. Chem. Phys.* **2003**, *80*, 348–355.
20. Xiao, H. M.; Liu, X. M.; and Fu, S. Y.; *Compos. Sci. Technol.* **2006**, *66*, 2003–2008.
21. Rana, S.; Rawat, J.; Sorensson, M. M.; and Misra, R. D. K.; *Acta. Biomater.* **2006**, *2*, 421–432.
22. Lee, S. W.; Drwiega, J.; Mazyck, D.; Wu, C. Y.; Sigmund, W. M.; *Mater. Chem. Phys.* **2006**, *96*, 483–488.
23. Fu, W.; Yang, H.; Li, M.; Li, M.; Yang, N.; and Zou, G.; *Mater. Lett.* **2005**, *59*, 3530–3534.
24. Jiang, J.; Gao, Q.; Chen, Z.; Hu, J.; Wu, C.; *Mater. Lett.* **2006**, *60*, 3803–3808.
25. Beydoun, D.; Amal, R.; Low, G.; and McEvoy, S.; *J. Mol. Catal. A.* **2002**, *180*, 193–200.
26. Siddiquey, I. A.; Furusawa, T.; Sato, M.; and Suzuki, N.; *Mater. Res. Bull.* **2008**, *43*, 3416–3424.
27. Shukla, S.; Warrier, K. G. K.; Vaarma, M. R.; Lajina, M. T.; Harsha, N.; Reshmi, C. P.; PCT Application No. PCT/IN2010/000198.
28. Lajina, T.; Shereef, A.; Shukla, S.; Pattelath, R.; Varma, M. R.; Suresh, K. G.; Patil, K.; Warrier, K. G. K.; *J. Am. Ceram. Soc.* (No. doi: 10.1111/j.1551-2916.2010.03949.x).
29. Shukla, S.; Varma, M. R.; Suresh, K. G.; and Warrier, K. G. K.; Magnetic dye-adsorbent catalyst: a "Core-Shell" nanocomposite. In: Proceedings of anoTech Conference and Expo, TechConnect World Summit Conferences and Expo 2010, Anaheim, California, U.S.A., Vol. 1, pp. 830–833, June 21–24; **2010**.
30. Varma, P. C. R.; Manna, R. S.; Banerjee, D.; Varma, M. R.; Suresh, K. G.; and Nigam, A. K.; *J. Alloy Compd.* **2008**, *453*, 298–303.
31. Lee, S. W.; Drwiega, J.; Mazyck, D.; Wu, C.Y.; and Sigmund, W. M.; *Mater. Chem. Phys.* **2006**, *96*, 483–488.
32. Kasuga, T.; Hiramatsu, M.; Hoson, A.; Sekino, T.; and Niihara, K.; *Langmuir* **1998**, *14*, 3160–3163.
33. Kasuga, T.; Hiramatsu, M.; Hoson, A.; Sekino, T.; and Niihara, K.; *Adv. Mater.* **1999**, *11*, 1307–1311.
34. Bavykin, D. V.; Parmon, V. N.; Lapkina, A. A.; and Walsh, F. C.; *J. Mater. Chem.* **2004**, *14*, 3370–3377.
35. Sun, X.; and Li, Y.; *Chem. Eur. J.* **2003**, *9*, 2229–2238.
36. Harsha, N.; Ranya, K. R.; Shukla, S.; Biju, S.; Reddy, M.L.P.; and Warrier, K.G.K.; *J. Nanosci. Nanotech.* (in press)
37. Baiju, K.V.; Shukla, S.; Sandhya, K. S.; James, J.; Warrier, K. G. K.; *J. Phys. Chem. C.* **2007**, *111*, 7612–7622.

CHAPTER 19

SOLID POLYMER FUEL CELL: A THREE-DIMENSIONAL COMPUTATION MODEL AND NUMERICAL SIMULATIONS

MIRKAZEM YEKANI, MEYSAM MASOODI, NIMA AHMADI, MOHAMAD SADEGHI AZAD, and KHODADAD VAHEDI

19.1 INTRODUCTION

Fuel cell is a device that converts chemical energy directly into the electrical energy. The main components of the fuel are Cell Electrolyte, anode electrode and cathode electrode. The electrochemical reactions performing around the electrodes and generating the electric potential lead to electric current. Oxygen reduction reaction is three time slower than hydrogen oxidation reaction. In fuel cell system with theoretical viewpoint, electrical energy can be generated regularly while the oxidant and fuel are injecting into the electrodes. Depreciation, corrosion, and unsuited operation occur due to loss of the life time of the fuel cell. In this system, fuel, this is mostly hydrogen, flows in anode side and oxidant gases such as air or pure oxygen pass in cathode side.

The most attractive choice among variety of fuel cell is Polymer electrolyte membrane fuel cell (PEMFC). In PEMFC, there is a solid polymer electrolyte conducting protons between two platinum impregnated porous electrodes, the present electrodes are cast as thin films and bonded to the membrane. William T. Grubbs [1] originally performed the use of organic cation exchange membrane polymers in fuel cells in 1959. One of the best electrolytes used in PEMFC is Nafion. Water is produces as liquid instead of steam in a PEFC. The first three-dimensional model was presented by Dutta et al. in 2000 [4]. Berning, Djilali, and Li et al. have also developed steady-state, three-dimensional, no isothermal models for PEMFCs recently [5]. PEMFC have been extensively suggested for future power supply in the vehicles, by developing the power generation systems and electronically applications. Although research and investigations in fuel cell systems have been developed a lot, but these systems and applications are still very expensive and complicated so they are not suitable for commercial uses. Complete fuel cell systems have been demonstrated for a number of transportation applications including public transit buses and passenger automobiles. One of the most important goals of recent development has been cost reduction and high volume manufacture for the catalyst, membranes, and bipolar plates. This issue will come off by ongoing research to increase power density, improve water management, operate at ambient conditions, tolerate reformed fuel, and extend stack life. In recent years various research and experiments on PEMFC, by various geometries have been developed. One of the main requirements of these cells is maintaining a high water content in the electrolyte to ensure high ionic conductivity.

During the reactive mode of operation water content in the cell will be determined by the balance of water or its transport. Contributing factors to water transport are the water drag through the cell, back diffusion from the cathode, and the diffusion of any water in the fuel stream through the anode.

Water transport is function of cell current and the characteristics of the membrane and the electrodes. Water drag refers to the amount of water that pulled by osmotic action along with the proton. Water management has a noticeable impact on cell performance, because at high current densities mass transport issues associated with water formation and distribution limit cell output. Without sufficient water management, an imbalance will happen between water production and evaporation within the cell. Against effects, include dilution of reactant gases by water vapor, flooding of the electrodes, and dehydration of the solid polymer membrane. If dehydration occurs the adherence of the membrane to the electrode also will be adversely affected. As there is no free liquid electrolyte to form a conducting bridge so intrinsic contact between the electrodes and the electrolyte membrane is important. If water was exhausted more than produced, thus it is important to humidify the incoming anode gas. If there is too much humidification, however, the electrode floods, which causes problems with diffusing the gas to the electrode. A smaller current, larger reactant flow, lower humidity, higher temperature, or lower pressure will result in a water deficit. A higher current, smaller reactant flow, higher humidity, lower temperature, or higher pressure will lead to a water surplus. There have been attempts to use external wicking connected to the membrane to either drain or supply water by capillary action in order to control the water in the cell. The ionic conductivity of the electrolyte is higher when the membrane is fully saturated, and this offers a low resistance to current flow and increases efficiency. Operating temperature has a significant influence on PEFC performance. If the temperature increases the internal resistance of the cell will decrease, mainly by downfall of the ohmic resistance of the electrolyte. In addition, mass transport limitations get reduced at higher temperatures. The overall result is an improvement in cell performance. Experimental data [1, 2] suggest a voltage gain in the range of 1.1 mV to 2.5 mV for each degree (°C) of temperature increase. The other advantage of Operating at higher temperatures is reducing the chemisorptions of CO as this reaction is exothermic. Improving the cell performance through an increase in temperature, however, limited by the high vapor pressure of water in the ion exchange membrane. This is due to the membrane's susceptibility to dehydration and the subsequent loss of ionic conductivity. Operating pressure also influences cell performance. An increase in the oxygen pressure from 3 to 10.2 atmospheres produces an increase of 42 mV in the cell voltage at 215 mA/cm^2. According to the Nernst equation, the increase in the reversible cathode potential that expected for this increase in oxygen pressure is about 12 mV, which is considerably less than the measured value. These results demonstrate that an increase in the pressure of oxygen results in a significant reduction in polarization at the cathode [1, 2]. Performance improvements due to increased pressure must be balanced against the energy required to pressurize the reactant gases. The overall system must be optimized according to output, efficiency, cost, and size. Operating at pressure above ambient conditions would most likely be reserved for stationary power applications. In this work, a three-dimensional, steady-state, mathematical model was developed. Because of no material found by zero permeability, some researches and investigations have been done by mathematical and numerical modeling of fuel cell, which respect bipolar plates as another parts of PEMFC as gas diffusion layers. Gas diffusion layer that currently made from carbon fiber or carbon cloth and functions to wick away liquid water, transport reactants of H_2 and O_2 and conduct electrons. Its thickness is normally between 200 and 300 μm. [7, 8] Design of a flow channel is very complicated and difficult to optimize geometry, shape, and size of flow fields like that the gas flow channels and bipolar plates since there are many parameters which affect the fuel cell operation such as different materials use in cathode and anode bipolar plates [8] channel/rib ratio [9], and channel path length [10] Catalyst layer which is the region where the membrane and the electrodes overlap and the H2 oxidation or O_2 reduction reaction occur. It allows electron and ion conduction at the same time. There are other parameters, which affect cell performance such as operating temperature, pressure, and humidification of the gas in the cell. It is necessary to understand these parameter and their effects on cell performance. For this reasons we understand it is a point to set up these design factors at optimum values in order to increase the PEMFC operation performance. In this model, major transport phenomena in PEMFC and the effects of prominent gas diffusion layers on cell performance and output cell voltage were studied. The modeling data for base case validate with experimental data.

19.2 EQUATIONS

19.2.1 MATHEMATICAL MODEL

Figure 19.1 shows a schematic of a single cell of a PEMFC. It is made of two porous electrodes, a polymer electrolyte membrane, two-catalyst layer and two-gas distributor plates. The membrane is sandwiched between the gas channels.

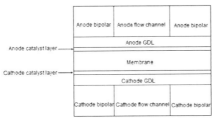

FIGURE 19.1 Schematic of a single straight channel of PEMFC.

19.2.2 GOVERNING EQUATIONS

In this numerical simulation, a single domain model formation was used for the governing equations. These governing equations consist of mass conservation, momentum, and species equations, which can be written as:

$$\left(\nabla.\rho u\right) = 0 \tag{19.1}$$

$$\frac{1}{\left(\varepsilon^{\text{eff}}\right)^2} \nabla.\left(\rho u u\right) = -\nabla P + \nabla.\left(\mu \nabla u\right) + S_u \tag{19.2}$$

$$\nabla.\left(u C_K\right) = \nabla.\left(D_K^{\text{eff}} \nabla C_K\right) + S_K \tag{19.3}$$

$$\nabla.\left(\kappa_e^{\text{eff}} \nabla \Phi_e\right) + S_\Phi = 0 \tag{19.4}$$

In Eq. (19.1) ρ is the density of gas mixture. According to model assumption, mass source and sink term neglected. ε is the effective porosity inside porous mediums, and μ is the viscosity of the gas mixture in the momentum equation is shown as Eq. (19.2) The momentum source term, Su, is used to describe Darcy's drag for flow through porous gas diffusion layers and catalyst layers [11] As:

$$S_u = -\frac{\mu}{K} u \tag{19.5}$$

Solid Polymer Fuel Cell

K is the gas permeability inside porous mediums. D_k^{eff} In the species equation shown as Eq. (19.3), is the effective diffusion coefficient of species k (e.g., hydrogen, oxygen, nitrogen, and water vapor) and is defined to describe the effects of porosity in the porous gas diffusion layers and catalyst layers by the Bruggeman correlation [12] as:

$$D_K^{\text{eff}} = \left(\varepsilon^{\text{eff}}\right)^{1.5} D_K \tag{19.6}$$

Transport properties for species are given in Table 19.1

TABLE 19.1 Transport properties [13]

Property	value
H2 Diffusivity in the gas channel, D0H2	1.10×10^{-04} m²/s
O2 Diffusivity in the gas channel, D0o2	3.20×10^{-05} m²/s
H2O Diffusivity in the gas channel, D0H2o	7.35×10^{-05} m²/s
H2 Diffusivity in the membrane, DmemH2	2.59×10^{-10} m²/s
O2 Diffusivity in the membrane, Dmemo2	1.22×10^{-10} m²/s

Additionally, diffusion coefficient is function of temperature and pressure [13] by next equation:

$$D_K = D^{\circ}{}_K \left(\frac{T}{T_{\circ}}\right)^{\frac{3}{2}} \left(\frac{P_{\circ}}{P}\right) \tag{19.7}$$

The charge conservation equation is shown as Eq. (19.4) and K_e is the ionic conductivity in the ion metric phase and has been incorporated by Springer et al. [14] as:

$$\kappa_e = \exp\left[1268\left(\frac{1}{303} - \frac{1}{T}\right)\right] \times (0.005139\lambda - 0.00326) \tag{19.8}$$

Moreover, in recent equation, λ is defined as the number of water molecules per sulfonate group inside the membrane. The water content can be assumed function of water activity, a, is defined according to experimental data [15]:

$$\lambda = 0.3 + 6a\left[1 - \tanh(a - 0.5)\right] + 3.9\sqrt{a}\left[1 + \tanh\left(\frac{a - 0.89}{0.23}\right)\right] \tag{19.9}$$

Water activity, a, is defined by:

$$a = \frac{C_w RT}{P_w^{sat}} \qquad (19.10)$$

The proton conductivity in the catalyst layers by introducing the Bruggeman correlation [16] is defined as:

$$\kappa_e^{eff} = \varepsilon_m^{1.5} \kappa_e \qquad (19.11)$$

In recent equation ε_m is the volume fraction of the membrane phase in the catalyst layer. The source and sink term in Eqs. (19.3) and (19.4) are given in Table 19.2. Local current density in the membrane can be calculated by:

$$I = -\kappa_e \nabla \Phi_e \qquad (19.12)$$

Then the average current density is calculated as follow:

$$I_{ave} = \frac{1}{A} \int_{A_{mem}} I dA \qquad (19.13)$$

Where A is the active area over the MEA.

19.2.3 WATER TRANSPORT

Water molecules in PEM fuel cell are transported via electro-osmotic drag due to the properties of polymer electrolyte membrane in addition to the molecular diffusion. H$^+$ protons transport water molecules through the polymer electrolyte membrane and this transport phenomenon is called electro-osmotic drag. In addition to the molecular diffusion and electro-osmotic drag, water vapor is also produced in the catalyst layers due to the oxygen reduction reaction.

Water transport through the polymer electrolyte membrane is defined by:

$$\nabla \cdot \left(D_{H_2O}^{mem} \nabla C_{H_2O}^{mem} \right) - \nabla \cdot \left(\frac{n_d}{F} i \right) = 0 \qquad (19.14)$$

TABLE 19.2 Source/sink term for momentum, species, and charge conservation equations for individual regions

	Momentum	Species	Charge
Flow channels	$S_u = 0$	$S_K = 0$	$S_\Phi = 0$
Bipolar plates	$S_u = -\frac{\mu}{K} u$	$S_K = 0$	$S_\Phi = 0$

GDLs	$S_u = -\dfrac{\mu}{K} u$	$S_K = 0$	$S_\Phi = 0$
Catalyst layers	$S_u = 0$	$S_K = -\nabla \cdot \left(\dfrac{n_d}{F} I\right) - \dfrac{S_K j}{nF}$	$S_\Phi = j$
Membrane	$S_u = 0$	$S_K = -\nabla \cdot \left(\dfrac{n_d}{F} I\right)$	$S_\Phi = 0$

Where n_d and $D_{H_2O}^{mem}$ are defined as the water drag coefficient from anode to cathode and the diffusion coefficient of water in the membrane phase, respectively.

The number of water molecules transported by each hydrogen proton H⁺ is called the water drag coefficient. It can be determined from the following equation [15]:

$$n_d = \begin{cases} 1 & \lambda < 9 \\ 0.117\lambda - 0.0544 & \lambda \geq 9 \end{cases} \qquad (19.15)$$

The diffusion coefficient of water in the polymer membrane is dependent on the water content of the membrane and is obtained by the following fits of the experimental expression [17]:

$$D_w^{mem} = \begin{cases} 3.1\times 10^{-7} \lambda \left(e^{0.28\lambda}-1\right) e^{\left(\frac{-2346}{T}\right)} & 0 < \lambda \leq 3 \\ 4.17\times 10^{-8} \lambda \left(1+161 e^{-\lambda}\right) e^{\left(\frac{-2346}{T}\right)} & \text{Otherwise} \end{cases} \qquad (19.16)$$

The terms are therefore related to the transfer current through the solid conductive materials and the membrane. The transfer currents or source terms are nonzero only inside the catalyst layers. The transfer current at anode and cathode can be described by Tafel equations as follows:

$$R_{an} = j_{an}^{ref} \left(\dfrac{[H_2]}{[H_2]_{ref}}\right)^{\gamma an} \left(e^{\alpha_{am} F \eta_{an}/RT} - e^{-\alpha_{cat} F \eta^{an}/RT}\right) \qquad (19.17)$$

$$R_{cat} = j_{an}^{ref} \left(\dfrac{[O_2]}{[O_2]_{ref}}\right)^{\gamma cat} \left(-e^{\alpha_{am} F \eta_{cat}/RT} + e^{-\alpha_{cat} F \eta^{cat}/RT}\right) \qquad (19.18)$$

According to the Tafel equation, the current densities in the anode and cathode catalysts can be expressed by the exchange current density, reactant concentration, temperature, and overpotentials according to the Tafel equations. Where the surface over potential is defined as:

The difference between proton potential and electron potential.

$$\eta_{an} = \varphi_{sol} - \varphi_{mem} \tag{19.19}$$

$$\eta_{cat} = \varphi_{sol} - \varphi_{mem} - V_{oc} \tag{19.20}$$

The open-circuit potential at the anode is assumed to be zero, while the open-circuit potential at the cathode becomes a function of a temperature as:

$$V_{oc} = 0.0025T + 0.2329 \tag{19.21}$$

The protonic conductivity of membrane is dependent on water content, where σ_m is the ionic conductivity in the ionomeric phase and has been correlated by Springer et al. [23]:

$$\sigma = (0.005139\lambda - 0.00326)\exp\left[1268\left(\frac{1}{303} - \frac{1}{T}\right)\right] \tag{19.22}$$

Energy equation given by Eq. (19.23):

$$\nabla \cdot (\rho u T) = \nabla \cdot (\lambda_{eff} \nabla T) + S_T \tag{19.23}$$

Where, λ_{eff} is the effective thermal conductivity, and the source term of the energy equation, S_T, is defined with the following equation:

$$S_T = I^2 R_{ohm} + h_{reaction} + \eta_a i_a + \eta_c i_c \tag{19.24}$$

In this equation, R_{ohm}, is the ohmic resistance of the membrane, $h_{reaction}$, is the heat generated thorough the chemical reactions, η_a and η_c, are the anode and cathode overpotentials, which are calculated as:

$$R_{ohm} = \frac{t_m}{\sigma_e} \tag{19.25}$$

Here, t_m is the membrane thickness.

$$\eta_a = \frac{RT}{\alpha_a F} \ln\left[\frac{IP}{j_{0_a} P_{0_{H_2}}}\right] \tag{19.26}$$

$$\eta_C = \frac{RT}{\alpha_c F} \ln\left[\frac{IP}{j_{0_c} P_{0_{O_2}}}\right] \tag{19.27}$$

Where, α_a and, α_c are the anode and cathode transfer coefficients, $P_0 o_2$, is the partial pressure of hydrogen and oxygen, and, j0 is the reference exchange current density.

Solid Polymer Fuel Cell

The fuel and oxidant fuel rate, u, is given by following equations:

$$u_{in,a} = \frac{\xi_a I_{ref} A_{mem}}{2C_{H_2,in} FA_{ch}} \qquad (19.28)$$

$$u_{in,c} = \frac{\xi_c I_{ref} A_{mem}}{4C_{O_2,in} FA_{ch}}$$

In present equation, I_{ref} is the reference current density and ξ is a stoichiometric ratio, which is defined as the ratio between the amount supplied and the amount required of the fuel based on the reference current density. The species concentrations of flow inlets are assigned by the humidification conditions of both the anode and cathode inlets.

19.2.4 BOUNDARY CONDITION

Equations (19.1) to (19.4) form the complete set of governing equations for the traditional mathematical model. Boundary conditions are dispensed at the external boundaries. Constant mass flow rate at the channel inlet and constant pressure condition at the channel outlet, the no-flux conditions are executed for mass, momentum, species, and potential conservation equations at all boundaries except for inlets and outlets of the anode and cathode flow channels.

19.3 RESULTS AND DISCUSSION

19.3 1MODEL VALIDATION

To validate the numerical simulation model used in this study, the performance curves of voltage and current density for base case compared with the experimental data of wang et al [18].

Fuel cell operating condition and geometric parameters are shown in Table 19.3.It is used fully humidified inlet condition for anode and cathode. The transfer current at anode and cathode can be described by Tafel equations.

Polarization and power density curve of present model is shown in Figure 19.2 this curve signifies the good agreement between present model (for base case) with experimental data.

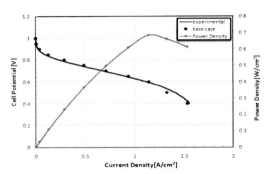

FIGURE 19.2 Comparison of polarization curve of model with experimental data and power density curve at 1.5 (A/m2).

TABLE 19.3 Geometric Parameters and operating condition of PEMFC [18].

Parameter	value
Gas channel length	7.0×10^{-2} m
Gas channel width and depth	1.0×10^{-3} m
Bipolar plate width	5.0×10^{-4} m
Gas diffusion layer thickness	3.0×10^{-4} m
Catalyst layer thickness	1.29×10^{-5} m
Membrane thickness	1.08×10^{-4} m
Cell temperature	70° C
Anode pressure	3 atm
Cathode pressure	3 atm

According to Figures 19.3 and 19.4, oxygen mass fraction is high at the entrance and then decrease along the fuel cell length due to consumption of oxygen by water formation. On the other hands, water increase along the cell length, this increase of water mass fraction was related with the fact that the water was formed by electrochemical reaction along the channel and water was transported from anode side by electro-osmotic drag coincidently.

Solid Polymer Fuel Cell

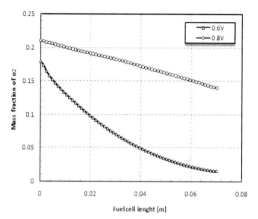

FIGURE 19.3 Comparison mass fraction of oxygen at the interface of cathode GDL and cathode catalyst layer for two different cell voltages, along the cell.

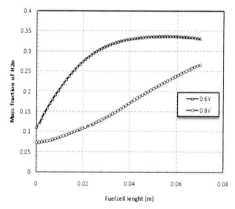

FIGURE 19.4 Comparison mass fraction of water at the Interface of cathode GDL and cathode catalyst layer for two different cell voltages, along the cell.

In addition, it can be anticipated that the mass fraction of oxygen and water will be decreased and increased respectively, by increasing of cell voltage due to increasing the electrochemical reactions.

The current density on the catalyst layer is shown in Figure 19.5. The current density at the inlet was the highest and decreased along the cell. The highest value of current flux density at inlet is probably because of the high concentration of hydrogen and oxygen and high electro-osmotic mass flux at the inlet region.

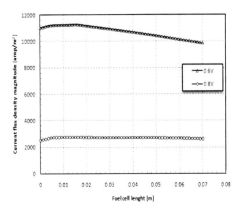

FIGURE 19.5 Comparison current flux density at the Interface of cathode GDL and cathode catalyst layer f.or two different cell voltages, along the cell

In this work, also the effect of prominent gas diffusion layers on cell performance and efficiency was studied and compared with base case (Figure 19.6).

Geometrical specification of case with prominent GDLs given in Table 19.4.

Figures 19.7 and 19.8.compare polarization curve and power density curve of two numerical case respectively.

Figure 19.9 shows the cross sectional schematic of base case which validated with experimental data.

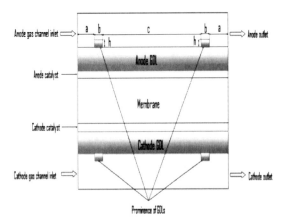

FIGURE 19.6 Cross sectional schematic of case with prominent GDLs.

Solid Polymer Fuel Cell

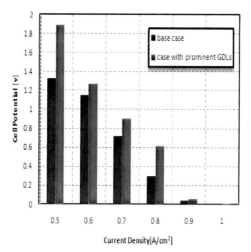

FIGURE 19.7 Comparison polarization curve of two numerical models.

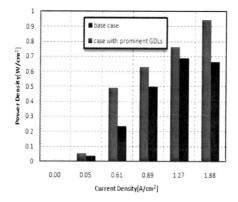

FIGURE 19.8 Comparison power density curve of two numerical models.

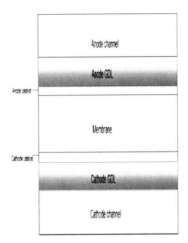

FIGURE 19.9 Cross sectional schematic of base case.

TABLE 19.4 Geometrical specification of case with prominent GDLs

Symbol	Value
a	15mm
b	5 mm
c	30 mm
h	0.25mm

It is notice that case with prominent GDLs produces more current density than base case at same cell voltage. Figure 19.10 illustrates the comparison of distribution of velocity at cathode gas channel. It is clear that the case with prominent GDLs yields a notable increase in velocity. Prominence of GDLs increases the velocity by decreasing the cross sectional area of gas flow at gas channels. This increase of velocity provides the reactant gases to the catalyst layers thus the efficiency of catalytic reaction increases and therefore the performance of PEMFC improves. Also Prominences of GDLs improve the flow of reaction hence reduces the membrane drawing effect. Thus the performance of fuel cell especially at the higher current densities improves. Figures 19.11 and 19.12 show the contours of velocity distribution for two case and confirm the mentioned reasons. The comparison of protonic conductivity at membrane and cathode catalyst layer interface for entry region is shown in Figures 19.13 and 19.14.

Solid Polymer Fuel Cell

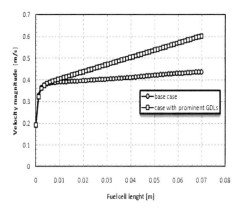

FIGURE 19.10 Comparison velocity magnitude of two numerical cases at cathode gas flow channel along the fuel cell.

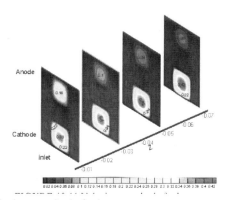

FIGURE 19.11 Velocity magnitude for base case.

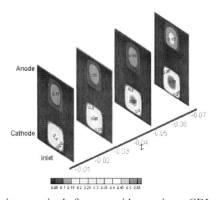

FIGURE 19.12 Velocity magnitude for case with prominent GDLs.

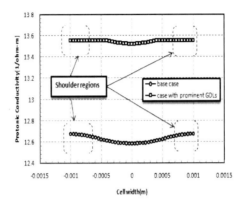

FIGURE 19.13 Comparison protonic conductivity at the Interface of membrane and cathode catalyst layer for two different numerical model at entry region(L/L$_0$ = 0.1428).

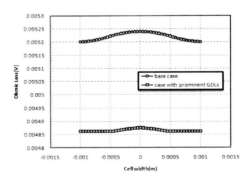

FIGURE 19.14 Comparison ohmic loss at the Interface of membrane and cathode catalyst layer for two different numerical model at entry region(L/L0=0.1428).

Figure 19.15 illustrates the distribution of oxygen mass fraction at membrane-cathode catalyst interface, for two different cases at the entry region of fuel cell. In case with prominent GDLs since the current density generative the cell higher than base case therefore high current density due to high accelerate in the electrochemical reaction rate and oxygen consumption. Lack of oxygen at the shoulder region of cell causes to higher mass fraction losses.

Solid Polymer Fuel Cell

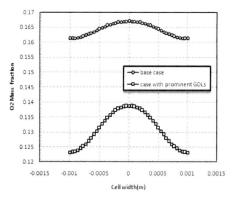

FIGURE 19.15 Comparison mass fraction of oxygen at the Interface of membrane and cathode catalyst layer for two different numerical model at entry region(L/L0=0.1428).

For downstream region of channel high mass fraction losses becomes worse due to diminution of the reactant with moving downstream which shown in Figure 19.16.

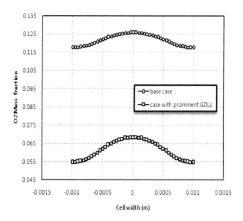

FIGURE 19.16 Comparison mass fraction of oxygen at the Interface of membrane and cathode catalyst layer for two different numerical model at exit region(L/L0=0.8571).

The contour of oxygen mass fraction at interface of membrane and cathode catalyst layer is shown in Figure 19.17. This fig shows oxygen mass fraction decreases gradually along the flow channel due to the consumption of oxygen at the catalyst layer. At the catalyst layer, the concentration of oxygen is balanced by consuming the oxygen and the amount of oxygen that diffuses toward the catalyst layer, driven by the concentration gradient. The lower diffusivity of the oxygen along the flow with the low concentration of oxygen in ambient air results in noticeable oxygen diminution along the fuel cell.

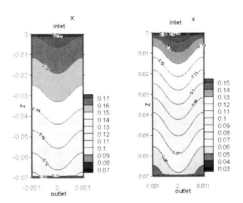

FIGURE 19.17 Mass fraction of oxygen at the Interface of membrane and cathode catalyst layer for base case (*left*) and case with prominent GDLs (*right*).

Hydrogen at the anode provides a proton, releasing an electron in the process that must pass through an external circuit to reach the cathode. The proton, which remains solvated with a certain number of water molecules, diffuses through the membrane to the cathode to react with Oxygen and the returning electron. Water successively produced at the cathode.

Comparison of mass fraction of water at the cathode side is shown in Figures 19.18 and 19.19. Respectively for entry and exit region of cell for two cases. Water concentration at the cathode membrane and catalyst layer interface increases along the flow channel. This increase of water concentration associates with the phenomenon that the Water composes by electrochemical reaction along the channel and transports from anode side by electro-osmotic drag coincidently. In case with prominent GDLs higher current density due to higher reaction rate. Then the mass fraction of water for base case lower than case with prominent GDLs. Figure 19.20 shows the distribution of water along the cell at membrane and cathode GDL interface.

Solid Polymer Fuel Cell

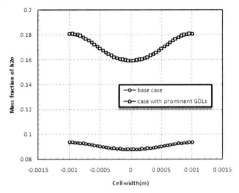

FIGURE 19.18 Comparison mass fraction of water at the Interface of membrane and cathode catalyst layer for two different numerical model at entry region(L/L0=0.1428).

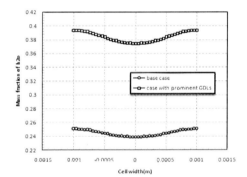

FIGURE 19.19 Comparison mass fraction of water at the Interface of membrane and cathode catalyst layer for two different numerical model at exit region(L/L0=0.8571).

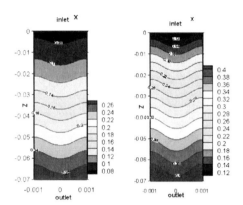

FIGURE 19.20 Mass fraction of water at the Interface of membrane and cathode catalyst layer for base case (*left*) and case with prominent GDLs (*right*).

19.4 CONCLSIONS

In this paper a full three-dimensional computational fluid dynamics model based on finite volume method of a proton exchange membrane fuel cell (PEMFC) has been developed to study the effect of prominent gas diffusion layers on cell performance such as cell current density, power density and efficiency of fuel cell. In this work we repose the prominence of GDLs at entry and exit region of gas flow channel both anode and cathode side. By comparison the result of this case such as polarization curve, power density curve, velocity distribution and species mass fraction by result of base case it can be seen that the prominences of GDLs increase the flow velocity which it is due to improvement of the gas flow along the channel which provides the reactant gases to the catalyst layers. Also gas flow improvement is due to reduce the membrane drawing effect thus the efficiency of catalytic reaction increases and more current density achieves with comparison by the base case at same cell potential. Therefore the performance of PEMFC improves. Thus it is clear that using the GDLs with prominence in PEMFC can be useful in increasing the efficiency of the cell.

KEYWORDS

- **PEM fuel cell**
- **Geometrical configuration**
- **Modeling and simulation**
- **CFD, catalyst layer**

6 NOMENCLATURE

a	Water activity
C	Molar concentration (mol/m^3)
D	Mass diffusion coefficient (m^2/s)
F	Faraday constant (C/mol)
I	Local current density (A/m^2)
J	Exchange current density (A/m^2)
K	Permeability (m^2)
M	Molecular weight (kg/mol)
n_d	Electro-osmotic drag coefficient
P	Pressure (Pa)
R	Universal gas constant (J/mol-K)
T	Temperature (K)
t	Thickness
$\vec{u}$	Velocity vector
V_{cell}	Cell voltage
V_{oc}	Open-circuit voltage
W	Width
X	Mole fraction

GREEK LETTERS

α	Water transfer coefficient
ε^{eff}	Effective porosity
ρ	Density (kg/m^3)
μ	Viscosity (kg/m-s)
σ_e	Membrane conductivity (1/ohm-m)
λ	Water content in the membrane
ζ	Stoichiometric ratio
η	Over potential (v)

λ_{eff} Effective thermal conductivity (w/m-k)

ϕ_e Electrolyte phase potential (v)

SUBSCRIPTS AND SUPERSCRIPTS

a	Anode
c	Cathode
ch	Channel
k	Chemical species
m	Membrane
MEA	Membrane electrolyte assembly
ref	Reference value
sat	saturated
w	Water

REFERENCES

1. Fuel Cell Handbook EG&G Services, Parsons, Inc. Science Applications International Corporation.
2. Amphlett, J. C., et al., The Operation of a Solid Polymer Fuel Cell: A Parametric Model. Royal Military College of Canada.
3. Ledjeff K., et al., Low Cost Membrane Fuel Cell for Low Power Applications. Fraunhofer-Institute for Solar Energy Systems, Program and Abstracts, Fuel Cell Seminar (**1992**).
4. Dutta, S.; Shimpalee, S.; and Van Zee, J. W.; Three-dimensional numerical simulation of straight channel PEM fuel cells. *J. Appl. Electrochem.* **2000**, *30*, 135–146.
5. Gurau, V.; Liu and Kakac, H. S.; Two-dimensional model for proton exchange membrane fuel cells. *AICHE J.* **1998**, *44*(11), 2410–2422.
6. Seung, Chi Lee and Chul, S. Y.; *Korean J. Chem. Eng.* **2004**, *21*, 1153–1160.
7. Chi Seung Lee, Chang Hyun Yun, and Byung Moo Kim, Se Chan Jang and Sung Chul Yi, *J. Ceram. Proc. Res.* **2005**, *6*(2), 188–195.
8. Kumar, A.; and Reddy, R. G.; Power, J.; *Sources.* **2003** *114* 54–62.
9. Shimpalee, S.; and Van Zee, J. W.; *Int. Hydrog. J. Energy.* **2007**, *32*(7), 842–856.
10. Shimpalee, S.; Greenway, S.; and Van Zee, J. W.; *J. Power. Sour.* **2006**, *160*, 398–406.
11. Garau, V.; Liu, H.; and Kakac, S.; *AIChE J.* **1998**. *44*(11), 2410–2422.
12. Meredith, R. E.; and Tobias, C. W.; Advances in Electrochemistry and Electrochemical Engineering 2, (Tobias, C. W., Ed., Interscience Publishers, New York, **1960**.
13. Byron Bird, R.; Warren, E. Stewart; and Edwin, N. L.; Transport Phenomena. John Wiley & Sons, Inc.; **1960**.
14. Springer T. E.; Zawodzinski, T. A.; and Gottesfeld, S.; *Electrochem J. Soc.* **1991**, *138*, 2334–2342.
15. Kuklikovsky, A. A.; *Electrochem J. Soc.* **2003**, *150*(11), A1432–A1439.

16. Meredith, R. E. and Tobias, C. W.; In Advances in Electrochemistry and Electrochemical Engineering 2, Tobias, C. W., Ed. Interscience Publishers: New York, 1960.
17. Yeo, S. W.; and Eisenberg, A.; *Appl J. Polym. Sci.* **1997,** *21,* 875.
18. Wang, L.; Husar, A.; Zhou, T.; and Liu, H.; *Int. J. Hydrog. Energy.* **2003,** *28*(11), 1263–1272.
19. Jung, Ch. Y.; Kim, J. J.; Lim, S. Y.; and Yi, S. Ch. *J. Ceram. Proc. Res.* **2007,** 8(5), 369–375.
20. Patankar, S. V.; Numerical Heat Transfer and Fluid Flow. Hemisphere Publishing Corporation; **1980.**
21. Ahmed, D. H.; Sung, H. J.; *J. Power. Sour.* **2006,** *162,* 327–339.
22. Maher, A. R.; and Albaghdadi, S.; *Renewable Energ.* **2008,** *33,* 1334–1345.
23. Springer, T. E.; Zawodinski, T. A.; and Gottesfeld, S.; *J. Electrochem. Soc.* **1991,** *136,* 2334.

INDEX

A

Aarony-Stauffer rule, 23
  natural nanocomposites, 20
Ab initio approaches, 250
  molecular simulation methods, 250
Acetic acid, 105, 106, 300
Acid dye, 256, 298, 300, 303, 305
Acid red, 298
Acrylonitrile-butadiene rubber, 81, 89, 92
Activation energy, 186, 195, 200, 202, 204
Activation entropy, 200, 202, 204
Active surface area, 51, 79
Adhesive agents, 248
Adsorption method, 324
Aerosil 200
Agglomerates, 56, 98, 216, 220, 221, 229
Agricultural research, 324
Air-gas mixture, 104
Air-ozone mixture, 104–106
Alkali-earth metals, 212
Aluminum hydroxide, 127, 128, 132, 133–136
Aluminum sheet, 262
Amine-terminated hyperbranched polymer, 299
Amine-terminated poly(amido amine), 307
Amorphous glassy polymers, 16, 2, 5, 10, 12, 14, 16, 20, 46, 47
Analysis of variance, 265, 266, 268
Analytical techniques, 79
  Raman spectroscopy, 79
  X-ray diffraction, 79
  X-ray fluorescence, 79
Anode electrode, 339
Applied voltage, 262, 264, 271–273, 276
Arc-discharge method, 280
Armco iron counterface, 113
Armco iron specimen, 115
Aromatic rings, 85
Arrhenius equation, 187
Arrhenius' coordinate, 199
Artificial neural network, 261, 262, 265, 266, 269, 276
As-received natural graphite flakes, 55
Association (dissociation) processes, 10
Atomic force microscopic, 310
Atomic-scale characteristics, 241, 252, 255
Atomistic modeling, 250
Atomistic simulation, 254
Avogadro number, 6

B

Back-propagation algorithm, 265
Band intensity, 301
Basal dislocations, 58
  see, dislocations
Beads, 174, 245, 253, 254, 271, 307, 313–319
Bentonite, 212
  bentonite clays, 212
Biomedical agents, 262
Blend components, 248
Block-on-ring friction couple, 114
Boron oxide, 125, 127, 128, 137
Boron oxide particles, 128
Boundary conditions, 251
Brabender Plasticorder, 82
Brabender Plasticorder laboratory, 92
Bright-field TEM images, 282
Brownian dynamics, 242
Brownian movement, 182
Bruggeman correlation, 343
Bulk modulus, 44
Butadiene-styrene copolymer, 104
Butyl-rubber, 104

C

Cable industry, 130
Cadmium sulfide, 324
Calcareous sediments, 51

Calcined kaolin, 127, 128, 133–136
Calibration, 148
Carbide, 52, 139, 140–143
Carbonate, 52, 146
Carbon-carbon (C-C) cross-links, 81
Carbon-carbon bonds, 58
Carboxyl and carbonyl terminal groups, 108
Cartilage, 307
Catalytic dye-agent, 323, 327
Cathode electrode, 339
Cationic modifiers, 213
Cationic UV curing, 250
Cell electrolyte, 339
Cell performance, 250, 340, 341, 350, 358
Central composite design, 262, 276
Ceramics, 223
Ceramizable composites, 125, 126, 128, 138
Ceramizable silicone, 125
Ceramization process, 126, 127
Channeling catalyst particles, 54
Chark, 142
Chemical vapor decomposition, 280
Chemical vapor deposition, 261
Chinox 1010 stabilizer, 151
CIELAB method, 297
Clay hydrophobic, 213
Clay surface, 213, 214, 228
Close-up agglomerated flake, 55
Close-up of sieved flake, 56
Cluster model, 3, 6, 9, 13, 14, 16, 26, 34, 37, 38, 40, 41, 45, 46, 48, 49
Clusters, 2, 8, 22, 49
Coal tar, 72
Coater system, 294
Coating materials, 145
Colloid-chemical properties, 105, 107, 108
Computer simulation, 241
Concentrated solutions, 174
Conducting polymer, 155
Contact angle (CA) measurements, 261
Contaminated natural graphite flake, 62
Continuous Configurational Boltzmann Biased, 254
Continuum mechanics approaches, 250
Contour plots, 272
Convergent methodology, 243
Convex, 279, 287, 288, 294

Core-shell, 323, 325, 328, 329, 337, 338
Correlation length index, 44
Corundum pots, 158
  see, instrumental
Coupling reaction, 307–319
Covalent bonding, 58
Critical mode, 5
Cross-link density, 82–84, 86–88, 112
Cross-link structure, 82–86
Cross-linking accelerator, 82, 86, 88
Cross-linking efficiency, 85
Crystal lattice, 57, 140, 212
Crystal planes, 159
Crystallite, 2, 5, 40
Crystallization process, 330
Cyclic voltammetry, 159, 165
Cylindrical geometry, 129
Cylindrical surface, 31, 32

D

Darcy's drag, 342
Dark dots, 287, 288
Dark-field electron microscopy method, 14
David Bridge, 82
Degenerated" nanocluster, 9
Degraded edge structure, 76
Degree of branching, 244, 245, 257
Degree of polymerization, 244, 245, 246
Degrees of branching, 253, 255
Dekabrominediphenyloxide, 231, 232
Dendrigraft (arborescent) polymers, 243
Dendrigraft polymers, 242, 243, 297
Dendrimers, 241–246, 253, 254, 256–258, 260, 297, 307, 308, 311, 313–319
Dendritic architectures, 241–244, 248, 249, 252, 254–256
Density functional theory, 250
Description theories, 14
Design of experiment, 264
Design-Expert software, 264
Dibenzothiazole disulfide, 82
Differential scanning calorimetry, 46, 145, 153
Differential thermal analysis, 166
Dilute solutions, 310
Dimetyldioctadecylammonium, 218
Dipole-dipole attraction, 251

Index

Dislocations, 52, 58, 80
  basal dislocations, 58
  nonbasal edge dislocations, 58
  prismatic edge dislocations, 58
  prismatic screw dislocations, 58
Divergent methodology, 243
DNA delivery, 249
Dogadkin's theory, 81
Doping agent, 162, 165–167, 169
Doping–dedoping process, 165
Dose rate, 81
Downstream region, 355
Dried-sample, 333, 334
Drug delivery, 249
Dumbbell specimen, 84
Dumbbell shape, 94
Dye absorbance behavior, 302
Dyeability, 248, 249, 258, 297, 298, 302, 305
Dye-adsorbent catalyst, 323–325, 329–338
Dyeing process, 300
  acid red, 300
  cold water, 300
  dyeing profile, 300
  occasional stirring, 300
Dye-receptive additives, 248
Dye-removal, 323, 325, 333, 337
Dynamical properties, 201

E

Earthquakes, 11
EBID process, 279
Ebonite, 112, 113, 115, 116, 118–122
EDX spectra, 162
Elastic component, 175
Elasticity modulus, 14–16, 20, 29, 33–35, 38, 39, 40, 42, 43–45, 47, 48, 211
Electric arc furnaces, 51
Electrochemical deposition, 261
Electrodes, 92, 105, 159, 165, 339, 340, 341, 342
Electron beam, 83, 89, 94, 279–291
Electron paramagnetic resonance EPR-spectroscopy, 12
  see, spin probes
Electron-beam irradiation fluences, 282
Electron-beam-induced deposition, 279

Electro-osmotic drag, 344, 348, 356, 359
Electrospinning, 261–265, 271–273, 276, 277
  aluminum sheet, 262
  metal needle, 262
  PAN powder, 262
  power supply, 262
  rectangular grounded collector, 262
Electrospinning apparatus, 262
Electrospun fiber mat, 262–266, 268–276
Elongation at break, 95, 99, 129, 131, 136
Energy of electrons, 158
Energy of ionizing radiation, 81
Epoxy polymer, 35, 37, 38, 43
EPR data, 14
Erratic edge, 63
Ersteds, 12
Ethylenepropylene-diene rubber, 125
Ethylenoxide grouping, 214
Euclidean space, 44
Evaporation, 200, 223, 261, 280, 340
Evernox 10 stabilizer, 150
Evonik Industries, 82
Exfoliated structure, 216
Exfoliation, 330
Experimental inhibition, 84, 85
External magnetic field, 323, 325, 327, 331, 332, 337
Extractive distillation, 249
Extrusion, 37, 38, 39, 41, 42, 44, 130, 223
Eye vitreous humor, 307

F

Fabrication methods, 280
  arc-discharge method, 280
  chemical vapor decomposition, 280
  evaporation, 280
  hydrothermal reactions, 280
FEGSEM resolution, 53
Fiber diameter distribution, 264, 273, 274
Fiber spinning systems, 261
  melt spinning, 261
  wet spinning, 261
Filtration, 262
Fire retardant additives, 51
Flory ball, 180
Flory-Huggins interaction, 83

Fluxing agent, 125, 127, 137
Food technology, 324
Fourier transform infrared spectroscopy, 298
Fractal (multifractal) structure, 2
Fractal-like tree morphology, 287
Friction force, 118
Frictional component, 178, 182–186, 193, 194, 196, 201, 203–208
Fritsch GmbH, 128
FTIR spectroscopy, 170
FTIR spectrum, 298
Fuel cells, 339
Full cell, 262

G

Gas diffusion layers, 341–343, 350, 358
Gas separation, 249
Gene carrier, 249
Gene delivery, 249
Globular structure, 243, 255, 297
Gold-sputtered electrospun fibers, 263
Governing equations, 250, 342, 347
Gradiently dependent value, 176
Graphic method, 22
Graphite atoms, 58
Graphite crystal, 51, 57, 59
Graphite intercalation compounds, 51
Grasselli machine, 141
Grinding process, 128

H

Hair-coloring, 324
Halogen treatment, 52
HBP treatment, 297
Hexagonal crystal lattice, 54
Hexagonal edge structures, 57
Hostguest encapsulation, 248
Hyaluronic acid, 307, 319
Hybride polymer systems, 2
Hydrodynamic flow, 173–175, 177, 183, 184, 189, 190, 197, 198
Hydrogen bonds, 91, 214
Hydrostatic extrusion method, 37
Hydrothermal reactions, 280
Hydrothermal treatment, 326, 329, 330, 335
Hydroxylation, 103
Hyperbranched polymer, 241, 242, 244–246, 248, 253, 256–258, 260, 298–306

I

Impurity analysis, 52
Independent variables, 264, 273
    applied voltage, 273
    solution concentration, 273
    spinning distance, 273
    volume flow rate, 273
Indian rubber, 103
Individual catalyst particle, 54
Inert gas (argon), 148
Inherent surface area, 79
Inhibiting effect, 84
Inhibiting particles, 68
In-lens detection system, 53
Inlet region, 349
Intercalation capacity, 51
Intercomponent bonds, 30
Interfacial interaction, 160
Interlayer space, 212
Intermolecular forces, 205, 251
Intermolecular interaction, 252
Instrumental, 158
    see, scanning electron microscopy
IR spectroscopy, 297
Irganox 1010 antioxidant, 149
Irradiation process, 82
Iso-methyltetrahydrophthalic anhydride, 37
Isostearic acid, 113
Isothermal DSC scans, 149

J

Jet movement, 261
Jet-milling, 71

K

Kerner equation, 17
Kubelka–Munk single-constant theory, 298
Kuhn's wire model, 183

L

Laboratory furnace, 129
Large-scale interactions, 252

Latex epoxidation peroxyformic acid, 104
Leather tanning, 324
Lennard-Jones (LJ) potential, 251
Light-harvesting arrays, 324
Linear polymer, 253
Linoleic, 106
Linseed oil, 113
Liposomes, 307
Liquid monomer, 184
Liquid phase, 65, 248, 298, 305
Liquid rubbers, 145
Lithium ion batteries, 51
Lithium polyisoprene, 103
Local density approximation, 250
Local order domains, 10, 40
   see, nanoclusters
Low-temperature plasma, 91

M

Macromolecular building blocks, 243
Macroscopic characteristics, 9
Magnetic ceramic particle, 323, 325, 326, 328–334, 337
Magnetic dye-adsorbent catalyst, 323, 330–333, 335, 337
Magnetic photocatalyst, 323
Maleic anhydride, 103
Maleinization, 103
Martin Marietta Magnesia Specialties, 127
Material testing, 153
Mathematical model, 341
Maxwell's equation, 175
M–ball rotation, 185, 189
Mechanical properties, 224
Mechanodegradation, 115
Melt spinning, 261
Melt-polycondensation reaction
Metal ion extractant, 249
Metal oxide, 52
Metamorphosed siliceous, 51
Methylene blue, 323
Metropolis Monte Carlo algorithm, 354
Mica flakes, 132
Michael addition reaction, 300
Microcomposite models, 20
Minelco minerals, 127
Mineral fillers, 128

Modified montmorillonite, 127, 130–134, 136
Molecular dynamics, 242, 250–260
Molecular simulation methods, 250
   atomistic modeling, 250
   continuum mechanics approaches, 250
Monomeric link, 180, 184, 203–205
Monte Carlo method, 250
Monte Carlo simulation, 251
Montmorillonite group, 212
Montmorillonites, 226, 227
Mooney viscosity, 107
Mrozowski cracks, 75
Multifunctional membranes, 262

N

N,N-dimethylformamide, 223
Natural nanocomposites reinforcement, 12
   chaotic distribution, 12
   dipole-dipole interaction, 12
   EPR-spectroscopy, 12
   Ersteds, 12
   polymer matrix, 12
   spin probes, 12
Nanoclusters, 3, 5–19, 21–23, 25–46
Nanodendrites, 279, 280, 284, 285, 287–294
Nanofiller, 2, 12, 14–16, 29, 32, 34–36, 43
   see, nanoclusters
Nano-microworld objects, 11
Nanoparticle, 2, 5, 6, 10, 211, 292, 293, 294
   see, nanoclusters
Nanopyramid, 68
Nanostructural systems, 2, 37
Nanotechnology, 236, 260, 276, 279, 296, 337
Nanotree structures, 283
Nanowire arrays, 279, 294
National Science Centre Poland, 100
Natural graphite flakes, 51
Natural nanocomposites structure, 2, 12
Natural rubber, 112
Necklace model, 174
Needle coke, 72, 77
Newton's equations, 251
Nitrile rubber, 89
Nonbasal dislocations, 58
Nonbasal edge dislocations, 58

Noncovalent-bonded complexation, 319
Nonequilibrium molecular dynamics, 253
Nuclear grade natural graphite, 70
Nuclear grade synthetic graphite, 72
Nuclear reactors, 51
Nylon-6, 224, 229–231, 234

O

Ohm's law, 165
Oleic, 106
Oligomers, 104, 145, 146, 223
Onset temperatures, 78
Optimization method, 178
Organic dye-removal mechanism, 325
Organic peroxide, 81
Organic red pigment, 146
Organoclay, 29
Oscillatory motion, 228
Otsuka Electronics, 309
Oxidation induction time, 145, 148, 152, 153
Oxidation onset temperature, 145, 153
Oxidative reactivity, 51, 52, 79
Oxidized natural graphite flake, 57
Oxidized pitch particle, 76
Oxygen plasma treatment, 94, 100
Oxygen reduction reaction, 77
Ozone electrosynthesis, 105
Ozone-air mixture, 104
  see, air-ozone mixture
Ozonization degree, 339
Ozonization reaction, 104

P

Paper production, 324
Parameter B, 197
Pauli repulsion, 251
PC radicals-probes, 12
PEM fuel cell, 344
Percolation relationship, 13, 15–17, 33, 44
Percolation system, 16, 44
Percolation theory, 44, 49
Percolation threshold, 44
Petroleum, 72
Phosphonium ions, 213
Photo-electrochemical cells, 324

Photosensitizers, 155
Piperylen, 145
Plasticization, 42, 107, 109, 130, 136
Platinum sheet, 159
Poisson's ratio, 17
Polar compounds, 103
  lithium polyisoprene, 103
  maleic anhydride, 103
Polarization curve, 348, 350, 351, 358
Poly (methyl methacrylate), 13
Poly(ethylene terephthalate), 248, 256, 298, 305
  PET fabrics, 248
Poly(propyleneimine) dendrimers, 254
Polyacrylate, 235
Polyaddition, 243
Polyamide-6, 248, 258
Polyamidoamine, 253
Polyaniline, 156, 158–170
Polyarylate, 4, 35, 37
Polybenzoxasene, 235
Polybutyleneterephtalate, 225, 226, 235
Polycarbonate, 3, 34, 45, 235
Polycondensation reaction, 297
Polyethylene, 125, 138, 153, 234, 236, 239, 257, 258
Polymer chains, 3, 32, 81, 85, 87, 209
Polymer electrolyte membrane fuel cell, 339
Polymer irradiation, 85
Polymer matrix, 16, 12, 16, 17, 20, 21, 27, 35, 45, 88, 167
Polymer rings, 114
Polymeric ball, 174, 175, 185, 190
Polymeric chain, 174, 175, 177, 181–185, 191, 199, 200, 203, 207, 213, 223, 225, 228, 229
Polymeric materials, 16, 2, 37, 45, 47, 49, 235, 236
Polymeric nanocomposite, 211, 215, 216, 222, 223
Polymeric polystyrene, 232
Polymethylmetacrylate, 211, 235
Polyolefins, 145, 148, 153, 256, 305
Polypropylene, 25, 26, 90, 153, 229, 231, 234, 235, 248, 249, 256, 257, 297, 305
Polystyrene, 90, 138, 173, 176, 180, 185, 186, 191, 193, 194, 195, 196, 199, 200,

201, 202, 203, 205, 206, 207, 208, 209, 218, 229, 231, 232, 233, 234, 235, 238, 257
Polystyrene's melt, 191
Polysulfide cross-links, 85, 86, 87, 111
Polysulfide rubber, 112, 113, 115, 118, 119, 121, 122
Polysulfone, 13
Polysulphide cross-link, 83
Polysulphide rubber, 112
Polysulphone, 112, 113, 115–119, 121, 122
Polyurethane elastomers, 145
Potassium hydroxide, 105, 108
Potato-shaped graphite, 71
Powders, 91, 347, 348, 350, 351, 358
  black, 91
  carbon, 91
  kaolin, 91
  silica, 91
Power density curve, 347, 348, 350, 351, 358
Prismatic edge dislocations, 58, 60
Prismatic screw dislocations, 58
Protective clothing, 262
Pseudoorthorhombic structure, 159
Pure oxygen, 53, 77, 339
Pure polystyrene, 232
Purified Natural Graphite, 51
Pyramid like structures, 67
Pyrolysis process, 140

Q

Qualitative variation, 333
Quantum (wave) aspect, 11
Quasi-equilibrium state, 13
Quaternary ammonium montmorillonite, 232

R

Radiation cross-linking, 81
Radiation energy, 85
Radical polymerization, 243
Radius of gyration, 246, 253, 254
Raman spectroscopy, 115
Rayonet photoreactor, 328
Rectangular grounded collector, 262

aluminum sheet, 262
Reflectance function, 298
Refractory filler, 125–127, 130, 136, 137
Reptational mechanism, 175
Resinous acids, 106
Resonance energy dissipation, 85
Response surface methodology, 261, 262, 264, 276, 277
Returning electron, 356
Reverse Monte Carlo, 255
Rheology modifiers, 248
Rhombic sulfur, 82
Rolling mills, 107, 109
Room temperature, 83, 113, 129, 131, 132, 133, 134, 135, 136, 157, 262, 263, 281, 291, 299, 310
Rotary movement, 182, 183
Rotation motion, 205, 207
Rounded nuclear graphite particle, 70
Rouze-Zimm hydrodynamic, 254
Rubber mixes, 82, 83, 91–93, 113
Rubber technology, 91
Rubber vulcanizates, 89, 91, 92, 94, 97–112, 140

S

SAED pattern, 327
Sawtooth edge formation, 67
Sawtooth structures, 67
Sawtooth-like edge formation, 66
SBR vulcanizate, 93
Scanned beam, 83
Scanning electron microscope, 94, 130, 141, 220, 263
Scanning tunneling, 218
Scorch time, 130
Screw dislocations, 52
Secondary electron signal, 94
Segmental motion, 194, 196–207
Self–avoiding walks statistics, 176
Semicrystalline polymer, 2
Serpinsky carpet, 44
Shear modulus, 18, 44, 176, 182, 187
Silane, 82
Silicate (organoclay), 35
Silicone rubber, 125
Sliding friction, 111

Small edge dislocation, 61
Sodium hyaluronates, 307, 319
Sodium poly-L-glutamate, 307
Solid-body synergetics, 3, 4, 5
Solid-state component volume fraction, 44
Solution concentration, 262, 264, 271, 272, 273, 276
Songnox 1010 stabilizer, 150
Spherical edge particles, 69
Spin probes, 12, 13, 47
Spin-coated film, 313, 314, 318
Spinning distance, 271, 272, 273
Stabilizer, 145, 146, 147, 149, 150, 151, 152
Standard free energy, 204
Stearic acid, 82, 83, 112
Stirring time, 333–336
Strings, 307, 313–319
Styrene–butadiene rubber, 111
Sulfur cross-linking system, 81, 82, 88
Sulfur vulcanization, 90, 111
Superposition principle, 183, 205
Suprasegmental (nanocluster) structure, 42
Surface chemistry, 91
  silica, 91
Surface free energy, 92, 94, 95, 96, 100, 101
Surface-adsorption mechanism, 325, 333, 337
Surface-cleaning treatment, 324, 333
Synovial fluid, 307
Synthetic graphite, 51
Synthetic isoprene rubber, 103

T

Tafel equations, 345
Takayanagi model, 17
Tear strength, 129, 131, 136
Technological revolution, 241
TEM micrographs, 285
Tensile strength, 87, 88, 95, 99, 129, 131, 136, 224, 228, 229
Tensometric method, 106
Tensor of gyration, 246, 253
Test compounds, 140
Tetrahedron, 140
Textile dyeing, 249
Thermal aging, 81, 149
Thermal stability, 145

Thermal-oxidation aging, 145
Thermodynamical state, 204
Thermogravimetric (TG) curves, 168
Thermogravimetric analysis, 211, 226, 236
Thiol—amine analysis, 83, 86
Tissue engineering scaffolds, 262
Titanium (IV) oxide, 170
Tixotropic properties, 212
TOF-SIMS, 115
Toluene, 156
Toyota, 211
Translational motion, 201, 206
Transmission electron microscope, 279, 323, 327
Triethylenglycole dimethacrylate, 200
TTI furnace, 52
Tumbler reactor, 92
Twinning angle, 60
Twinning band edge, 60
Two-stage glass transition process, 40

U

Ultrahigh resolution field-emission microscope, 53
Ultraviolet, 249, 323, 325
UV-radiation exposure, 328

V

Vacuum chamber, 92, 280
Vacuum drier, 157
Van der Waals bonding, 58
Van der Waals forces, 251
  see, intermolecular forces
Verlet methods, 251
Verse 3D, 141
Vietnam, 104
Virgin filler, 100
Virgin latex, 106
Viscosity, 104
Volume flow rate, 262, 264, 271–273, 276
Vulcanization kinetics, 129

W

Waste-water, 324, 325
Water drag coefficient, 345
Water transport, 340

Index

Wet spinning, 261
Wiener index, 244
Witten-Sander clusters, 22
W-nanodendrite structures, 281
Wollastonite, 92, 94–100, 127, 128, 131, 133–137
Wound dressing, 262

X

X-ray analysis, 216
X-ray diffraction, 327
X-ray fluorescence, 79

XRD broad-scan spectra, 328

Y

Yellowish gel, 326
Yield tooth, 38
Young Scientists' Fund, 88
Young's modulus, 17

Z

Zinc sulfide, 324
Zwick 1435, 84, 94